# Quantitative Biology of Metabolism

# Quantitative Biology
# of Metabolism

Models of Metabolism, Metabolic Parameters, Damage
to Metabolism, Metabolic Control

3rd International Symposium,
Biologische Anstalt Helgoland
September, 26—29, 1967

Organizers
A. Locker and F. Krüger

Editor
## A. Locker

With 66 Figures

Springer-Verlag  Berlin · Heidelberg · New York 1968

ISBN 978-3-540-04301-0      ISBN 978-3-642-51065-6 (eBook)
DOI 10.1007/978-3-642-51065-6

Library of Congress Catalog Card Number 68-55620

Title No. 1527

# Opening Address

Ladies and Gentlemen,

This is the third time that I have the honour to welcome the participants of the *International Symposium on Quantitative Biology of Metabolism* here on Helgoland. I cordially welcome you to this big rock of red sandstone in the North Sea, and to the Marine Station of the Biologische Anstalt Helgoland! We are very glad that this symposium series has proved a valuable platform for the exchange of information and ideas concerning quantitative aspects of metabolism, and that many leading scholars representing a wide variety of biological and medical disciplines have come again in order to present and to discuss new scientific results. This third symposium has attracted the largest number of participants from English-speaking countries yet — an indication, it seems, of the growing international reputation of this conference.

Please let me say a few words about the history of this symposium series. The idea to initiate biennial international and interdisciplinary symposia devoted to the fast developing field of quantitative biology was put forward in 1962 by Drs A. Locker (Österreichische Studiengesellschaft für Atomenergie GmbH, Wien) and F. Krüger (Biologische Anstalt Helgoland, Zentrale, Hamburg). I did not hesitate to support their idea and to offer the assistance of the Biologische Anstalt Helgoland as sponsoring institution. Thus the first two symposia of this series were organized by Locker, Krüger und myself, and held in September 1963 and 1965, respectively, on Helgoland. All papers and discussions have been published in volumes 9 and 14 of the "Helgoländer wissenschaftliche Meeresuntersuchungen".

From the beginning, I intended to devote myself to getting the symposium off the ground. Now that it is flying, I shall restrict myself to making available, as before, the assistance of our institution. As some of you may know, I am a marine biologist and experimental ecologist; as such I should no longer exert influence on an important field which progresses rapidly beyond my professional horizon. It is also felt that the proceedings should be published, from now on, in a special publication in order to do justice to their importance, not only to marine biologists but also to physiologists, biochemists, radiation biologists, geneticists, biomathematicians, theoretical biologists and physicists. Certainly this will make it easier to reach the great diversity of institutions and individuals interested in the papers and discussions of this conference. We are grateful that Springer-Verlag has agreed to act as publisher, and are convinced that the proceedings of these international symposia will continue to assist all colleagues interested in quantitative aspects of metabolism, as well as to encourage further studies in this exciting new scientific frontier.

I am glad to notice that some of the "old-timers", who attended the first and second symposia, have returned and equally happy to recognize several "newcomers"—well-known scholars in their scientific fields—who have come for the first time thus providing a highly welcome source of "fresh blood".

May this third symposium enlarge and advance our knowledge on quantitative aspects of metabolic processes as did the two preceding ones! I wish all of you rich hours of receiving and giving, and herewith open the *Third International Symposium on Quantitative Biology of Metabolism*.

O. KINNE

Biologische Anstalt Helgoland,
Zentrale, Hamburg

# Preface

The development of the life sciences may be said to have effected a gradual transition from a more or less intuitive prescientific approach based on crude observation, via a more refined type of observation to experimentation and hence to the level of formal theories. Quantitative methods are introduced at the second level; they comprise: (a) quantitative design of experiments; (b) regrouping of experimental results; (c) evaluation of results by means of mathematical or special statistical techniques.

The last step implies the introduction of theoretical concepts, but we are not justified in speaking of theoretical science unless true theoretical considerations—models or hypotheses—precede experiment, and this is then followed by an attempt to link results with theory so as to verify the theory. Biology at present seems to lie somewhere between the second and third level mentioned above, not yet having achieved the status of a theoretical science in all its branches. Thus, though the need for quantification and mathematical formulation is widely recognized, many biologists still believe that e.g. general systems theory is too abstract to be of use in handling concrete problems. Those, however, who look critically at the present state of affairs cannot adopt this attitude.

Here I would like to pay may respects to the founders of theoretical biology: A. J. Lotka in his *Elements of Physical Biology* laid the groundwork for the theoretical approach to almost all the problems of the life sciences, especially those of metabolism and related phenomena, such as occur in ecology; N. Rashevsky, in several now classical monographs, in *Mathematical Biophysics*, and through the *Bulletin of Mathematical Biophysics*, which he founded, made a decisive contribution to the subject, and his efforts are now culminating in a universal theory of "Relational Biology"; L. von Bertalanffy used the term "Theoretical biology" in several monographs, and we owe to him the concept of steady-state (Fließgleichgewicht) which, generalized as an open system, led him to the principal postulate of a General Systems Theory. The late H. Quastler deserves a special mention for having convinced biologists that information theory may usefully be applied to biology—and psychology. As a result of this pioneering work, a world-wide research movement was set off—though the centre of gravity lies in the United States. The time is now ripe for attempts to bridge the gap which remains in biology between steps two and three as defined above.

The chief aim of the *Symposia on Quantitative Biology of Metabolism*, held every other year at the Biologische Anstalt, Helgoland, is to promote recognition by producing acceptable results. The first (publ: *Helgoländer wiss. Meeresunters.*, Vol. **9**, 1964) comprised 36 papers, most of which dealt with experimental problems, and included a long discussion on the problems of metabolism under adaptive conditions. But even in 1963, the need to define and delimit the subject produced some theoretical contributions. The second (publ: *Helgoländer wiss. Meeresunters.*, Vol. **14**, 1966) included 41 papers, many considering the general implications of models and systems, while several outlined models for metabolic regulation and control based on differential equations or binary logic.

The lines of investigation pursued at the first two Symposia are found again at the third, some of the contributions achieving an even higher level of formalization. Four interrelated themes, chief among them the application of models to metabolism, can now be distinguished. The section on models is prefaced by a general review of their role in human cognition. The two main types of model capable of representing metabolism and the reactions of biosystems in general are, depending on the mathematics involved, either discrete and continuous or deterministic and stochastic; examples of both are found here. They show how fruitful purely theoretical considerations can be when applied to otherwise intractable problems. Certain limitations inherent in the design of computers exclude mere representation of all possible reactions in a given system; however, computer efficiency can be improved by the omission of certain terms. It is interesting that, for certain systems of differential equations, the solutions are shown not be to elementary.

Several more or less experimental trends in the quantitative biology of metabolism are illustrated by papers on the effects of temperature, activity stage, adaptation to temperature and season, growth and aging. Some of these problems run parallel in animals and plants. Among noxious influences, ionizing radiations play a leading role, and the quantitative aspects of the mechanisms responsible for protection and damage repair are of great interest. For models of metabolic control, the optimization of reactions is of fundamental importance and may be based on irreversible thermodynamics, as well as on the theory of automata. The link with time is of great importance in the theory of metabolic control and the topic of biological rhythmicity is also involved. Control mechanisms based on coordination of elementary reactions display oscillatory features at the cellular, intercellular and systemic levels; the steady state in the central nervous system is a good example of the power of such coordination. The field of thermoregulation requires a special model.

Thus we may say that these papers represent a cross-section through several of the main lines of advance of quantitative (theoretical) biology;

it is for the reader to decide to what extent they constitute a positive contribution to an ever-growing field of great scientific topicality. This Symposium, like its predecessors, was a truly interdisciplinary occasion, bringing together experts from such diverse fields as mathematics, computer science, theoretical physics, biophysics, biochemistry, human, animal and plant physiology, pharmacology and psychology. The fundamental nature of the subjects of our common study enables certain general conclusions to be reached which are of philosophical significance.

The Symposium was held at the Biologische Anstalt Helgoland, Marine Station, through the kindness of the Leading Director, Prof. O. KINNE, who made available to us in a most hospitable manner the excellent facilities of his Institute. The local organization was in the capable hands of Prof. and Mrs. F. KRÜGER, who looked after the accommodation and welfare of the participants supremely well. I wish, too, to thank the Mayor of Helgoland and the municipality for the welcome they extended to us. Grants from the Federal Ministry of Scientific Research, Bonn-Bad Godesberg, for which I herewith express my profound gratitude, enabled us to reimburse some of the participants' travel expenses. The financial contribution made by the Österreichische Studiengesellschaft für Atomenergie GmbH, Vienna, greatly helped me in publishing the program of the Symposium; this Society also generously assumed all the expenses associated with the editorial work. My special thanks go to Dr. K. F. SPRINGER of Springer-Verlag, Heidelberg, for agreeing to publish the proceedings of the Symposium. Last but not least, Miss H. KRUMPHOLZ assisted me in editing and retyping some of the manuscripts.

In conclusion, may I express the hope that the material presented here will come into the hands of all who are interested in quantitative and theoretical biology.

Vienna, November, 1968. A. LOCKER

# Contents

## 3. Damage to Metabolism

## 4. Metabolic Control

# List of Participants

* ADAMIKER, D., Institut für Biologie und Landwirtschaft, Reaktorzentrum Seibersdorf, Lenaugasse 10, A 1082 Wien, Österreich

* AMMON, H. P. T., Pharmakol. Univ. Institut, Universitäts-Str. 22, D 852 Erlangen

* ARESE, P., Istituto di Chimica Biologica, Universita di Torino, Via Michelangelo 27, Torino, Italia

* ARLEY, N., Niels Bohr Institute, University of Copenhagen and Norsk Hydro's Institute for Cancer Research, Oslo, Norway

BARTHELMESS, Ilse, Institut für Vererbungs- und Züchtungsforschung, D 1 Berlin-Dahlem (now: Institute of Animal Genetics, University of Edinburgh, Edinburgh, Scotland)

* BARTHOLOMAY, A. F., Center of Research in Pharmacology and Toxicology, School of Medicine University of North Carolina, Chapel Hill/NC 27514/ USA

* BERGNER, P.-E. E., Oak Ridge Institute of Nuclear Sciences, P. O. Box 117, Oak Ridge, Tenn. 37830, USA

* BETZ, A., Institut für Molekulare Biologie, Biochemie und Biophysik, Mascheroder Weg 1, D 3301 Stöckheim/Braunschweig

BOROFFKA, Irene, Zoolog. Univ. Institut, D 8 München, Luisenstr. 14

* BREMERMANN, H. J., Department of Mathematics, University of California, Berkeley, Calif., 94720, USA

BULNHEIM, H. P., Biologische Anstalt Helgoland, Zentrale, Palmaille 9, D 2 Hamburg-Altona

* BURNS, J. A., Institute of Animal Genetics, University of Edinburgh, West Mains Road, Edinburgh 9, Scotland

CHRISTOPHERSEN, J., Landwirtschaftl. Hochschule, D 7 Stuttgart-Hohenheim

* COE, E. L., The Medical School, Northwestern University, Ward Memorial Building, 303 E. Chicago Avenue, Chicago, Ill., 60611, USA

* COLLOT, F., Société Internationale de Biologie Mathématique, 19, Rue Roger Salengro, F 92 Antony, France

* CRISTEA, A., Center of Radiobiology and Molecular Biology, Casuta Postala 160, Str. C. A. Rosetti 37, Raion 30 Dec., Bucarest 1, Romania

* DETCHEV, G., Central Biophysical Laboratory, Bulgarian Academy of Sciences, Sofia, Bulgaria

* FISCHER, A., II. Med. Univ. Klinik, Szentkiralyi utca 46, Budapest, Ungarn

* FISCHER, R., Department of Psychiatry, Division of Behavioral Sciences, Ohio State University, 410 West 10th Avenue, Columbus, Ohio, 43210 USA

* GELINEO, S., Starigrad na Hvaru, Yougoslavia

---

* active contributors

* GRAINGER, J. N. R., Department of Zoology, University of Dublin, Trinity College, Dublin 2, Ireland

* GRIFFITH, J. S., Department of Mathematics, Bedford College, Regent's Park, London N. W. 1, England. Now: Chemistry Department, Indiana University, Bloomington, Ind., 47401/USA

* HORWITZ, Barbara, Department of Physiological Sciences, University of California, School of Veterinary Medicine, Davis, Calif., 95616, USA

JAIN, K. V., Max-Planck-Institut für Biophysik, Kennedy-Allee 70, D 6 Frankfurt/M.

* JANKOWSKY, H.-D., Zool. Univ. Institut, Hegewischstr. 2, D 23 Kiel

* KACSER, H., Institute of Animal Genetics, University of Edinburgh, West Mains Road, Edinburgh 9, Scotland

* KIEFER, J., Universitäts-Institut für Biophysik, Südanlage 6, D 63 Gießen

KINNE, O., Biologische Anstalt Helgoland, Zentrale, Palmaille 9, D 2 Hamburg-Altona

* KIRK, J., Regional Physics Department, Western Regional Hospital Board, 9—13 West Graham Street, Glasgow C 4, Scotland

* KLAMERTH, O. L., Bundesanstalt für Lebensmittelforschung, D 75 Karlsruhe (now: Institut für Biologie und Landwirtschaft, Reaktorzentrum Seibersdorf, Lenaugasse 10, A 1082 Wien, Österreich)

KÖRNER, W. F., F. Hoffmann-La Roche & Co., A. G., CH 4000 Basel, Schweiz

* KRÜGER, F., Biologische Anstalt Helgoland, Zentrale, Palmaille 9, D 2 Hamburg-Altona

LANGER, H., Lehrstuhl für Tierphysiologie, Abteilung Biologie, Ruhr-Universität Bochum (dzt.: Zool. Univ. Institut, Röntgenring 10, D 87 Würzburg)

LEUENBERGER, H., Max-Planck-Institut für Biophysik, Kennedy-Allee 70, D 6 Frankfurt/M.

* LOCKER, A., Institut für Strahlenschutz, Reaktorzentrum Seibersdorf (Österr. Studiengesellschaft für Atomenergie GmbH), Lenaugasse 10, A 1082 Wien, Österreich

MEYER-DÖRING, H., D 2 Hamburg 52, Cranachstr. 49

* MILLER, A. T. jr., Department of Physiology, School of Medicine, University of North Carolina, Chapel Hill, N. C., 27515, USA

* MÜLLER, E., Institut für Biochemie der Pflanzen, Deutsche Akademie der Wissenschaften, DX 401 Halle/Saale, Weinbergweg

PANDIAN T. J., University of Madras, India (now: Biologische Anstalt Helgoland, Hamburg)

* PAYNE, T. J., Department of Human Nutrition, School of Hygiene and Tropical Medicine, University of London, Keppel Street, London W. C. 1, England.

* RAO, K. P., Zoology Department, Bangalore University, Bangalore 1, India

RAUTENBERG, W., W. G. Kerckhoff-Institut, D 6350 Bad Nauheim

* ROSEN, R., Center of Theoretical Biology, State University of New York at Buffalo, Health Sciences Center, 4248 Ridge Lea Road, Amherst, N. Y., 14226, USA

* ROSSINI, L., Istituto di Farmacologia, Universita di Camerino, Camerino, Italia

* Schoen, H. R., Chirurg. Univ. Klinik, Klinikstr. 37, D 63 Gießen

* Simon, Z., Universitate de Timisoara, Bd. V. Pervan Nr. 4, Timisoara, Romania.

* Smith, R. E., Department of Physiological Sciences, School of Veterinary Medicine, University of California, Davis, Calif., 95616, USA

* Stolwijk, J. A. J., John B. Pierce Foundation Laboratory, 290 Congress Avenue, New Haven, Conn., 06519, USA

* Strubelt, O., Pharmakol. Institut, Medizinische Akademie, Ratzeburger Allee 160, D 24 Lübeck

* Sugita, M., Department of Physics, Hitotsubashi University, Kunitachi/ Tokyo, Japan

  Vergroesen, A. J., Unilever Research Laboratory, P. O. Box 114, Vlaardingen, Nederland

* Voss, R. E., Chirurg. Univ. Klinik, Klinikstr. 37, D 63 Gießen

* Walter, Ch., Department of Biochemistry, University of Tennessee, Memphis, Tenn., 38103, USA

  Wenzl, H., Lehrstuhl für Tierphysiologie, Abteilung für Biologie, Ruhr-universität Bochum, dzt., D 89 Würzburg

* Wheeler, Erica F., Department of Human Nutrition, School of Hygiene and Tropical Medicine, University of London, Keppel Street, London W. C. 1, England

* Woodstock, L. W., United States Department of Agriculture, Market Quality Research Division, Beltsville, Md., 20705, USA

* Zerbst, E., Physiol. Institut, Freie Universität, D 1 Berlin 33, Arnimallee 22,

  Zinkler, D., Lehrstuhl für Tierphysiologie, Abteilung für Biologie, Ruhr-universität Bochum, dzt., D 89 Würzburg

# The Epistemological Significance of Models in Science

A. LOCKER

*Abstract*

The usefulness of models for science is associated with the role they play in human thinking in general. The main features of models as well as of systems—the objects mapped by models and called originals in this respect—may be described in terms of the set theory based on certain pre-suppositions; the original model relationship may be formalized. A special class of systems or models is represented by automata, the elaborate theory of which for certain kinds of them allows a more profound application to biology.

## Introduction

The term and concept of a model, now a familiar phenomenon in all branches of science as well as in everyday life, nonetheless requires a careful theoretical study with respect to its scientific meaning and its epistemological implications, i.e. its relationship to human thinking. In other words, what is required is not only a careful study of the possibilities man has in obtaining knowledge, but also an analysis of the basic conditions for carrying out this task. The present use of the term model indicates a new turning point in present-day thinking, the opening of a scientifically orientated era in which, however, science can no longer—as it did in the past—neglect the philosophical background.

In order to understand what a model actually does represent, it would be desirable to start with an object that is only vaguely conceived and of which a model has to be built. An object of this kind may be called a system—provided we bear in mind that systems too do not exist *per se* but, on the contrary, are products of the creative or constructive capability of the human mind. Assuming the critical position of KANT's gnoseology—which is by no means a theory suspended in midair but rather a doctrine well founded in ontology [2]—we have to establish its constructive nature, the well known "productive power of imagination", which in its transcendental synthesis lends to the concept its image, although the concept as such can only be formed by means of perception. This interrelated creative activity of reason and perception—analogous to intellect and ideas—is obviously what was meant by MILLER [9] when he studied the implications of systems theory for biology. He distinguishes between "conceived" systems, "concrete" and "abstract" systems [9]. Since conceiving systems originates in the human mind and is inevitably determined by decisions made previously in order to specify the purpose of a system, it may be concluded that between such ambiguously conceived systems on the one hand and models, as their still more or less vaguely conceived representatives on the other, no sharp distinction can be made—at least not in the initial phase of the process of cognition. Only when systems have been "concretized"—by ascribing certain properties to them—are we able to infer the existence of certain properties as existing in other objects or systems as well; for various reasons, e.g. convenience, these are called models.

The relation between original (i.e. system) and model with which we are faced here may ultimately be formalized and this puts us into the position of making statements about properties which are characteristic of models in general. It is the merit of STACHOWIAK [12] to have thoroughly analyzed the main traits occurring here and to have emphasized—besides having introduced a suitable terminology—the distinction between the two possible main types of adjustment of the model to its relevant system or original. If the adjustment of the model to the original is based on structural features alone, then in its maximum degree of perfection the adjustment may be termed isomorphism; in the other instance where this adjustment is based on qualitative characteristics alone, its maximum degree of perfection leads to isohylism. However, between the two extreme instances, so to speak, an intermediate one may be conceived if, for a certain prescribed structural adjustment, a well-defined qualitative adjustment (or even an isohylic adjustment) is assumed; this clearly represents a case of analogy and the model is thus becoming an analog model. It is precisely this range to which all the models science—including quantitative biology—is concerned with, belong, e.g. simulation models programmed on computers, electronic circuits, hydrodynamic analogs and any other physical realization. The term "partially analog model" has also been proposed [12] in order to characterize models of this range, mapping time-dependent processes. These types of adjustment between models and relevant originals hold true, irrespective of the role models play in the process of cognition; in this process models may induce further cognition or they may merely exemplify by demonstration what is already known.

To characterize models appropriately, we have to take into consideration the following three characteristics which reveal the functional nature of models [12] or their "pragmatic" significance: (1) The mapping function deserves prime attention since here a principal feature may be conveniently formalized. Mapping depends on the degree of adjustment previously achieved between the model and the original. (2) The reducing function implies—which is what makes a model so useful—that certain traits on which a pre-decision has been made will be retained in the relationship between model and original—whereas other traits will have to be abandoned.

It has to be emphasized, however, that in order to know which of the traits should be neglected and which retained, we inevitably require a complete knowledge of all the features of the model. This requirement, of course, can only be satisfied by artificial objects; of these we have to regard rather arbitrarily the one as a system and the other as a model. Therefore, whenever an exact comparison between models and naturally occurring originals seems possible, this is truly an "artefact of perception" [12]. (3) The third function of the model, its subjectivization, expresses the fact that models cannot exist in their own right but rather that they are designed and applied in order to satisfy certain needs which arise only in relation to certain subjects.

## Models in Epistemology

In raising the question of the position and function of models in epistemology, we are confronted with the modern theory of cognition [7]. This theory states that science consists of two parts, linked by a third or intermediate one. The first part, concerned with the immediate observation of facts, represents the base and may be described in appropriate terms $L_O$ (language of observations) or $L_{O'}$ (language of measures); the second part is the apex as represented by $L_T$ (language of theories). Between the two parts we may insert the intermediate one $L_C$, consisting of rules of correspondence between base and apex. Thus it seems to be beyond any doubt that models are indeed tools for effecting such correspondence. In particular, we may identify the object (or system) "conceived" with $L_O$, together with the more or less "concretized" object (system) as introduced by MILLER [9]. On the basis of the general role of a model, we may now endeavor to search for an approach to $L_T$, i.e. the "abstract" system. Since the two (or three) levels of language do not express a unidirectional principle of construction, e.g. from base to apex, but rather are more or less interdependent, we can assume that the model may be applied both to obtain knowledge (by ascending from $L_O$ to $L_T$) and to verify it (by descending from $L_T$ to $L_O$). Moreover, the three main functions of a model, as compared with the respective original, viz. mapping, reducing and subjectivization, appear to be equally valid in both directions.

By inference from their intermediate position (pertaining to the level of correspondence), it becomes obvious that models represent only the sphere of signs—provided we accept the division of "reality" into (a) a sphere of objects, (b) a sphere of signs and (c) a semantic sphere in which the signs attain a certain meaning [10].

Therefore, models need to be distinguished in terms of possible signs and hence at least two types may be delineated [4]: (a) iconic models, preserving some perceptual elements, and (b) symbolic models, in which even the remnants of a perceptive element have completely vanished. Iconic models, for instance, are block diagrams, flow diagrams, etc., whereas symbolic models are purely mathematical descriptions.

In modern scientific usage the term model is increasingly displacing the term hypothesis; indeed, the use of the word model tacitly implies a sort of explanation and expresses a certain level of knowledge already achieved. Hypothesis, however, not being clearly separable from the term model, rather emphasizes the pre-suppositions to be accepted in order to attain new knowledge.

### The Concept of a System (in Verbal and Formal Respects)

To keep in touch with the epistemological nature of the model, it becomes necessary to establish the position for constructing the original, i.e. the system; this position is reflected in the model. However,—without reference to the manner of attaining a model—the general definition of system and model is completely identical. The following is an example of a verbal definition of a general system [5]: a system is a set of objects, of the relations between these objects, and of their properties. The position of the objects within the system (or model), being determined by their relations, shows (1) how fundamentally, even in such a preliminary verbal definition, the structural or relational aspect prevails and (2) how impossible it is to define a system (or model) without previously clarifying the objects and relations entering into such a definition. One has, therefore, to consider as fundamental what was casually mentioned above, namely, the resolution level as an expression of a position accepted by the perceiving, conceiving and operating subject [6]. The level of resolution depends (a) on the quantities or properties being assigned to an entity (system or model), (b) on the accuracy with which these properties are expressed (or measured) and (c) on the time intervals at which this should be done. In approaching this problem, we are faced with certain terms which are necessary in order to arrive at an appropriate definition, i.e. terms like activity, behavior, structure and organization as determinants for activity or behavior. No system exists by itself, it is permanently connected with an environment, either described as a system (supersystem in this respect) or represented by the perceiving and conceiving subject. From a position taken outside a given or created system, several quantities entering the definition must appear as observables.

A formal definition may be attempted in the following manner, somewhat simplified and starting with such observables [6]:
Let the set of external observables be:

$$X = \{x_1, x_2, \ldots, x_n\} \tag{1}$$

Relations between the quantities of the set $X$ at resolution level $u$ will be $r_i(X)_u$, the set of all relations being

$$R(X)_u = \{r_1(X)_u, r_2(X)_u, \ldots, r_m(X)_u\} \qquad (2)$$

The system $S$ is the set of the relations between the external quantities:

$$S = \{R(X)_u\} \qquad (3)$$

This very simplified definition of a system may be brought into accord with other definitions [8], all on the basis of set theory. Because the definitions of systems (or models) are based on the set theory, it should be mentioned that a system is distinguishable from the pure set by its structure. Moreover, an important feature of a system, namely its "wholeness", may also be described in terms of set theory.

A definition similar to set theory may in principle be specified on the basis of the logic of predicates and classes [12]. This latter definition becomes necessary if one is dealing with more than a first predicate logical step, viz. with properties of properties, properties of relations, relations of relations, etc. We may distinguish between a predicate logical one-step system $S^{(1)}$ and a predicate logical $k$-step system $S^{(k)}$; the first of which may be formalized as follows:

$$S^{(1)} = A \cup 1\, A^1 \cup 1\, A^2 \cup, \ldots, \cup 1\, A^n \qquad (4)$$

$A$: basic set of elements; $k\, A^n$: class of $n$-occupancy predicates of the $k$-th step.

## Models in Formal Respect

With the formalization of a model, the three main functions of models with respect to systems, namely mapping, reducing and subjectivization, have, of course, to be included. The first of them, mapping, may be described as follows [3]: Given are 2 sets; of these $B$ with predicates $Q_1, \ldots, Q_k$ is a model of a set $A$ with predicates $P_1, \ldots, P_k$ if there exists a mapping $\varphi \mid A \to B$ of $A$ into $B$ and a mapping $\varphi \mid P_j \to Q_j$ of $P_j$ into $Q_j$ such that $\varphi\, P_j(x_1, \ldots, x_n) = Q_j(\varphi\, x_1, \ldots, \varphi\, x_n)$ for $j = 1, \ldots, n$ and all $x_1, \ldots, x_n \in A \times, \ldots, \times A$. In this definition $\varphi$ becomes a homomorphism with respect to the structure $\mathfrak{S}$ which will be in both sets ($A$ and $B$) defined as a collection of predicates of one or several variables ranging over the set. A homomorphism with respect to the structure $\mathfrak{S}$ is called a mapping $\varphi$ if for every predicate $P$ on $A \times, \ldots, \times A$ a predictae $\varphi P$ defined on $B \times, \ldots, \times B$ is associated.

Reduction may be considered in a model definition in which the uni-unique mapping refers to predicates which comprise only subclasses of the two respective entities, viz. model and system, wherein the complexity of the "abundant" properties of the latter exceeds that of the former. Subjectivization may finally be formalized as follows: If two homomorphic sets are given, one of them is called model if at least one human exists for whom during the time $T$ with respect to certain perceptive and operative functions $\eta$ the system $S_1$ may be replaced by $S_2$ [12]. Formally, the definition may be written as follows:

1st step: statement of the formalized conditions in which $S_2$ is a homomorphic mapping of $S_1$:

$$\mathrm{hm}(S_2, S_1) = \mathrm{def}. \; \exists \, U_1 \, \exists \, U_2 \, | \, U_1 \subset S_1 \wedge U_2 \subset S_2 \rightarrow \mathrm{bij.\ m.} \; (U_2, U_1) \quad (5)$$

bij. m. $(U_2, U_1)$: $U_1$ is bijectively mappable in $U_2$.

2nd step: statement of the conditions under which the homomorphic mapping equals a model:      (6)

$$\mathrm{mod.}\; (S_2, S_1) = \mathrm{def.\ h.\ m.}\; (S_2, S_1) \wedge \exists \, K \, \exists \, \eta \, \exists \, T \; \mathrm{repl.}\; (S_2, S_1, K, \eta, T)$$

$K$: "kybiac"-organism (comprising natural and artifical organism); repl.: replacing function of $S_2$ with respect to $S_1$ and with respect to $K, \eta, T$.

Of course, since the definition of a model, like that of a system, may be based on time-invariant and time-dependent features, it should be possible to propose other definitions of models, too [6]. However, it should be pointed out that whatever a formal definition of a model may be like, its mapping, reducing and subjectivization functions must be preserved.

## Automata as Models

As mentioned above, a system may be considered only in conjunction with its environment—especially with regard to epistemology—; the system which has to be considered as open [1], in its response to the environment maps the stimulus received [3]. If the stimulus-response relationship is of a rigid and prefixed nature—as determined by a program—we are dealing with an automaton, especially if its internal structure is not known or arbitrarily neglected, so that it may be regarded simply as "black box". Among automata we may distinguish between those with internal states and those with internal memory [6].

To begin with, for automata in which the number of internal states is finite or the memory is finite, it should be pointed out that the former constitutes a more general concept; whereas, in an automaton with finite memory, the response is derived from the stored parts of the stimulus together with the activity already stored in the automaton, an automaton with a finite number of states stores the past of its activity as an internal state which, in connection with the instantaneous stimulus just acting, determines the response. The internal state is recursively defined in as much as it depends on the previous internal state as well as on the instantaneous stimulus. The fundamental difference between the two types of automata lies in the fact that the automaton with a finite memory has an explicitly expressed dependence of its response on the previous stimuli and responses; in the automaton with a finite number of states, the response is expressed implicitly—by means of the internal states [6].

It is certainly not without general (epistemological) significance that models, systems and automata have several features in common with respect to (1) the mathematics inherent and (2) to their behavior—if we define behavior as the relationship between stimuli and responses or activity and reactivity to environment. Besides deterministic and stochastic automata—with completely different onsets—additional classes of automata, namely prospective and retrospective ones, may be mentioned as examples. With the exception of some still problematic attempts at continuous automata, the theory of automata is at present entirely based upon time-discreteness. Three particular forms of automata are of special interest for biology, i.e. self-organizing, growing and self-reproducing automata.

## Self-Organizing Automata

Self-organization means that the structural order increases. A presupposition is, of course, that the system (automaton) is open to its environment, a fact which allows it to adapt to prevailing situations and to attain an improved organization. One may also state that in those automata the variety of responses produced may under certain circumstances change and, in general, must diminish [6]. Practically the same holds true for the learning automaton which improves its efficiency with respect to a certain task that is to be fulfilled. An automaton is also considered to be learning if it changes in such a manner as to require different quantities for its description at different times.

## Self-Reproducing Automata

In considering automata with an infinite number of states, we are confronted with the TURING machine named after the famous British mathematician, A. M. TURING. It consists of an indefinitely long tape, divided into squares, along which a reading head moves; the latter is able to write or erase symbols in the squares and moves only one square per time interval. The movement is governed by standard code instructions; finally, there is also provision for recording the current state of the machine. However, this device represents only a basic one, called the TURING automaton by von NEUMANN, so as to distinguish it from the TURING machine proper where the device is provided with a tape of infinite length in both directions; this latter has been contrived for purely mathematical purposes so as to prove the computability of certain functions. The Turing machine is capable of simulating other machines since code instructions may be developed by means of which the first machine can be induced to imitate another, and, as a matter of fact, any other machine. Also, a third type of Turing machine may be envisaged, the so-called universal Turing machine which imitates the action of a Turing automaton [13].

Growing automata, from the viewpoint of Turing machines, are therefore those which can write symbols on their own tapes and thus induce their enlargement and growth. Self-reproducing automata, in their simplest form, are those which write down their own coding programs, provided a universal Turing-machine is at hand.

Examples of more elaborate self-reproducing automata may be presented: (1) the kinematic model, (2) the tesselation model (both of them proposed by von Neumann) and (3) the string-processing automaton proposed by Stahl [13]. The kinematic model, however, may be shown to contain an intrinsic paradox [11]. The tesselation model, in which self-reproduction is related to a configuration (and not to a concrete entity) has not been successfully solved to this day and, therefore, there remain only a few words to say on the string-processing automaton which is not merely a product of the mind but has actually been realized by a computer [13]. Herein the cell is considered as a string-processing enzyme automaton which accepts or reads not only single symbols but whole strings of symbols and may process such strings in a flexible manner, e.g. carry out enzyme synthesis, etc.

It should be mentioned that the application of the automata theory to biology introduces into the latter certain paradoxes and logical limitations such as those arising in a given axiomatic system of mathematics. Examples of unsolvable cases have been reported, e.g. if a cellular computational automaton (like a self-organizing one) is designed to synthesize a new gene, then the automaton is unable to determine whether the newly formed cell with the new gene can undergo self-reproduction or not. Several other paradoxes are associated with the problems of self-applicability, self-referencing and self-describing (cf. [13]).

## Concluding Remarks

In this brief and rather sketchy outline I have tried to trace a few possible aspects of the nature of models in general, of the application of models to science, and biology in particular, and I have also tried to throw some light on the importance of the gradually maturing theory of models for the accomplishment of which the fundamental investigations of Stachowiak [12] have been valuable contributions. Since the constructing of models is most intimately associated with the process of human knowledge, the evolution which the theory of cognition has recently undergone certainly also underlines the importance of applying models in a manner which is both governed by reason and continuity and directed towards the aim of universal recognition. The freedom to invent models, which nevertheless remain useful for recognizing "reality", supports the assumption ("hypothesis" in the true sense of the word) expressed by critical philosophy that things-in-themselves are only comprehended insofar as they are established in the human mind.

## References

1. BERTALANFFY, L. VON: Br. J. Phil. Sci. **1**, 134 (1950).
2. BLAHA, O.: Die Ontologie Kants. Salzburg-München: Pustet 1967.
3. BREMERMANN, H. J.: The evolution of intelligence, Techn. Rep. No 1, Contr. Nonr 477(17), Dept. Math., Univ. Washingt. Seattle 5, Wash.: July 1958.
4. FREY, G.: Synthese **12**, 213 (1960).
5. HALL, A. D., and R. E. FAGEN: Gen. Syst. **1**, 18 (1956).
6. KLIR, J.: Gen. Syst. **10**, 29 (1965).
7. LEINFELLNER, W.: Struktur und Aufbau wissenschaftlicher Theorien. Wien Würzburg: Physica 1965.
8. MESAROVIC, M. D.: In: MESAROVIC, M. D. (Ed.), Views on General Systems Theory, p. 1. New York: John Wiley & Sons 1964.
9. MILLER, J. G.: Perspect. Biol. Med. **9**, 107 (1965).
10. PEIRCE, C. S.: cited from: BENSE, M., Grundl.studien Kybern. Geisteswiss. **4**, 12 (1963).
11. ROSEN, R.: Bull. Math. Biophys. **21**, 387 (1959).
12. STACHOWIAK, H.: Studium Generale **18**, 432 (1965).
13. STAHL, W. R.: Perspect. Biol. Med. **8**, 373 (1965).

*Discussion*

ROSEN:

A complete development of VON NEUMANN's tesselation models for self-reproduction has been made by ARTHUR BURKS, in a recently published book. This has been modified by a number of authors, especially APTER and ARBIB, to give the beginnings of a theory of automata in which development is possible, reminiscent of embryological development, at least in a metaphorical sense.

GRIFFITH:

In connection with self-replicatings systems, I should like to draw the attention of the symposium to the very simple mechanical models put forward by PENROSE (L. S.) and PENROSE (R), Nature **179**, 1183 (1957). I showed recently (GRIFFITH, J. S., Nature **215**, 1043 (1967)), that a logically very similar kind of model could be constructed using protein subunits as components.

R. FISCHER:

I am most grateful to Dr. LOCKER for the stimulation he has provided with his masterful lecture. Within our Aristotelian, binary (two-valued) ontology, logic and language, there are only objects and subjects but no specific designation exists for *image* and *thought*. The need for them becomes especially apparent when we have discuss the nature of models. A model is an "as if object" created in imagination and thought. This Pygmalion, our creation, is then repeatedly mirrored and reflected — with the aid of observation and measurement—within the complexity of our own organization until it becomes alive as an "as if subject" at the highest level of resolution of our own nature (which we call physical reality). Thus we, the system, were modeling another system until the model became real; so real that at the moment of understanding we could exclaim (when talking to this formerly iconic model): „Ich bin im Bilde" (I am in the picture, i. e. in the image).

LOCKER:

You correctly touch upon the crucial point that the so-called objective being can only be attained within the medium of awareness. Models thus constitute an attempt at unifying subjective thinking and objective being.

# Causality, Complexity and Computers

H. KACSER, and J. A. BURNS

With 6 Figures

*Abstract*

Biological systems containing many components and complex interactions between them require a different approach than is customary when dealing with simpler situations. In particular, analytical procedures reveal mainly the structure of the system. The behaviour of the system, however, is equally dependent on the quantitative values of the parameters which may determine the type of response. Experimental and theoretical questions must therefore take account of the system as a whole. Some of the difficulties and possible solutions are discussed.

The problem of complexity has been facing biologists right from the beginnings of the scientific study of their subject. The recent advances in the understanding of some of the molecular mechanisms operating in organisms has, in some circles, tended to the view that we are nearing a solution. It is our view that, on the contrary, a close examination has revealed far greater complexity than was evident before. More importantly, it has become clear that a new methodological and conceptual approach is necessary if we are to make progress beyond the mere cataloguing of an even greater number of biochemical details.

## Structure and Quantity

To explain the difficulties which we have in relating the properties of an organism to its fundamental biochemical events it is instructive to analyse very briefly how we proceed in our studies. We can resolve our activities broadly into a series of steps. The first step is essentially analytical. It consists of the specification and sometimes the 'discovery' of the elements in terms of which the investigation is to proceed. Thus, the anatomist 'dissects' the organism into organs, the physiologist into tissues, the geneticist into characters, the chemist into molecules. Naturally there is considerable overlap between the fields but each field, or perhaps more accurately, each investigator, deals with a chosen set of elements which he considers appropriate.

The specification of the elements is followed by and often closely linked up with the establishment of the relations between them. The establishment of which elements are coupled to each other and what the nature of this relationship is, constitutes a large part of everyday scientific pursuit. We thus establish biochemical pathways, nerve and muscle connections, circulatory and transport systems, cell and tissue organisation and so on. When, as is usual, a large number of elements are involved, the method of choice is to take these two at a time. This method of isolation is sometimes achieved by studying the elements actually isolated from the system. When this is not possible or inconvenient, the rest of the system is immobilised, held constant or uncoupled and the relations are studied. The elements may be *in* the system but not *of* it.

By a series of such isolation procedures under different conditions it is, in principle, possible to establish the relationship between all the elements even if each element stands in some relation to many others.

The net result of such investigations is the establishment of the *structure* of the system. We usually represent this diagrammatically as a map, such as anatomical, biochemical or neurological maps, and this is helpful in some simple cases to understand and predict the behaviour of the system. In particular, a linear sequence of interactions is easily grasped. Elaborate hierarchies of causes can thus be established.

Difficulties, however, begin to appear when it is found that an 'effect' is connected by a feed-back to a 'prior cause'. Which is 'prior' will depend on where you started to investigate the system. This is a very general problem but we would like to demonstrate it by reference to systems with which we are more familiar. For example, in certain enzyme systems of organisms there is a phenomenon known as repression which shows that the level of a substrate 'determines' the quantities of the enzymes. These quantities in turn 'determine' the level of the substrate (Fig. 1). Both are

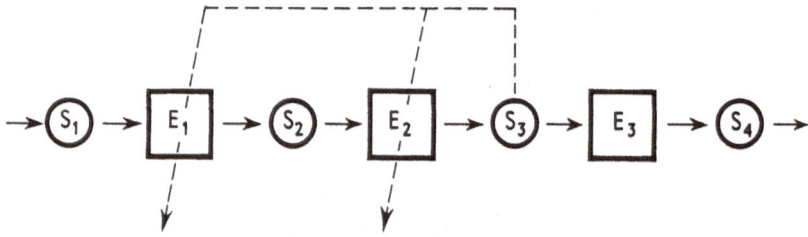

Fig. 1. Enzyme sequence with repression. The substrate $S_3$ acts in some way as an antagonist of enzyme synthesis and hence enzyme quantities are negative functions of the value of $S_3$. Enzymes $E_1$ and $E_2$ catalyse the production of $S_3$ and its value is therefore a positive function of the enzymes

mutually dependent on each other and 'cause and effect' language, useful as it may be in discussing binary relations, must be abandoned in such cases for a mathematical description of the functional relations. In other words to obtain the system's behaviour a synthesis in quantitative terms is required of the separately established properties of its parts. We must therefore substitute for the question "What cause?" the question "How much?"

Let me give as a simple example a slight expansion of the enzyme-substrate relation referred to above (Fig. 2). The substrate $S_3$ forms a

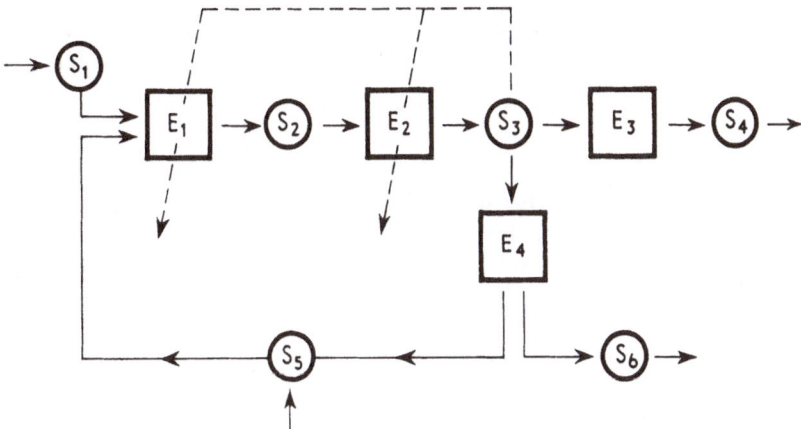

Fig. 2. As in Fig. 1 with the addition of a loop. The first reaction is now bi-molecular

negative feed-back loop on the enzymes $E_1$ and $E_2$ which catalyse the flux for the production of $S_3$. Substrate $S_5$, on the other hand, derived from $S_3$, is a partner in the first reaction and therefore has a positive influence on the level of $S_3$. The net outcome of these opposing interactions cannot be predicted by inspection of the structural map. Nor is the result immediately obvious even if we know the quantitative relationships of each part of the system such as the repression function, the enzyme equations etc.

When we allow all interactions to proceed simultaneously, the behaviour of a particular element within the system will depend, in general, on all other elements. For example, depending on the relative and absolute quantities of the individual parameters, the system may either oscillate or come to a steady state, $S_3$ may respond positively or negatively to changes in the activity of enzyme 4 and so on. In other words, systems with the same structure may display different properties depending on the values of the quantitative relations between their elements. In more complicated systems 'switches' between different states may occur [5, 7] and whether such a switch is possible or whether it occurs in a given circumstance will, again, depend on the quantitative values and not only on the structure.

The first important conclusion, then, is that the quantitative formulation is not simply an added refinement of the structural one but is the essential basis for our understanding. In fact the quantitative formulation contains within it the structural information but not *vice versa*. The set of simultaneous differential equations which describes the whole system is the only means of answering the functional questions which we can pose.

## System Limits and Master Reactions

The carrying into practice of this desirable prescription, however, meets with two formidable difficulties. The one concerns the method of obtaining the necessary quantitative information, the second the problem of solving the equations which this information makes possible.

Our first concern must be to define our system. This is a necessary first step both to determine what measurements to make and what equations to set up. In the example in Fig. 2 we have arbitrarily defined it by a small number of elements. In any real organism, however, this represents only a part and is connected, perhaps in many ways, to other elements. Must we include these too? And if we do, must we not include further elements which, in turn, are connected to the bigger system? Does this reasoning not lead us to the conclusion that we must consider the whole organism even if we are to ask questions about only a part of it? This is a fundamental methodological problem and we must enquire whether there are any *a priori* or empirical reasons which allow us to escape from this dilemma. Before attempting some kind of answer I would like to discuss an aspect of measurement which is basic to this problem.

As was mentioned before, the investigator can choose his elements of analysis or, put in another way, the measurements he wishes to make. Furthermore he can make measurements both on the system and on chosen parts of it. In doing so, he will find that certain classes of measurement can be made in both situations, on the system and on part of it. For example, he can weigh an organism, then dissect it—if necessary down to the molecular level—and weigh each part. (If, in addition, he knows the law of additivity of mass, he will come out with the same answer). But the main point is: he can determine weight on any part of the system.

There is, however, another class of measurement, possible on the system but not on the parts. These are properties which arise only because of the *interactions* of the parts and which therefore naturally 'disappear' in the process of elementary analysis. In this class are most aspects of growth and many of the homeostatic properties, variously described as buffering, canalisation, integration, etc. Certain aspects of morphogenesis—the development of specific forms in growing organisms—also fall in this category.

The ability of an organism to maintain, say, constant internal temperature against external fluctuation is not measurable when the analysis is carried out at the cellular or molecular level but is a property of the system as a whole—or at least a considerable part of it. It is precisely because many variables change within the system that the temperature does not change. The temperature behaviour of each part is uninformative of the behaviour of the whole. The whole, if not greater, is at least different from the sum of its parts. These systemic properties can never be understood as a hierarchy of causes. Their mathematical formulation usually involves non-linear and often higher degree differential equations for which no explicit solutions exist. They are, however, in many ways the most interesting and important properties of organisms.

It has often been argued that, complex as biological phenomena might be, there is usually some 'master-reaction' (see e.g. [1, 3, 6]) on which all others depend in some simple manner. The phenomenon is said to reflect the underlying single process. Confirmation of this view is sometimes sought by demonstrating unitary behaviour, that is to say that some simple model adequately describes the situation. This is, however, rarely a sound procedure. In e.g. the increases in weight of embryonic tissues one often finds 'logarithmic' growth. This time/weight correlation may, on analysis. be found to be a compound of increase in cell numbers, increase or decrease in cell mass and increase in water content—none of them changing at the same rate. The model of an 'autocatalytic' mechanism is clearly inappropriate for anything but descriptive and possibly predictive purposes. This is not to deny that in *some* cases a simple overriding 'master-reaction' will govern the behaviour of a complex dependent system. But in biological phenomena it may be safer to assume that simple effects have rarely simple causes. Those who use Occam's Razor without due caution are likely to find they have cut off more than their beard. The apparent unit process behaviour is often due to a *system behaviour*, that is to the simultaneous response of many interacting elements. In such cases the simple behaviour of the system is not attributable to a simple master process on which the rest of the complex system depends, but on the contrary, the behaviour is the result of the *irreducible* interactions of the whole. Our studies, both theoretical and experimental, of linked enzyme systems, have shown that some of their properties, which of necessity arise from the interactions of many components, behave remarkably like single processes. Although not attributable to any one entity, a model based on such an assumption would be a consistent one. It appears that there is a kind of saturation of complexity, that the addition of more elements or more interactions beyond a certain level, merely generates redundancy so far as *types* of behaviour are concerned. Every now and then novel properties may arise, such as oscillations, switchings and so on. But it appears that these behaviour types are small in number. The reasons for the apparent loss of complexity will require discussion.

We may not have come much nearer to answering the question as to how many elements must be considered in defining a system but at least we are now aware that the search for the master reaction may be a futile one and we must find other criteria to assess the importance of any component.

How difficult this can be is perhaps demonstrated by some of the work by Mr. G. Bulfield in our laboratory.

In Fig. 3 we show two mice, a small one and a large one. Both were born in the same litter and brought up in the same cage. If we enquire into the differences between these two we find a considerable number. (Table 1). No doubt with more work we will find more differences. How are to understand these? Which are causes, which effects? Some of the observed differences could be used with certain assumptions, to 'explain' the size difference, others seem to point to the opposite of the observed effect. How large is the system? Which are the functionally important elements? Perhaps all, perhaps none of them.

In a way we have a kind of explanation: The difference between the two mice is due to a single gene difference located on the 11th chromosome. In that sense this gene is the cause of the syndrome. In another sense this is no explanation at all. It is perhaps like the opinion that the bullet which killed Archduke Ferdinand at Sarajevo was the cause of the first World War. Any serious student of history or biology wants to understand the many forces that come into play as result of such an event, some tending

Fig. 3. Litter mates seven weeks old from a cross in which the 'obese' gene segregates

Table 1

| Character | Obese litter mate | Normal litter mate |
|---|---|---|
| 1. Body weight | 60 gm. | 30 gm. |
| 2. % Fat | 33 | 12 |
| 3. Fatty acid composition | normal | normal |
| 4. Pancreas | increased insulin secretion | normal |
| 5. Resistance to cold | drastically reduced | normal |
| 6. Blood glucose | hyperglycaemic | normal |
| 7. Mating behaviour | absent | normal |
| 8. Cholesterol on fasting | increased | normal |
| 9. Liver glycogen turnover | high | normal |
| 10. Retention of steroid hormones | high | normal |
| 11. Ketone levels | slightly high | normal |
| 12. Enzyme levels | | |
|    (a) Citrate cleavage | 300 | 100 |
|    (b) Isocitric dehydrogenase | 120 | 100 |
|    (c) Pyruvate kinase | 180 | 100 |
|    (d) Malic enzyme | 200 | 100 |
|    (e) Monoglyceride lipase | 20 | 100 |
|    (f) Pancreatic lipase | 70 | 100 |
|    (g) Malic dehydrogenase | 100 | 100 |
|    (h) Lactate dehydrogenase | 100 | 100 |
|    (i) Fumarase | 100 | 100 |

References 1—11 inclusive from: JEAN MAYER, Amer. Jour. of Clin. Nutr. **8**, 712 (1960).

to escalate the situation, some tending to stabilise it, some making no impact on the net result. A complete enumeration of changes does not allow us to make a decision.

Can we at least eliminate certain components or parts of our system? We can e.g. propose that all components whose values have not changed need not be considered as part of the system for the purpose of our calculations. That this is not invariably a sound procedure is shown by some of the work[1] in our laboratory on the arginine pathway in the fungus *Neurospora crassa*.

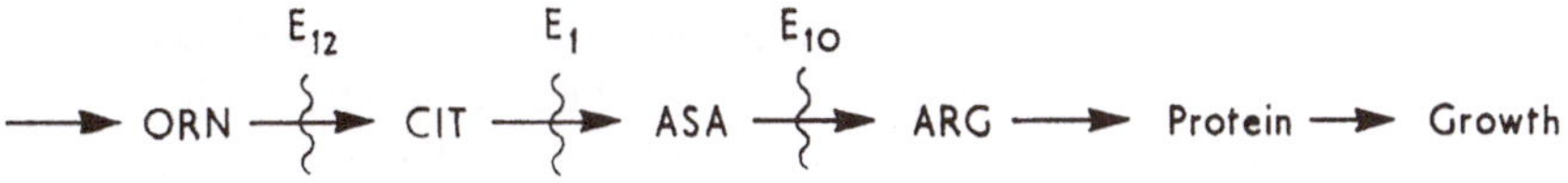

$$\longrightarrow ORN \xrightarrow{E_{12}} CIT \xrightarrow{E_{1}} ASA \xrightarrow{E_{10}} ARG \longrightarrow Protein \longrightarrow Growth$$

Fig. 4. Some amino acid transformations in the arginine pathway

[1] Those engaged in this work have been: Drs. D. W. DONACHIE, I. R. BROWN, C. F. CURTIS, O. GILLIE, Miss C. CHILCOTT and Mr. B. WATSON.

Three successive enzymes, which we have simply numbered, form part of this system. All of them are essential enzymes in the sense that growth ceases if any one of the activities is zero. We can, however, make defined changes in the enzyme activities by genetic and other means and as a result of this the following picture emerges.

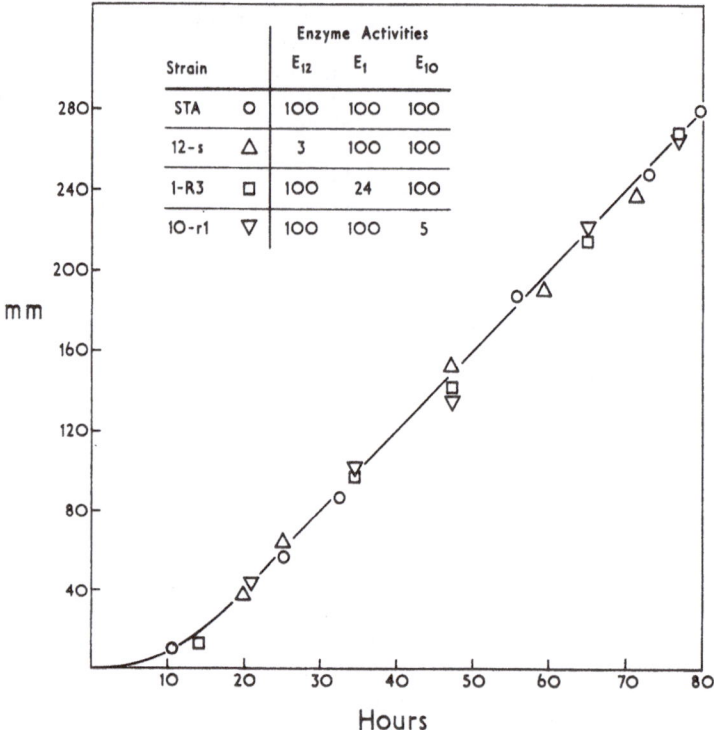

Fig. 5. Comparison of the growth rates on solid medium of four strains of *Neurospora crassa*. They differ from each other by single genes specifying three enzymes in the arginine pathway

The linear growth rate of three strains is compared to that of a standard strain. Each strain has, as indicated, suffered a considerable reduction in the activity of one of the enzymes. Without going into any of the complications, we shall only point out that the identity of the growth rates is accompanied in each case by different adjustments of the system. These adjustments involve more components than shown in Fig. 4 but the important point is the demonstration of the remarkable buffering capacity which may be concealed by the apparent constancy of some of our measurements. It is precisely the interaction of the enzymes and their substrates which generates the invariance of growth rate for the changes under discussion. They are therefore very much 'in the system'. A proper understanding of the system must therefore take account of these phenomena and explain rather than ignore them.

So we come to the somewhat unsatisfactory conclusion that there are no general rules for defining the limits of a system. The analysis has, however, pointed the way we must go. It is one of our aims to give a quantitative description of enzyme systems of this kind and in the following we shall describe how we have approached the problem.

## The Sensitivity Coefficient

We are faced with the situation that we have fairly detailed knowledge of the structure of enzyme systems as well as information on the properties and relations of the individual (isolated) enzymes and their substrates. But since there are many interactions—sequential, competitive, inhibitory and activating,—we must devise a method of studying each enzyme when it is acting and interacting in the whole system.

We have already pointed out that the total elimination of enzyme activity (by e.g. the classical genetic block) usually leads to elimination of the measured property and sometimes to the elimination of the whole organism. This kind of experiment therefore yields mainly structural information and gives no quantitative answers as to the role of the enzyme. (Fig. 6a). There is, however, a related experiment which immediately gives quantitative information. This consists of 'modulating' the activity of one enzyme and measuring the result on as many systemic properties such as products, fluxes (and other enzymes!) as we may choose. We can e.g. introduce into an organism a mutant enzyme with altered activity (as in the strains in Fig. 5) or we could alter the quantity of enzyme by controlling the nuclear dose, or we could inhibit it in some specific manner.

Let us consider the investigation of, say, a flux $F$ through a pathway in a system consisting of many enzymes. We shall assume (and may have evidence) that the system is in steady state. I shall not here go into the technical difficulties (which are not small) of measuring the flux of molecules in a particular part of an organism but shall assume that we have some means of doing so. Similarly I shall assume that we can estimate the quantity or activity of a chosen enzyme $E$. If we now compare, say, a different mutant with enzyme activity altered by a step $\Delta E$ we can ask what change in the flux, $\Delta F$, has resulted. (Fig. 6b). It is best to express these as percentage changes or as fractional changes, i.e. as $\dfrac{\Delta E}{E}$ and $\dfrac{\Delta F}{F}$ since this eliminates scale effects. As the flux change depends on the magnitude of the enzyme change, the ratio, $R$, of these two quantities, i.e. $\dfrac{\Delta F}{F} / \dfrac{\Delta E}{E} = R$, is some

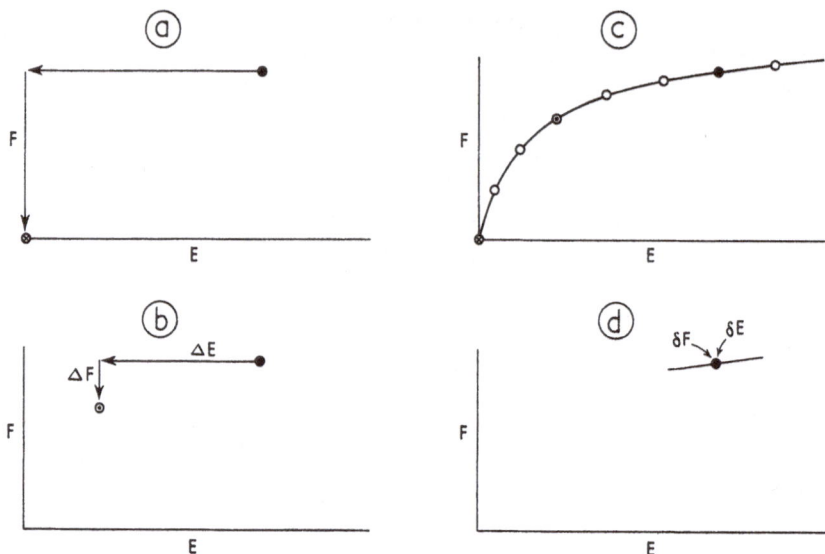

Fig. 6. Alteration of one enzyme in a system and its effect on one flux

a) Reduction of enzyme activity from its given value to zero. The flux becomes zero.
b) A finite change in enzyme activity, $\Delta E$, results in a finite change in flux, $\Delta F$.
c) A series of enzyme activities with their associated fluxes defines the system's response to the change.
d) An infinitesimal change at one point defines the slope and hence the response at that point.

measure of the importance of the enzyme to the flux since it measures the response to an imposed alteration. However, in general, the new position we reach will depend in a non-linear manner on the step we have taken. (Fig. 6c). The value of the ratio will therefore vary with step length and hence it is not particularly informative. If we make the change exceedingly small, however, it becomes nearly independent of step length and will reflect the 'local' response at the position. (Fig. 6d). In the limit this is, of course, a differential when the ratio of the two changes will give the slope at the point.

$$\frac{dF}{F} \Big/ \frac{dE}{E} = \frac{dF}{dE} \times \frac{E}{F} = \text{Slope} \times \frac{E}{F}$$

Experimentally we can obtain the shape of the curve by a series of measurements with different mutants or other means and are thus in a position to estimate the slope for any value of the enzyme.

The ratio of the differential changes we have called the 'sensitivity coefficient'. Thus

$$\frac{dF_j}{F_j} \Big/ \frac{dE_i}{E_i} \equiv C^{F_j}_{E_i}.$$

It will be noted that the value of the coefficient varies over the whole range of enzyme activities because of the non-linearity. This means that the response of the system to the same fractional change will be different at different absolute values of the enzyme.

We can now see clearly what meaning can be attached to the term 'controlling' enzyme or 'master reaction'. Obviously this must be the one whose coefficient is numerically equal to 1, that is, a reaction where, say, a 1% change in the enzyme produces a 1% change in the dependent flux. The same enzyme may be rate-controlling at the lower end of its range but quite insensitive at high values. The non-linearity is the result of all the interactions of the system and hence depends, in general, on the values of all the other components. The coefficient is therefore a systemic property depending not only on the one enzyme but on all others, as well as on certain external parameters such as fixed substrate sources. What applies to one enzyme applies to all others. The coefficients of all enzymes therefore depend on each other and will change if any one of them changes (by e.g. a movement to a new position). It is therefore seen that many enzymes contribute to the flux *and* to its control. In a simple pathway, for example, no single enzyme may have a coefficient of 1 but all may have small values and therefore none of them will be master reactions (compare Fig. 5). Flux control is therefore seen to be dependent not simply on the enzyme itself, nor on its structural position nor on the structure as a whole. The absolute and relative values of all parameters are equally important. It is a truly systemic property, being the response characteristic of the whole system at one point of its phase space.

Coefficients can be determined for a large number of properties with respect to any changes in the parameters under experimental control. Thus it is possible to determine the coefficient $C_{X_j}^{F_i}$ of a certain flux $F_i$ with respect to a controlled food source $X_j$. In general all systemic properties, i.e. dependent variables, have coefficients with respect to all independent parameters, $C_P^V$. In principle all such coefficients are determinable and a matrix of these displays in quantitative terms the role each element plays when acting within the system. Although these are 'properties' of the elements, they are dependent (and sometimes critically dependent) on other elements. We shall therefore find that this property may change, (change in the value of the coefficient) if e.g. the organism is exposed to different environments (change in 'external' parameters) or if the genetic background ('internal' parameters) is changed. The element, e.g. an enzyme, having undergone no change is nevertheless seen to have altered its role. Although by no means restricted to biological systems, it is in organisms that these concepts are essential to our understanding.

## Predictability and Understanding

We now have at our disposal values for the non-systemic, elementary parameters (enzyme quantities, Michaelis constants etc.) and means of altering these experimentally as well as values for the systemic variables (pools, fluxes, coefficients, etc.). The well-known enzyme equations, established by enzymology, together with the structural information of our metabolic maps enables us to write down the set of equations in their differential form which represent the whole system. These will solve for us the questions for which we had no simple logical answer. The truth of the matter is, however, that none but the simplest set of equations is soluble. Our only way out of this dilemma is to obtain *particular* solutions. With the advent of computers, provided one has sufficient funds to use them, this has become a practical possibility. A system of any complexity can now be handled and the computer can be made to run through as many sets of data as you wish to supply it with. This is indeed the usual use of computers in simulation (see e.g. Garfinkel [4]). In this way we are able to calculate the fluxes and sensitivity coefficients. We are able to show that most coefficients are likely to be small, thus confirming the remarkable buffering capacity in the arginine pathway already referred to (see [2]).

The computer can be used to confirm or deny suggested mechanisms or quantitative assumptions; it can suggest experiments and predict their outcome.

By such means we may regain predictability which we had lost because of complexity. But we have not regained understanding. We have, so to speak, delegated the observing of the processes to the computer and must rest content with observing the outcome of its arithmetic. We are in fact making empirical observations on a system—which is what we started off by doing in our biological experiments. Of course, the computer system is constructed in a known manner with specified elements and specified relations. But then, provided we have made a correct and exhaustive analysis of our 'real system', we had this knowledge before we programmed the computer. The two systems are believed to be isomorphic and their behaviour *is what it is*. We are left with what we might call the 'transparent box' problem. Although we see everything in the box, how things are connected and what each part is, the behaviour of the whole is largely unknown,—except after the event. The status of the observations on the computer is exactly the same as that of observations on the system which it simulates. We have no insight or expectations beyond those obtained by the empirical observations of either system. As with bird-watching, computer watching has its fascinations and advantages. In particular, it has been fashionable for some time to claim that prediction is the only aim of scientific endeavour. I think this view is mistaken. I think that prediction is but one part of what we, as scientists, are trying to do. The problem of complexity should not be relegated to a transparent box which we trust but do not understand.

We are therefore led to consider complexity as a problem in its own right apart from the simulation of a particular system under review.

There are three aspects of a system which can be usefully considered as influencing its behaviour: (1) the number of elements, (2) the structure, (3) the values of the parameters. By constructing computer models and varying any of these three classes, we can ascertain their influence on systemic properties. But since the computer gives only particular solutions, we are using it as a source of empirical data from which empirical generalisation must be made. By starting with relatively simple systems we might hope to build up an understanding of systemic behaviour which we are, as yet, lacking. We must begin to learn what our evolutionary or educational past has apparently not forced upon us: namely, to think in non-linear and quantitative terms. I do not think that our inability to comprehend complex situations is due to the 'structure of the mind' or other such limitation but essentially due to unfamiliarity with the behaviour of such systems. A good chess player, e.g. can understand situations which completely baffle most of us. By making a large number of observations on computer results and relating these to structure and quantities, it should be possible to obtain a direct and intuitive understanding of 'standard' types of complex systems. We then might build up a 'dictionary of complexity' which could become as familiar as the rules of logic are alleged to be.

Computer technology, far from solving the problems of complexity, has high-lighted the necessity to approach it in a different way. The computer is not only a powerful tool for simulation and prediction, but if properly used can also aid our understanding. But we must deliberately set out to use it in this way—as a 20th century abacus—not only for calculation but for the training of our unsophisticated minds.

## References

1. BÜCHER, T., u. W. RÜSSMANN: Angew. Chem. **75**, 881 (1963).
2. BURNS, J. A.: 3rd Symposium on Quant. Biol. of Metabolism, p. 75 (1968).
3. BURTON, A. C.: J. Cell Comp. Physiol. **14**, 327 (1939).
4. GARFINKEL, D.: Biochem. Soc. Symp. **26**, 81 (1966).
5. KACSER, H.: in: Biological Organisation at the cellular and supercellular Level, p. 25. Academic Press 1963.
6. KREBS, H. A., u. H. L. KORNBERG: Ergebn. Physiol. biol. Chem. exp. Pharmakol. **49**, 212 (1957).
7. ROSEN, R.: Int. Revs. Cytology, **23**, 25 (1968).

# Discrete and Continuous Representations of Metabolic Models

R. ROSEN

*Abstract*

A wide variety of important biological systems admit both a continuous and a discrete representation. For example, the central nervous system can be described in terms of both a continuous two-factor theory and a discrete theory of neural nets; biochemical control systems admit a continuous theory of chemical kinetics and a discrete theory of "biochemical automata". Each of these representations possesses an intrinsic character not shared by the other; for example we may talk about transient behavior and stability in continuous systems, and about computability in discrete systems, but we cannot at present transfer such concepts readily from one sphere to the other. Nevertheless, from the fact that both kinds of theories describe the same systems, there must be some kind of "correspondence principle" by means of which transfer of concepts can be made between discrete and continuous models. We shall be concerned with the development of such a principle, and the implications of this principle for the description of metabolic control and regulation.

## I. Introduction

The behavior of biological systems simultaneously admits a number of independent types of description. For instance, it is possible to construct models for the central nervous system in terms of quantities which are continuously variable in time. The kinetic properties of such systems are then described by a system of ordinary differential equations (see e.g., [10]). This type of description allows us to speak meaningfully of intensity gradations of stimuli and responses, and of the stability or instability of system behavior. On the other hand, the same central nervous system may be *independently* described in terms of a mathematically discrete theory [8]; here, ideas of continuity, continuous variability and stability are meaningless, and are replaced by entirely new concepts: computability, regularity [7] and ease of computation [3, 4]. An exactly analogous situation has arisen in the theory of cellular control and regulation (see [12] for an elaboration of the analogous features). Here too, we have both continuous descriptions [5, 6, 15] and independent discrete descriptions [16, 18, 19]. Again, the discrete descriptions are characterized by the fact that continuity and stability have no intrinsic meaning within these descriptions, while on the other hand the basic concepts of the discrete description, such as computability, have no meaning in the continuous description.

It is most important to find a way of reconciling and comparing the discrete and continuous descriptions of biological activity and indeed of systemic activity in general. A variety of attempts at such a reconciliation have already been proposed of which we may briefly mention several here.

## 1. Sampled Data Systems

In this case we really do not have two independent descriptions. We begin with a continuous description and simply discretize it by standard relaxation methods. In particular, every state of a sampled data system is already a state of original system. The time scale and the structure of the state space of the sampled data system, together with all observable properties of that system, are inherited from the continuous description directly. We shall not consider sampled data systems further in this paper, although a particular class of such systems will emerge from our analysis.

## 2. Hybrid Systems

SUGITA [20] has proposed the consideration of hybrid systems, part of which can be described only discretely, and part only continuously. Here again, we do not have independent descriptions of overall system activity; although hybrid systems may prove convenient for important applications, it is to be expected that a hybrid description will be superseded by purely digital and continuous descriptions which may then be directly related.

## 3. Homologies between System Descriptions

A number of authors [1, 2, 22] have noted that the continuous and discrete descriptions are related by formal homologies, and have on this basis attempted to construct a single formalism which could encompass both kinds of description. These attempts have a number of fruitful consequences, but from our point of view suffer from one basic difficulty: they are of an external and formal character, and do not relate the independent digital and continuous descriptions *of any particular system*. We have shown [13] that the formal homologies on which these studies are based arise out of theoretical necessities governing all types of systems descriptions and are thus incapable of providing direct insight in the question we have raised.

What is really needed, then, is a kind a *correspondence principle* which will allow us to relate these two kinds of descriptions in terms of the intrinsic properties of any particular system being described, rather than to simply exhibit formal homologies between the descriptions. We would expect such a principle to exist from the simple fact, that independent alternate descriptions of the same system are possible. To be satisfactory, such a principle must allow us to express the basic concepts of each description in terms of the other in an effective manner. The purpose of the present note is to suggest such a principle and to examine some of its consequences for the description of a variety of biological systems. The arguments which follow represent a formalization of ideas implicit in many discussions of discrete and continuous systems, put into a general system theoretic perspective. For the purpose of this exposition, we omit mathematical details; the reader should consult Rosen [13] for notation and terminology.

## II. Continuous and Discrete Descriptions

The continuous description of system activity is invariably formulated within the framework of dynamical (or control) system theory. In this theory, the set of states of the system forms a manifold $S$, and the forces imposed on the system are related to the kinetic properties of the system through equations of motion. These equations describe the rate of change of appropriately chosen state variables in terms of the variables and the forces. Each possible kinetic activity of the system is represented by a trajectory in this state space. The state space is further contoured into regions of stability and instability in terms of the equations of motion in a well-known way.

The discrete description, on the other hand, generally is expressed in the form of an abstract sequential machine. Here, the set of states forms a discrete (usually finite) set. According to the homologies we have mentioned, the forces that can be imposed on the system are represented by abstract alphabet symbols, and the role of the equations of motion is played by the nextstate function. The time variable for sequential machines ranges over an abstract set of *instants*, which in general has no relation to any kind of real time (in sampled-data systems, on the other hand, the set of instants is directly specified in terms of real time).

Now let us introduce some convenient terminology. Any attribute of a continuous or discrete description will be denoted by prefixing that attribute with the letters *c-*, *d-* respectively. Thus, e.g. a *d*-state will refer to a state occurring in a discrete description, etc.

Suppose that we are given a system $\mathfrak{S}$, and two independent descriptions of its activity, one of which is continuous and one of which is discrete. How can we hope to establish a general correspondence between these descriptions, using nothing more than the fact the descriptions refer to the same system? The key observation to be made here is the following: the crucial property of the *c*-description is the fact that *any property whatsoever* of the system $\mathfrak{S}$ must necessarily be specified in terms of the observables of that description. In other words, if $S$ represents the manifold of *c*-states of $\mathfrak{S}$, then any property of $\mathfrak{S}$ is necessarily specified in terms of the values assumed on $S$ by a (finite) set of real-valued functions on $S$. Hence, in particular, it follows that the $d$-states themselves must necessarily be so specified. Let us denote the $c$-observables in terms of which the $d$-states are defined by $v_1, v_2, \ldots, v_r$.

Obviously, any two $c$-states $\sigma_1, \sigma_2 \in S$ such that $v_i(\sigma_1) = v_i(\sigma_2)$, $i = 1, \ldots, r$ will not be distinguished as separate states by the $d$-description. Hence the $d$-states can be regarded as inducing a decomposing of $S$ into equivalence classes, in such a way that the set $S_d$ of $d$-states is isomorphic to $S/R$, where $R$ is the equivalence relation on $S$, defined by writing $\sigma_1 R \sigma_2$ iff [1] $v_i(\sigma_1) = v_i(\sigma_2)$, $i = 1, \ldots, r$.

We can get further information regarding the equivalence classes in $S/R$ by looking at the kinetic properties of the $d$-description. These kinetic properties are embodied in statements of the form

$$\delta(s,a) = s'$$

where we know that $s, s' \in S_d$ correspond to equivalence classes of $c$-states in $S$. We do not yet know to extend this correspondence to the $d$-inputs $a$. But we can observe that, in the $d$-description, the system $\mathfrak{S}$ does not leave a $d$-state until it is presented with an input symbol at one of the abstract time instants of the system. Retreating to the $c$-description, this means that *any c-state transitions occurring in real time* (i.e. in $c$-time) after the system has assumed the $d$-state $s \in S_d$ but before it has received a new input symbol $a$ *must be restricted to the equivalence class corresponding to s.*

Moreover, *the application of the input symbol a to the system must correspond in the c-description to some activity which will be capable of causing a c-state transition from one equivalence class to another.*

If we think about these two statements for a moment, we will realize that they must be interpreted in the following manner: The $d$-states of $\mathfrak{S}$, corresponding to individual equivalence classes of $c$-states of $\mathfrak{S}$, are simply what Thom [21] has called the *basins* of stable orbits of the $c$-equations of motion of $\mathfrak{S}$, and hence are completely determined by the $c$-kinetic properties of $\mathfrak{S}$. The $d$-inputs $a$ must correspond to *transient forcings* or *controls.*

Let us make this last remark more precise. It is intuitively to be expected that, since the $d$-states of $\mathfrak{S}$ correspond to *equivalence classes* of $c$-states, the $d$-inputs will correspond likewise to equivalence classes of forcings of $\mathfrak{S}$, rather than to individual such forcings. This turns out in fact to be the case. In greater detail, let the basins of the $c$-description be enumerated as $B_1, B_2, \ldots, B_r$ (we may assume that the $c$-description has a finite set of basins, since otherwise an independent finite $d$-description could not exist,

---

[1] "iff" means: "if and only if", a notation originally introduced by P. R. Halmos (Measure Theory, Princeton, N. J. 1964).

contradicting our hypothesis that such a description has been given). Let $\sigma \in B_i$ denote a $c$-state, lying in the $i$-th basin. Let $F(x, t)$ denote the set of all transients which can be applied to the system. We call two transients $G_1, G_2$ *equivalent* if, for all $\sigma \in B_i$, both $G_1(\sigma, t)$, $G_2(\sigma, t) \in B_j$ for all sufficiently large $t$ (i.e. if $G_1, G_2$ cause the same $d$-state transitions). Denote by $F_j(x, t)$ the totality of all such equivalent transients; we shall in general have one such class for every basin $B_j$. Now consider the set

$$\bigcup_{x \in B_i} F_j(x, t).$$

This is the set of all perturbations which, from an initial $c$-state in $B_i$ cause a transition to a $c$-state in $B_j$.

Thus we see that any independent $d$-description of $\mathfrak{S}$ must be related to the $c$-description in the following way: The set of $d$-states corresponds to (perhaps a subset of) the set of basins of the $c$-description; the alphabet of the $d$-description corresponds to the sets of the form

$$\bigcup_{x \in B_i} F_j(x, t);$$

the next-state function of the $d$-description is simply a description of the transition properties of the $c$-description of $\mathfrak{S}$, reduced modulo the equivalence relation $R$.

Because we are dealing with equivalence classes of $c$-states and $c$-forcings, there is no relation between the set of instants of the $d$-description and the real time of the $c$-description. However, we may convert the $d$-description to real time by choosing particular representatives from the equivalence classes in question. By so doing, we obtain a real-time discrete system very much like a sampled-data system; we obtain a different such system for each set of representatives we choose. Moreover, if we try to convert an abstract sequential machine into a real-time system by positing the set of instants *ad libitum*, we see that not every such system will be *realizable* in dynamical terms; this will be the case iff appropriate representatives from the equivalence classes can be chosen. Indeed, it is not difficult to construct real-time sequential systems which cannot in this sense be realized in dynamical terms; this is because, heuristically speaking, flows on state-space manifolds only propagate with a finite velocity; we may posit real-times assignments in discrete systems which would require flows to propagate with an infinite velocity in order to be realized.

The above constructions are intrinsically determined by the dynamical properties of $\mathfrak{S}$, and do not simply exhibit a formal homology between descriptions of $\mathfrak{S}$ (although they exhibit these homologies in a new light). Nor are they like sampled-data descriptions (although we have seen that we can obtain such descriptions from our construction by choosing proper representatives from equivalence classes of forcings). In the next two sections, we shall show that these constructions allow us to interpret the basic concepts of both the continuous and the discrete descriptions, each in terms of the other.

Thus, our constructions satisfy the basic properties which we laid down in the preceding section, and which should be satisfied by a correspondence principle between discrete and continuous descriptions. Finally we may note that, according to these constructions, a dynamical system with a small number of very *deep basins* will have very strongly *digital characteristics*, while on the other hand, a dynamical system with a very large number of shallow basins will not. A system whose state space is partly contoured into a small number of deep basins and partly contoured into a large number of shallow basins will exhibit a hybrid character.

## III. Computability

Let us now sketch how the discrete concept of computability may be related to dynamical concepts in terms of the correspondence principle we have outlined above. Computability refers not to ordinary sequential machines, but to sequential machines with an appropriate feedback (i.e. memory); a sequential machine with such a feedback is generally called a Turing machine. More specifically, a Turing machine is a *pair* of interacting finite sequential machines, one of which serves as a memory of the previous activity of the other. This memory machine further chooses inputs to the first machine, to be supplied at subsequent instants as a function of this activity (cf. [13]). In the language of Turing machines, the first machine corresponds to a tape scanner and printer. The second machine corresponds to a tape scanner and mover, capable of moving the tape in either direction. The tape initially is regarded as having a finite number of input alphabet symbols printed upon it, all the rest of the tape being blank.

Problems concerning computability can always be translated into a halting problem for Turing machines, e.g. deciding whether a given Turing machine with a given input tape will halt its computation after a finite time. What does this mean dynamically? Using the principle we have enunciated above, it is possible to pass from the pair of interacting sequential machines to a pair of interacting dynamical systems whose stability properties are of the correct type. Call these dynamical systems $D_1$, $D_2$, respectively. Let $D_2$ correspond to the tape mover (or "memory" for $D_1$). Any transition induced in $D_1$ because of external forcings (corresponding to the alphabet symbols on the tape) will in general induce a similar transition in $D_2$, while any transition in $D_2$ will constitute a forcing of $D_1$. The halting problem in a Turing machine context obviously translates into the question whether the system $D_2$ will ever come to equilibrium. Conversely, any pair of dynamical systems interacting in this way will give rise to a TURING machine via our correspondence principle. Hence, the halting problem for TURING machines resolves itself into a purely dynamical problem concerning interacting systems of a particular type. All questions regarding computability in digital systems can now be reformulated entirely in dynamical terms (compare ROSEN [11]).

## IV. Continuity and Stability

We have remarked earlier, that because dynamical systems are governed by local laws, the set of states which may be occupied after a small time has elapsed, starting from any initial state under the influence of bounded forces, is itself bounded (like a spreading wave in an optical medium). Thus, in a dynamical system, two states are close if (roughly) they can be reached from one another in a small time interval by means of a (small) perturbation. This observation now allows us to define a measure of closeness in abstract sequential systems. This measure of closeness allows to define a metric, and hence a topology, on the state set of any abstract sequential machine in terms of any of its corresponding dynamical systems. These topologies are naturally closely related to the quotient topologies induced on the set of basins of these dynamical systems. The metric thus induced on the state set of an abstract sequential machine may differ, depending on the particular dynamical system we choose from the class determined by our correspondence principle, but for dynamical purposes these topologies may be regarded as equivalent.

Once the metric is determined, it may be specified for the discrete system without reference to any corresponding dynamical system, purely in terms of intrinsic properties of the machine. Thus, ideas of continuity, and with them, of stability, can be re-interpreted directly into intrinsic properties of sequential systems, as required by a suitable correspondence principle.

## V. Some Consequences

Discrete descriptions possess an algorithmic, constructive character which is not possessed by dynamical systems. If a digital activity can be specified at all, it can be described by the iteration of a (small) number of elementary procedures; or what is the same thing, any sequential machine can be replaced by an equivalent modular net. This explicit constructive character makes it possible to directly characterize the class of all digital systems which possess any particular functional property which we may specify.

Our correspondence principle suggests, therefore, a constructive approach to such problems as the elucidation of biochemical control pathways in differentiating systems. Using digital descriptions, it is possible to characterize the class of (digital) systems which will exhibit any pattern of differentiation which we may specify. In the present state of knowledge, however, it is impossible to use these digital descriptions to make detailed predictions concerning actual differentiating systems, because such predictions (and their verification on actual experimental systems) involve primarily dynamical aspects of system activity. However, by the judicious applications of a correspondence principle relating digital and discrete descriptions, we can translate any particular digital description into dynamical terms; this will permit us to make detailed predictions concerning real systems exhibiting the activities in terms of which the original digital systems were specified, and to design experiments to test these predictions on the real systems. This type of systematic investigation offers, to me, an attractive alternative to the trial-and-error simulations which presently comprise the bulk of our theoretical efforts directed to the understanding of differentiation and multi-cellularity.

In another direction, the principle we have sketched may allow us to directly attack some of the problems raised by STAHL [16] concerning the role of computability in development and evolution. The translation of computability into a purely dynamical concept, as we have suggested above, should enable us to clarify the role of computability in the activity of dynamical systems. It may well turn out [9, 11, 14] that the class of (realizable) dynamical systems is strictly larger than the class of systems which can be replaced by TURING machines, and hence that computability *per se* does not place any restriction on the developmental potentialities of organisms. In any case, however, the dynamical properties of a dynamical system which is capable of computing a nonrecursive function must be very complex topologically; the catastrophic set must be such as to preclude the effective construction of equivalence classes of forcings upon which our transition from continuous to digital descriptions was based; otherwise *a fortiori* the original continuous system would necessarily be computable.

## References

1. ARBIB, M.: J. SIAM Control Ser. A., **3**, 206 (1965).
2. — Automatica **3**, 161 (1966).
3. — and M. BLUM: Proc. Amer. Math. Soc. **16**, 442 (1965).
4. BLUM, M.: Quart. Prog. Rep. RLE, MIT **72**, 237 (1964).
5. GOODWIN, B. C.: Temporal Organization in Cells. New York: Academic Press 1963.
6. HEINMETS, F.: Analysis of Normal and Abnormal Cell Growth. New York: Plenum Press 1967.

7. Kleene, S. C.: In: C. E. Shannon, and J. McCarthy, (Eds.), Automata Studies, p. 1. Princeton University Press 1956.
8. McCulloch, W. S., and W. Pitts: Bull. Math. Biophys. **5**, 115 (1943).
9. Myhill, J.: AF. Technical Documentary Report, Contract AF 30 (602)—3754, 48 (1966).
10. Rashevsky, N.: Mathematical Biophysics, 3rd. edition. New York: Dover Publications 1960.
11. Rosen, R.: Bull. Math. Biophys. **24**, 375 (1962).
12. — J. Theor. Biol. **15**, 282 (1967).
13. — International Reviews of Cytology **23**, 25 (1968).
14. Scarpellini, A.: Z. für Math. Logik **9**, 265 (1963).
15. Simon, Z.: J. Theor. Biol. **8**, 258 (1965).
16. Stahl, W. R.: J. Theor. Biol. **8**, 371 (1965).
17. — and H. E. Goheen: J. Theor. Biol. **5**, 266 (1963).
18. Sugita, M.: J. Theor. Biol. **1**, 415 (1961).
19. — J. Theor. Biol. **4**, 179 (1963).
20. — J. Theor. Biol. **13**, 330 (1966).
21. Thom, R.: Stabilité Structurelle et Morphogenèse. Benjamin, to appear.
22. Zadeh, L., and C. A. Desoer: Linear System Theory — The State Space Approach. New York: McGraw-Hill 1963.

*Discussion*

Walter:

Could you describe again how this correspondence principle applies to transitions from limit cycles (which still may be time-dependent) rather than stable points. How is the proper "new basin" chosen?

Rosen:

The equivalence classes of transients which correspond to the discrete input symbols are defined pointwise on the phase space and the definitions work in all cases, including the one you cite.

Bremermann:

The abstract notion of "computability" is concerned with questions that have arisen in mathematical logic. Turing machines are ideal abstract machines unhampered by limitation of memory and time. Even in this setting there are considerable problems and uncomputable functions, etc.—There are problems that are trivial in this context, but uncomputable in a more restricted sense: for example, the game of chess. The problem to decide optimal moves is a finite one and hence trivial in abstract computability theory. However, the number of cases to be considered is so large that physical computation is impossible, while the solution through insight (that famous players seem to have) has not yet been achieved by computers.—The question to determine the maximal ease of computation of a given task may be undecidable.

Rosen:

I feel you are quite right in believing that computability plays at best a small role in the design of biological systems. As I said in my talk, it seems to me that criteria related to ease of computation are likely to be far more important. However, computability is a typical attribute of discrete systems, one which has no single continuous analog and which has been raised in the literature in connection with biological design. Therefore I thought it appropriate to use computability as an illustration of how the correspondence principle works in transferring ideas from the discrete to the continuous case.

# Interaction of a Digital (Genetic) System with a Continuous (Enzymatic) One in the Cell

M. Sugita

With 2 Figures

*Abstract*

According to certain assumptions concerning the interaction of a digital with a continuous system in the cell—represented by the coupling of an analog computer with a relay circuit—mathematical models for mitosis and DNA replication are proposed.

## Introduction

In several papers [5, 6] we discussed the problem of an applicability of continuous mathematics to cell metabolism, especially arising in the interaction of a continuous system with the so called molecular automaton [4]. Interactions of this kind can be observed e.g. in embryonic development, in cytodifferentiation and in mitosis. We attempt to simulate such an interacting system by applying a hybrid computing device and by taking as an example a mathematical model of cell division.

## Interaction of Continuous and Digital System

A DNA molecule carries genetic (i.e. base sequence of A, T, C, G) as well as temporal information (active or inactive, i.e. readable or not). The mRNA formation may correspond to reading and this digital information may be envisaged as a control tape. The so-called repression or induction may be considered as a feedback from the continuous to the digital system, including DNA. The total system constitutes a molecular automaton, whose internal states can be represented by a set of binary quantities, $g_1, g_2, \cdots$

Let us represent the kinetics of the continuous system by a set of simultaneous first order differential equations

$$\frac{d x_{lk}}{d t} = A_{lk} - B_{lk} \tag{1}$$

where $x_{lk}$ is the quantity of a metabolite $l$ in a compartment $k$, $A_{lk}$ and $B_{lk}$ are in- and outflux of $l$ and are functions of continuous variables $x_{1k}, x_{2k}, \ldots$ as well as of the binary quantities, $g_1, g_2, \ldots$ [5]. Let us consider a certain quantity $y_k$ being the function of $x_{1k}, x_{2k}, \ldots$ having a physiological meaning, and assume that a pulse $g'_k = 1$ is generated, when

$$y_k (x_{1k}, x_{2k}, \ldots) \geqq c_k \tag{2}$$

where $c_k$ is its threshold value, and that this pulse is fed back and affecting the internal state of the automaton. If, for example, $x_{1k}$ and $x_{2k}$ are the quantities of co-repressor and apo-repressor, then $y_k$ may be related to the repressor. In the relation (3.3) of a preceding paper [6] the symbols $r$ and $f(r)$ were used instead of $y_k$ and $g'_k$. Thus, the internal state of the automaton may be determined by the following logical function

$$g_k = G_k (g'_1, g'_2, \ldots; p_1, p_2, \ldots) \tag{3}$$

where $p_1, p_2, \ldots$ are exogenous digital inputs, e.g. of fertilization, virus infection, radiation induced mutation. The block diagram of such an interacting system is given elsewhere [5].

## Application of our Method to Cell Division

Our mathematical method can effectively be applied to some biological problems like development, cytodifferentiation, etc. During cell cycle some metabolites regulating mitosis or DNA duplication must fluctuate. Goodwin [1] transferred the ecological relation of prey—predator to oscillating metabolites. For this purpose, at least two kinds of metabolites, instead of one, are required. Let $x$ and $u$ be the metabolites regulating mitosis and DNA replication respectively, then $y$ and $v$ of the following kinetic equations may correspond to the predator. For simplicity let us assume the following form

$$\frac{dx}{dt} = \alpha (x,y,u,v) g_y - k_1 x, \qquad \frac{du}{dt} = \gamma (x,y,u,v) g_v - k_3 u$$

$$\tag{4}$$

$$\frac{dy}{dt} = \beta (x,y,u,v) g_M - k_2 y, \qquad \frac{dv}{dt} = \delta (x,y,u,v) g_D - k_4 v$$

where

$$
\begin{array}{llll}
g_M = 1 & \text{when } x \geqq x_c, & g_D = 1 & \text{when } u \geqq u_c \\
g_M = 0 & \text{when } x < x_c, & g_D = 0 & \text{when } u < u_c \\
g_y = 0 & \text{when } y \geqq y_c, & g_v = 0 & \text{when } v \geqq v_c \\
g_y = 1 & \text{when } y < y_c, & g_v = 1 & \text{when } v < v_c
\end{array}
\tag{5}
$$

$g_M = 1$ and $g_D = 1$ are signals of mitosis and DNA replication; $x_c, y_c$ etc. are threshold values. $\alpha$, $\beta$, $\gamma$ and $\delta$ are functions of $x, y, u,$ and $v$. However, we are considering the correlation between mitotic system and the system of DNA only qualitatively; therefore, we will simply assume that

$$\alpha\,(x,y,u,v) = \beta\,(x,y,u,v) = \gamma\,(x,y,u,v) = \delta\,(x,y,u,v) = 1 \qquad (6)$$

$k_1$ and $k_3$ are assumed to be constant (decay constant). For $k_2$ and $k_4$ let us assume the following step functions

$$k_2 = k_{20} + (g_i \cap \bar{g}_D \cup g_M)\,k_{21}$$
$$k_4 = k_{40} + (\bar{g}_i \cap \bar{g}_M \cup g_D)\,k_{41} \qquad (7)$$

where

$$g_i\,(t) = (g_i\,(t-\tau) \cup g_D) \cap \bar{g}_M \qquad (8)$$

and $\bar{g}_i$ or $\bar{g}_M$ is the negation of the respective quantities; $\cap$ and $\cup$ are the notations for logical product and sum. Yčas, Sugita and Bensam [7] have suggested the state of inactive DNA, which is one half of the DNA just replicated and inactivated during $G_2$ period. $g_i$ may correspond to such an inactive state.

The step function of (7) suggests "isozymes", i.e. enzymes catalyzing a certain metabolic pathway $i \to j$, however, being of several molecular origin and, thus, regulated by different genetic loci. The rate constant $k_{ij}$ can be represented by

$$k_{ij} = \sum_r (k_{ij})_r\, g_r \qquad (9)$$

where $g_r = 1$ shows that the locus $r$ is an open or readable state and $(k_{ij})_r$ is the rate constant catalyzed by the enzyme corresponding to $r$.

According to (7) $k_2$ and $k_4$ have the values shown by Fig. 1, in which the values of $g_M$, $g_y$, $g_D$ and $g_v$ are also listed; $g_M = 1$ during M period and

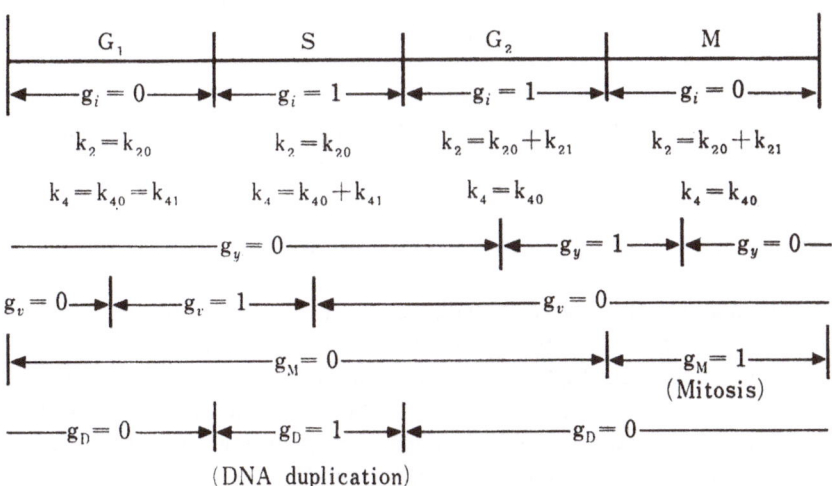

Fig. 1. Values of decay constants and binary parameters during cell cycle

$g_D = 1$ during S period. $k_2 = k_{20}$ during $G_1$ and S, and $k_2 = k_{20} + k_{21}$ during $G_2$ and M. If we assume that $k_{21} \gg k_{20}$, then $y$ (predator) decays strongly in $G_2$ and $x$ (prey) increases, i.e. mitosis is prepared in the period $G_2$. In the same way $k_4 = k_{40}$ during $G_2$ and M and $k_4 = k_{40} + k_{41}$ during $G_1$ and S. The possibility of DNA replication may be prepared during $G_1$ period.

Fig. 2 is a result of our hybrid computation. The constant values $k_1$, $k_3$, $k_{20}$, $k_{21}$, $k_{40}$, $k_{41}$, $x_c$, $y_c$, $u_c$ and $v_c$ are selected so as to present reasonable values for the interval of the periods $G_1$, S, $G_2$ and M. Looking at the curves we find the period M to be longer so that other sets of constant values have to be looked for.

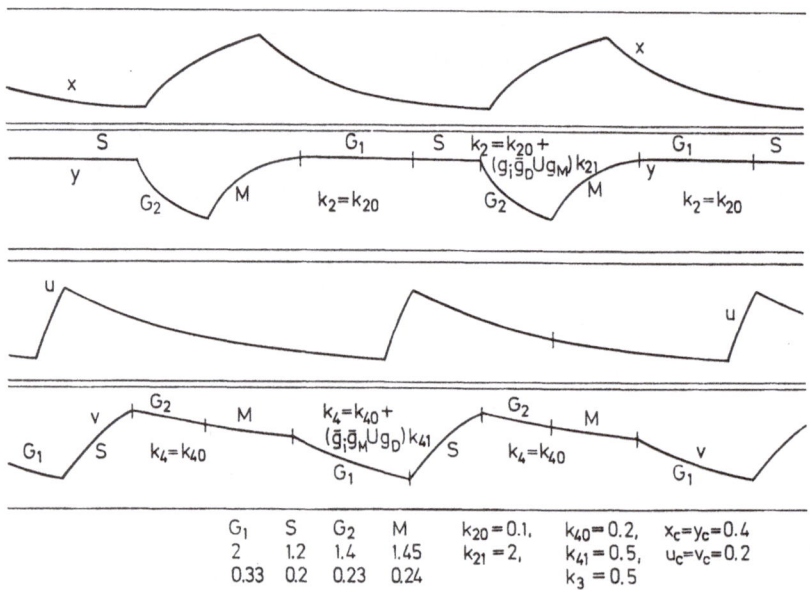

| $G_1$ | S | $G_2$ | M | $k_{20} = 0.1,$ | $k_{40} = 0.2,$ | $x_c = y_c = 0.4$ |
|-------|-----|-------|------|----------------|-----------------|-------------------|
| 2 | 1.2 | 1.4 | 1.45 | $k_{21} = 2,$ | $k_{41} = 0.5,$ | $u_c = v_c = 0.2$ |
| 0.33 | 0.2 | 0.23 | 0.24 | | $k_3 = 0.5$ | |

Fig. 2. A result of computation, assuming the following parameter values
$k_{20} = 0.1$, $k_{21} = 2$, $k_{40} = 0.2$, $k_{41} = 0.5$, $k_3 = 0.5$, $x_c = y_c = 0.4$, $u_c = v_c = 0.2$

The relative values of the periods are:

|  | $G_1$ | S | $G_2$ | M |
|------|-------|-----|-------|------|
|  | 2 | 1.2 | 1.4 | 1.45 |
| or | 0.33 | 0.2 | 0.23 | 0.24 |

The mathematical model considered here is only a preliminary one and the molecular nature of the signals of mitosis and DNA duplication must be clarified. Heinmets [2] has proposed an interesting but rather complicated model for cell division or DNA duplication. Therefore, it is not yet clear whether the method we described above is nothing but a simplification of Heinmets'. Quantitatively the assumption of (6) is also not accurate; that means that the validity of the selected constants is also limited.

*Acknowledgment*

Through kindness of the late Professor Jinbo the analogue computer of the Meiji University was used for this study. In the computer work the author was assisted by Dr. Ogawa and other members of the University. I wish to express to them my sincere gratitude for their cooperation.

## References

1. Goodwin, B. C.: Temporal organization in cells. New York: Academic Press 1963.
2. Heinmets, F.: Analysis of normal and abnormal cell growth. New York: Plenum Press 1966.
3. Sugita, M.: J. Theor. Biol. **1**, 415 (1961).
4. — J. Theor. Biol. **4**, 179 (1963).
5. — J. Theor. Biol. **13**, 330 (1966a).
6. — Helgoländer wiss. Meeresunters. **14**, 78 (1966b).
7. Yčas, M., M. Sugita and A. Bensam: J. Theor. Biol. **9**, 444 (1965).

### *Discussion*

Kacser:

I would like to ask Dr. Sugita whether he uses the switch in his models for ease of representation or whether he wants to make a point that there are in fact discontinuities in the system.

Sugita:

You want to know whether switching is only a computational convenience or whether it has some physical meaning for the model. In the analogy between the flux of chemical processes and the electrical current, I am considering the existence or non-existence of an enzyme as a correspondence to "make" or "break" of switches. Even though the quantity of an enzyme may vary continuously, only a step function approximation may lead to its complete efficiency. Thus, $(E) = ag + b\bar{g}$, that is, when $g = 1$, $(E) = a$; when $g = 0$, $(\bar{g} = 1)$, $(E) = b$. The binary quantity $g$ may be the state of an operator gene; when this one is repressed $(g = 0)$, the enzyme formation becomes $b$, i.e. very small.

Kacser:

In terms of observables we always find a small amount of enzyme. We have very little evidence that there is a zero; one situation in the organism.

Griffith:

In relation to Dr. Kacser's remark, I should like to point out that it is possible to have a simple cellular regulatory mechanism which admits of two stable states of activity, with an enzyme present in the one and absent in the other. Take the equations

$$\dot{M} = \frac{3S^2}{1 + 2S^2} - M \qquad \dot{S} = M - S$$

representing a cooperative inductive effect of a protein $S$ on the formation of its messenger $M$. These equations have stable stationary solutions $S = M = 0$ and $S = M = 1$ and an unstable solution $S = M = 1/2$.

Sugita:

I thank you for your valuable suggestions which are useful to support our own idea (or intuition) that in kinetics two possible states should be taken into consideration.

# A Model for Sustained Rhythmic Binary Logic in Biochemical Systems

CH. WALTER [1]

With 2 Figures

*Abstract*

A model chemical system is discussed as a possible basis for sustained rhythmic binary logic in biological systems. The model includes a "branch point" enzyme and two metabolic pathways. One pathway is a typical negative feedback scheme of the Yates-Pardee type; the other pathway involves a series of enzymes capable of displaying sigmoid relationships between activity and substrate concentration. Under the conditions described the concentration of one substance in the model is either very low ("off") or, except for a short interval, very high ("on"). The time of the "off" to "on" or "on" to "off" transition is small compared to the total period of each cycle. The "on"—"off" cycle is sustained for at least 200 periods.

## I. Introduction

Experimental results in biology often suggest the existence of rhythmic events which seem to either occur or not occur. Since the fundamental cause of all biological events stems from the constituent biochemical reactions, it seems likely that the cause of sustained rhythms of such biological "on-off" switches must also result from chemical reactions.

For a chemical model to serve as the source of sustained rhythmic events it is necessary for the model to be capable of sustained oscillation and to exhibit binary or "on-off" character. However, in most concepts of chemical reactions the process is considered to be essentially continuous. Furthermore, it is well known that chemical oscillations around an equilibrium state are thermodynamically impossible. These facts have served as the basis of criticism of suggestions that chemical systems can exhibit sustained oscillations or serve as the source of binary logic.

There are several reasons why this criticism is not justified in the case of chemical reactions in biological systems. First, experimental results clearly illustrate the existence of both oscillatory events and binary logic in biological systems. Second, the fact that chemical reactions are essentially

[1] Public Health Service Career Development Awardee, National Institutes of Health award number K 3-GM-11,237.

continuous does not preclude the possibility that in complex, interrelated chemical systems some function of the concentration of a particular chemical species is very nearly a step-function. Third, one of the important results obtained from the relatively new thermodynamics of "irreversible processes" (e.g. open systems) is that rotating chemical reactions are thermodynamically possible around non-equilibrium stationary states. Clearly, since biological systems are open and contain many complex, interrelated multi-enzyme pathways, one cannot exclude that rhythmic binary events are the result of constituent biochemical reactions.

## II. The Model

The model summarized in scheme 1 consists of two pathways for the metabolism of a substance $S_0$ which is supplied from an arbitrary source.

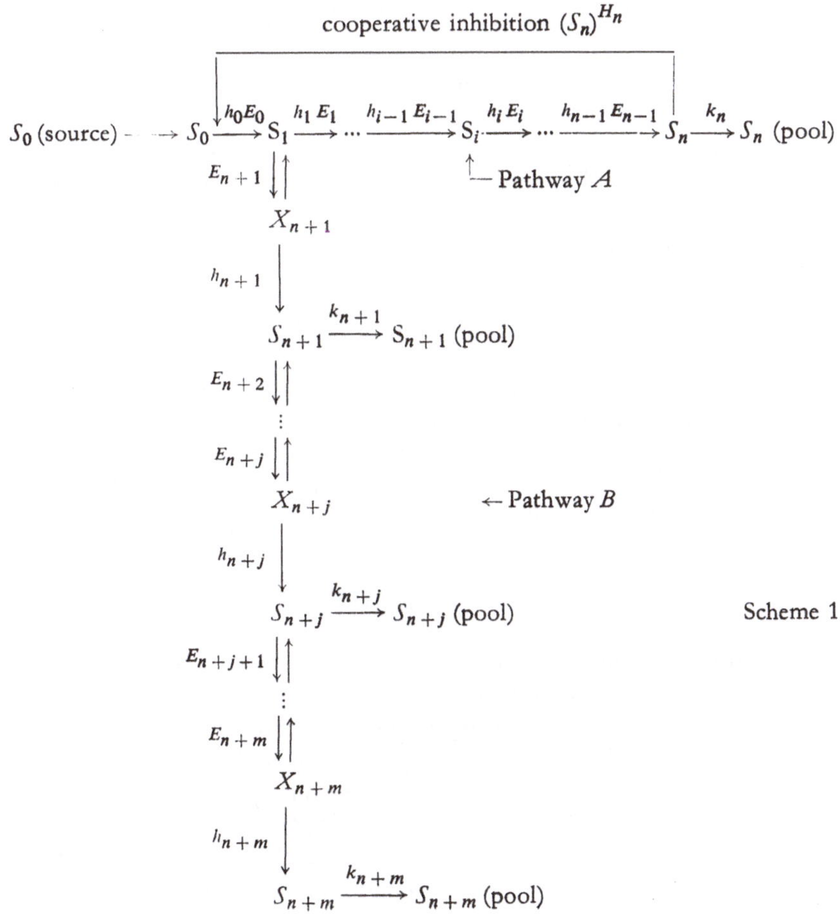

Scheme 1

$S_1$ is formed from $S_0$ by the catalytic action of a "branch point" enzyme, $E_0$. In pathway $A$ $S_1$ is metabolized via a series of consecutive reactions wherein the formation of substance $S_i$ from $S_{i-1}$ is catalyzed by enzyme $E_{i-1}$, and conversion to $S_{i+1}$ is catalyzed by enzyme $E_i$. The final substance in pathway $A$, $S_n$, exits from the system with rate constant $k_n$. $S_0$ is maintained constant from the external source and $E_0$ is inhibited cooperatively by $S_n$. Pathway $A$ is a typical negative feedback scheme of the Yates-Pardee [4] type.

In pathway $B$ $S_1$ is metabolized via a series of different consecutive reactions wherein formation of substance $S_{n+1}$ is catalyzed by enzyme $E_{n+1}$, and conversion to $S_{n+j}$ is catalyzed by enzyme $E_{n+j}$. Several of the enzymes in pathway $B$ are assumed to display a sigmoid relationship between activity ($F_{n+j}$) and substrate concentration (See equation 1). In each case it is

$$F_{n+j} = \frac{S_{n+j}\,\beta_{n+j}\,(1+S_{n+j}\,\beta_{n+j})^{H_{n+j}-1} + L_{n+j}\,S_{n+j}\,(1+S_{n+j})^{H_{n+j}-1}}{L_{n+j}\,(1+S_{n+j})^{H_{n+j}} + (1+S_{n+j}\,\beta_{n+j})^{H_{n+j}}} \qquad (1)$$

assumed that $S_{n+j}$ and $E_{n+j}$ are in quasi-equilibrium such that the rate of the $n+j$th step is determined by the fraction of enzyme sites occupied by substrate; furthermore each $S_{n+j}$ is assumed to exit from the system with rate constant $k_{n+j}$. Although the results that follow are independent of the mechanistic basis of the sigmoid relationship, an allosteric model of the type proposed by Monod, Wyman and Changeux [1] fits the needs of this system.

## III. The Question

The question to which we now address ourselves is: Given the general model described above, can the concentration of $S_{n+m}$ experience a sustained rhythmic sharp increase from nearly zero ("off") to the asymptote defined by equation 1 ("on") (and vice-versa) during an interval when $S_1$ changes by a relatively small per cent?

## IV. The Answer

An affirmative answer to the question posed above requires that one of the pathways of scheme 1 is capable of sustained oscillation and the other pathway exhibits essentially binary character; furthermore it is necessary that pathways $A$ and $B$ together are capable of operating in such a manner that the sustained oscillation is retained in the binary character.

It has recently been shown by Morales and McKay [2] that pathway $A$ is capable of exhibiting sustained oscillations of the concentrations of the $S_i$. Thus pathway $A$ fulfills the first condition for an affirmative answer.

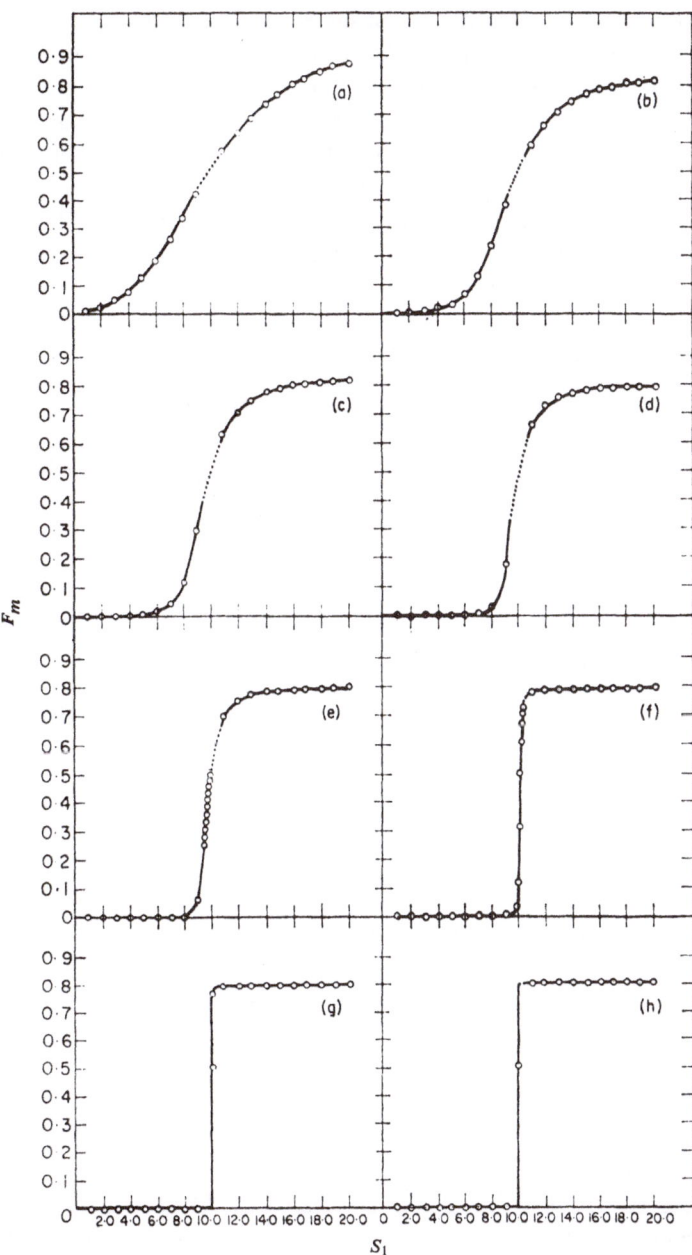

Fig. 1. Activity of the $n + m$th enzyme for $n = 0$ in scheme 1 (pathway $B$), $F_m$ is plotted versus $S_1$ for (a) $m = 1$; (b) $m = 2$; (c) $m = 3$; (d) $m = 4$; (e) $m = 5$; (f) $m = 10$; (g) $m = 15$; (h) $m = 20$. The parameters defined in equation 1 are identical to those used by MONOD, WYMAN, and CHANGEUX [1]: all $L_{n+j} = 10^{-4}$; all $\beta_{n+j} = 10^{-2}$; all $H_{n+j} = 4$,

That pathway $B$ is capable of exhibiting binary character can be seen in Fig. 1. Although the functions in Fig. 1 are continuous, it is not difficult to see that the relationships at larger $m$ could be accurately represented as a binary variable. A more complete discussion of the binary character of pathway $B$ is available elsewhere [3]. Finally, it can be seen from Fig. 2 that pathways $A$ and $B$ in scheme 1 can in fact operate in such a manner that the sustained rhythmic behavior is retained in the binary function. Thus $S_{n+m}$ ($n = 5$, $m = 4$) in scheme 1 exhibits a sustained rhythm of being very nearly

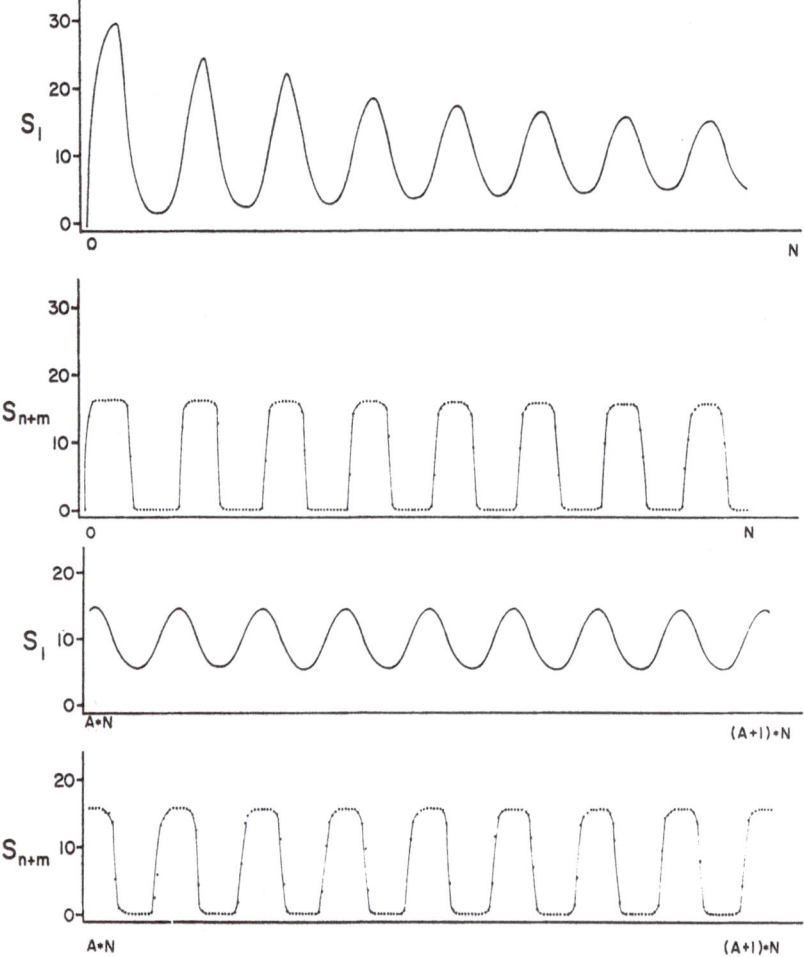

Fig. 2. Concentrations of $S_1$ and $S_{n+m}$ ($n = 5$, $m = 4$) in scheme 1 are plotted versus an arbitrary time scale. The $L_{n+j}$, $\beta_{n+j}$ and $H_{n+j}$ are identical to those used in Fig. 1 and $H_n$ (the stoichiometry of the feedback inhibition by $S_n$) is also 4. All $b_i$ ($i = 1$, $n - 1$) are 0.4; $k_n = .1795$; $b_{n+1} = 200$; $b_{n+2} = 400$; $b_{n+3} = 800$; $b_{n+4} = 1600$; $k_{n+1} = 10$; $k_{n+2} = 20$; $k_{n+3} = 40$; $k_{n+4} = 80$; $b_0 S_0 = 10.56$; $K$ (the feedback inhibition constant) is $1.0 \times 10^{-5}$; $A = 3$; $N = 200$

zero ("off") or very nearly 16 ("on"). The transformation from "off" to "on" or vice-versa occurs in a time interval that is short compared to the period of the rhythm.

At present it is a matter of speculation whether and to what extent such rhythmic step-functions play a role in biological processes. A possible role merits consideration however, since the main components of the model, enzymes which exhibit a sigmoid relationship between activity and substrate concentration and multi-enzyme systems which experience feedback product inhibition, are now well known. It is not unreasonable to expect that if a model like the one described above is useful, it could be readily provided by natural selection.

*Acknowledgement*

This work was supported in part by the National Science Foundation, grant GB 3625.

## References

1. MONOD, J., J. WYMAN and J. CHANGEUX: J. Mol. Biol. **12**, 88 (1965).
2. MORALES, M., and D. McKAY: Biophys. J. **7**, 441 (1967).
3. WALTER, CH., R. PARKER and M. YČAS: J. Theor. Biol. **15**, 208 (1967).
4. YATES, R. A., and A. B. PARDEE: J. Biol. Chem. **221**, 757 (1956).

*Discussion*

BETZ:

You certainly know the model proposed by HIGGINS to explain glycolytic oscillations which is based on PFK being inhibited by its substrate ATP and activated by the product FDP and by AMP, too, and, most efficiently, by F-6-P. Would this model be able to produce such step functions as yours does?

WALTER:

Any system describable by the VOLTERRA equations possesses the potential for oscillation. Although these models are considerably more complex than the YATES-PARDEE scheme (they require positive feedback) they could be used in place of pathway $A$ in scheme I. In fact, the source of the oscillatory character is irrelevant (WALTER, C. R. PARKER and M. YČAS: J. Theor. Biol. **15**, 208 (1967)).

BETZ:

In your model the existence of a net flux is not a sufficient condition for working as an oscillator or does it need some really irreversible enzymatic step?

WALTER:

If the $n-1$ chemical reactions of pathway $A$ were at equilibrium I would not expect the existence of a net flux to result in an oscillation. However, I do not think the oscillations depend on literal irreversibility of the chemical reactions.

GRIFFITH:

Have you any rigorous mathematical proof of the existence of sustained oscillations in any case? I have been investigating the equations

$$\dot{S}_1 = \frac{a_1}{1 + \beta S_n} - y_1 S_1$$

$$\dot{S}_i = a_i S_{i-1} - y_i S_i \qquad i = 2, \dots, n$$

Phase plane methods (BENDIXSON criterion) show the non-existence for $n = 2$, but although we have not obtained any undamped oscillations for $n = 3$, I have not proved rigorously that they cannot occur.

WALTER:

The equations for pathway $A$ are:

$$\dot{S}_1 = \frac{b_0 E_0 S_0}{1 + K (S_n)^{H_n}} - b_1 S_1$$

$$\vdots$$

$$\dot{S}_i = b_{i-1} S_{i-1} - b_i S_i \qquad i = 2, n-1$$

$$\vdots$$

$$\dot{S}_n = b_{n-1} S_{n-1} - k_n S_n$$

Your indication that for $n = 3$ and $H_n = 1$ oscillations may not occur is not surprising. I have not observed oscillations unless $H_n > 1$. I am attempting to determine analytically the necessary conditions for oscillation of these equations, especially with regard to minimum values for $H_n$, but at present these attemps are incomplete. — *Added at proof:* I now have vigorous proof that for $n = 2$ and $H_n \geq 0$, chemical systems described by the above equations cannot experience sustained concentration oscillations. For $n = 3$ and $n = 4$ the sustained nature of the oscillations suggested in [2] appears to arise only in analog computer simulations of the equations. However, since the source of the oscillator is irrelevant, the basic idea presented in this paper can be realized by altering pathway $A$ slightly.

# Some General Ideas on Deterministic and Stochastic Models of Biological Systems

A. F. Bartholomay[1,2]

*Abstract*

The paper discusses the formal expression in biomathematical terms of a general approach to the construction and analysis of mathematical models in biological systems. It points in the direction of a possible axomatization of such constructs and their associated methodologies.—The underlying general notion or "primitive concept" of biosystem $\mathfrak{S}$ $(C, S, E, F, t)$ is assumed to include: composition $C$, structure $S$, environment $E$, biological activities or functions $F$, and time relations $t$. In relation to $\mathfrak{S}$ a mathematical model is conceived as an associated mathematical construct $\mathfrak{M}$ such that there exists a correspondence $\mu\colon \mathfrak{S} \to \mathfrak{M}$ 1. between the biological objects in $\mathfrak{S}$ and the mathematical symbols in $\mathfrak{M}$ and 2. between the biological activities in $\mathfrak{S}$ and the allowable, biologically interpretable mathematical operations in the mathematical domains containing $\mathfrak{M}$.—Deterministic models are obtained by associating time-dependent, algebraic variables x with quantifiable aspects of $\mathfrak{S}$; and stochastic models by associating random variables x, usually defined with reference to a basic probability space corresponding to a discrete biological event. Stochastic and deterministic models for the kinetics of unimolecular processes are discussed in some detail as examples, and compared with each other.

## I. Introduction

The purpose of the mathematical biologist or biomathematician in constructing a mathematical model is the provision of a deductive or numerical overstructure defined to the biological process in terms of which the knowledge of the systems may be deepened and extended by taking advantage of (a) the formal deductive mathematical machinery inherent in the mathematical representation, and (b) the quantitative or statistical frame of reference which it provides for experiments.

The present paper begins with a general biomathematical-or molecular-set theoretic [4, 8, 9] (v.i.) formulation of the notion of biological

[1] The work was supported partially by Biomathematics Research Grant No. GM-10002 and Training Grant No. 5-Tl-GM-984 from the National Institutes of Health, Institute of General Medical Sciences, U.S. Department of Health, Education and Welfare; and by the Howard Hughes Medical Institute.

[2] Paper read at the Symposium by N. Arley.

system as a means of promoting the construction process of mathematical models and emphasizing their underlying biological bases. Its overall aims are: 1. the suggestion of a general formulation of the notion of mathematical model of a biological system, with a view toward further biomathematical development along axiomatic lines; and 2. an elucidation of the common features and associated scientific methodologies of mathematical models which, in the absence of specific operational algorithms for such constructions, may serve as a useful guide. Within this general framework a number of different approaches to the construction, and a variety of different classes of such models are discernible. The latter portion of this paper will confine itself to an elaboration of the two contrasting classes of deterministic and stochastic models.

The concepts from molecular-set theory transfer in obvious fashion to a large number of biological transformations by expanding the context of molecular sets $A$ to include appropriate sets of biological interacting units and material transformations $T$ governing the interconversions. In the next section, in fact, some of these extensions of molecular-set theory to obtain a biomathematical description of the notion of biological system are subsumed.

## II. Biological System or „Biosystem" $\mathfrak{S}$ $(S, C, E, F, t)$

Informally speaking, by "biological system" will be understood organized, dynamic collections of sets of material objects—either molecules, submolecular particles, cells, tissues, organs, organisms; or populations or mixtures of these units—which stand in certain physicochemical, mechanical, morphological or topological, and temporal relations to each other and to their environment, and with which are associated certain specific functions or sets of biological activities, evolving and ordered in some "natural" biological sense according to their relative times of occurence on interrelated time scales. The biological activities are associated with material transformations of the biological objects or with information processed by the system, which is assumed to be imbedded in an environment containing other systems with which it may communicate in this functional sense. The term biological system is thus meant to include any of the so-called systems studied in biology such as the circulatory system, the nervous system, the endocrine system, the cardiovascular system, the protein synthetic system, systems of growing and competing organisms, etc.

At the outset, it should be stressed that certain practical considerations, limitations and consequences of the present interpretation, and indeed of any notion of biological system itself, are well in mind. For example, any description or specification of a "biological system" is a relative or arbitrary matter depending, for example, on the current state of biological and personal knowledge; on what is regarded as the major or "target" level of biological abstraction being studied; and, where mathematical formulation is an aim, on the convenient selection of a minimum number of variables consistent with the biological integrity of the system and ultimately with mathematical tractability. These matters will be elaborated further in the sequel.

Along slightly more formal lines, partly in anticipation of the formation, ultimately, of an appropriate and comprehensive axiomatic framework, in this paper a biosystem $\mathfrak{S}$ $(C, S, E, F, t)$ will be taken as a kind of *primitive concept* entailing five major components, which are describable, using the biomathematical-set theoretic language and point of view, as follows:

a. *Composition C.* At a given instant the composition $C$ of the biological system is expressible as the biomathematical-set theoretic sum of subsets $C_1, C_2, \ldots, C_n$:

$$C = C_1 \cup C_2 \cup \ldots \cup C_n \tag{1}$$

where each $C_i$ is made up of a collection of well-defined chemical or biological elements or units, each possessing identical, or at least similar, biological and physicochemical characteristics. This partitioning is arbitrary in the sense that for purposes of a particular kind of study of system $\mathfrak{S}$ the component sets may include either macroscopic, microscopic, or submicroscopic elements or all of these. In other cases limited knowledge of $\mathfrak{S}$ may be the determining factor for the numbers and kinds of such compositional components. Thus, in population studies the $C$'s might be different interacting populations of whole organisms. In pharmacological studies using *in vivo* tracer kinetic techniques, the components might be "biological black boxes" containing whole "subsystems" of individual organisms. In Mendelian genetical studies the system might be specified down to the level of chromosomes or genes. In protein synthetic systems the $C$'s might be biological macromolecules, metabolites, ribosomal particles, etc.

It is, of course, understood that some or all of these components may be interconverting so that some may vanish or new ones arise, depending on the controlling conditions and "states" of the system at different times. For these reasons, it is more accurate to enlarge the set $C$ given in eq. (1) in the biomathematical mixed-Cartesian Product sense [4, 8] to include both a temporal set $I$ of biologically meaningful time instants $t$, as well as a collection $P^3$ of parametric sets $P_1, P_2, \ldots, P_m$ each consisting of different values of pressure, temperature, pH, etc., all of which would be included in the more extensive biomathematical-set theoretic statement format:

$$C = (C_1 \cup C_2 \cup \ldots \cup C_n) \times (P_1 \times P_2 \times \ldots \times P_m) \times I \tag{2}$$

---

[3] See also section 2 e.

b. *Structure S, or Form of a Biological System.* The term "structure" of a biological system may refer to a straightforward qualitative or quantitative description of the directly or indirectly visible parts or whole of the system. Or indeed, it may be simply a relational or topological-like statement of the interrelationship and organization of the component parts and units of composition $C$, or even of the "black boxes" (when the system is not geometrically or biologically resolvable into definite units of composition). In any case, some indication of the relation of the parts to the whole is required, since the "structure", or "form", or "organization" of a system we assume has to do with the way in which the compositional units are related each with the other, and with the media by which communication of all kinds between them and with other systems in the environment (v.i.) is accomplished and maintained. Moreover, the basic compositional decomposition (eq. 1 or 2) together with the other considerations mentioned would indicate that the structure $S$ may also be conceived as the sum of many substructures $S_1, S_2, \ldots, S_p$ subject to temporal changes and to sets of conditions $P$:

$$S = (S_1 \cup S_2 \cup \ldots \cup S_p) \times (P_1 \times \ldots \times P_m) \times I \qquad (3)$$

Just as the compositional aspect of $\mathfrak{S}$ is meant to be an arbitrary, or relative matter, so also is its structural $S$-aspect. For example, corresponding to different biological levels of interest in protein synthesis will be different structural conceptualizations and representations. Thus, for investigating simply the correlations of the most macroscopic subunits of protein molecules with those of the nucleic acids, the system $\mathfrak{S}$ has been conceived as consisting of amino acids and nucleotides, linearly arranged as "chains" in the corresponding total molecules. This arises in the "coding" problems of one-dimensional, ordinary information-theoretic variety. On the other hand, certain physicochemical studies would require that the typical helical conformations of the molecules be considered as the basic structures of the molecules of the system. Other physicochemical studies might require that the structure of molecules be specified down to known detailed stereochemical configurations of the molecules. Finally, for describing the genetic control process of protein synthesis the compositional elements might be simply the unanalyzed different amino acids, protein molecules, DNA and RNA molecules, ribosomal particles, etc. And the structure in which we would be interested would be carried by the schematic or symbolic diagrams, such as those of Jacob and Monod, which have to do with the overall topological description of the different biochemical pathways involved in the process.

At the more macroscopie levels of biological interest, in describing the human circulatory system, for example, the special requirements of a particular investigation might be met by taking for the $C'$s the various cells of the body and for the $S$'s certain geometric or analytic idealizations, obtained by analogy with simple physical or engineering systems, to represent the lungs, heart, arteries, capillaries, veins, etc.

c. *Environment E.* Conceptually, a distinction or separation between a biological system and its "environment", because of the open nature of biological processes and their activities, seems neither desirable nor possible. For this reason, we include it alongside the *composition* and *structure* of a system, as an additional component of a biosystem $\mathfrak{S}$. It is particularly a relative and arbitrary matter, depending on the degree to which it is biologically interpretable and reasonable in a given situation, to distinguish between, in effect, a target "subsystem" and a larger system which contains it. Perhaps the biological situation can be compared to certain mathematical situations. Thus, topology may be said to be concerned with global properties of surfaces (or "topological systems") of two kinds: 1. *intrinsic* topological properties entirely confined to a given surface and 2. *extrinsic* or imbedding properties that depend on the relationship of that surface to its spatial environment. Many biological systemic properties by analogy might be called "extrinsic" and hence dependent on their "environment" in any sense of the term.

In general, it seems that one could only safely say that included in environment $E$ of a system $\mathfrak{S}$ are all other biological, chemical and physical systems that can in some sense be regarded as secondary or supportive to those aspects of all living matter which have been arbitrarily included in the enumeration of parts $C$ und $S$ of a biosystem $\mathfrak{S}$. Thus for greatest clarity we might write out the notation $\mathfrak{S}$ $(C, S, E, F, t)$ for a biosystem more completely as $\mathfrak{S}$ $(C, S, F, t: E$ $(C', S', F', t'))$ to indicate that the total biosystem $\mathfrak{S}$ includes the "target system" $(C, S, F, t)$ and its associated environmental system $E$ $(C', S', F', t')$. Further remarks on the environment are included in section e.

d. *The Functions F, or Activities of a Biological System* $\mathfrak{S}$. It is also considered that associated with a given biological system $\mathfrak{S}$ is a set of biological, physical or chemical activities $F$ which, in molecular- or biomathematical-set theoretic terms, are representable as *transformations* $T$ with domains and ranges in the $C$-, $S$-, or $E$-sets of $\mathfrak{S}$. The set $F$ of such transformations includes simple and complex chemical and biochemical reactions, metabolic activities, the events of protein synthesis, cellular and other growth and development processes, struggles for existence between different populations of organisms, epidemic processes, etc.—in short, all the dynamic aspects of a system taking place within and between its various components and associated also with interactions and exchanges between that system and its environment.

As noted in previous papers [4, 8, 9], it would appear possible to infer an algebra of such functions and activities which translates directly into appropriate mathematical models.

e. *Time Scales t.* In connection with the compositional, structural and functional aspects of a biosystem, explicit attention has already been called to the importance of the time factor in defining these components of the system meaningfully. In all of these cases time is featured as the independent variable of any mathematical representation, in the following sense: given any measurable quantity connected with any component or activity of the system, it amounts to some kind of a function of time. Thus, anticipating the notion of mathematical model, if a deterministic approach is chosen, such a quantity becomes an ordinary analytic function of time; in the stochastic approach, it is treated as a random function of time as the independent variable. And in fact, in all models an ultimate mathematical aim is to deduce the exact nature of the variation of the variable with time as independent variable.

The importance of time in an elaboration of the environmental component has been accentuated particularly by Waddington [23] and taken up again somewhat later by Goodwin [14]. In fact, in their studies, time becomes the basis for classifying the different biological activities (v.i.) in such a way as to delineate between a system and its environment. Thus, three different magnitudes of time stretches or scales are differentiated with respect to which different orders of biological activities or functions are defined. The "shortest" phenomena are the biochemical and metabolic processes; so-called "epigenetic" or developmental processes taking place over much longer time stretches are next in the expanding order of classification; and finally the "longest" phenomena are the genetic and evolutionary processes. From this point of view then, given the appropriate time-scales over which the various activities of the target system transpire, related activities occurring over adjacent time-scales would lead inversely to the reconstruction of the collection of systems to be considered in the environment. The involvement of systems in the environment chosen in this way in a mathematical model representation of the biosystem could in many cases be limited simply to an inclusion in the set $P$ of parameters associated with the system. In this sense, the environmental component $E$ may be thought of as including the parametric sets $P$.

The preceding formulation in biomathematical terms of the concept of biosystem $\mathfrak{S}$ serves to call attention to specific aspects of the "domain" over which a mathematical model will be defined. It is therefore an integral part to the total conception of mathematical model, just as the precise specification of a probability space must precede the construction of a probability measure on a given set of random events. In fact this analogy is worth re-emphasizing for, as the sequel shows, the construction of a mathematical model of a biosystem is similar, conceptually, to imposing a "measure" on a well-specified space or collection of sets. In this case, the underlying space refers to some or all of the biomathematically described components indicated.

## III. The General Notion of Mathematical Model 𝔐
### of a Biological System 𝔖

Corresponding to the biomathematical formulation of biosystem 𝔖 just given, by the "mathematical model" 𝔐 of a biosystem 𝔖 will be meant an associated mathematical construct or structure such that: 1. the basic mathematical elements of that structure are mathematical symbols made to correspond with pertinent elements or properties of the biosystem 𝔖; and 2. the mathematical operations on these symbols, permitted by the identification of this construct with the appropriate domains of mathematics, accompany and formalize in some meaningful fashion the underlying biosystemic relations and transformations. It follows that in setting up a mathematical model, unless mere symbolic description is the sought-for end itself, it is necessary to enlarge the primary correspondence into a workable, deductive mathematical form. The utilization of such a representation will be discussed in greater detail presently, but first the concept itself will be examined more closely.

The relation between the model and the original system may be visualized, at first, simply as a mapping or association $\mu$ between the objects or elements in the biological domain defined by 𝔖 and the symbolic elements of 𝔐 in the mathematical domain:

$$\mu: 𝔖 \to 𝔐 \tag{4}$$

But, as already noted, more is required of $\mu$ than the fact that it establishes a correspondence or translating device between a biological real system and an abstract mathematical system: it must also set up a correspondence between inherent biological interrelationships and the formal mathematical operations allowed by the mathematical domains containing the symbolic system 𝔐. This kind of association may be compared to, say, a "homomorphism" $\mu: G \to H$ of an abstract mathematical group $G$ onto another group $H$. Homomorphisms allow the transfer of a problem in one mathematical structure to another, perhaps more computable structure. Similarly, the purpose of a mathematical model of a biological system is the translation of its quantitative or relational aspects into a mathematical form which then allows such properties to be pursued further along the lines of formal mathematical procedures, in much the same manner that one solves a mathematical problem or establishes a new theorem.

One cannot prescribe algorithms or operational procedures for passing formally from the biosystem to an appropriate mathematical model. First of all, in the description just stated of the latter concept, there was no implication of a unique *1:1* correspondence between biosystem and model[4].

---

[4] In fact, subsequent examples in this paper of deterministic vs. stochastic models as alternatives for a given biosystem will be seen to demonstrate this very point.

However, a careful biomathematical formulation of a given biosystem under study can greatly facilitate the process of model construction, referring now to the scientific truism that careful formulation of a mathematical problem leads one 50 % of the way toward its solution. An additional guide that may be offered is the delineation between two broad approaches to the construction of such models, as follows:

a. *The direct approach* refers to the direct search *de novo* for a mathematical system or discipline which matches the system $\mathfrak{S}$. It is particularly in this approach that the biomathematical formulation should be suggestive. For example, if the biological problem or system is one of kinetics or, more specifically, of the determination of an index for measuring rate of transformation of some material(s), then an obvious choice of corresponding mathematical domain would be the Calculus and the Theory of Differential Equations since these mathematical areas have to do essentially with the study of rates of change of mathematical functions (v.i.).

b. *The indirect or analogue approach.* The indirect approach amounts to finding another system, say, $\mathfrak{S}$ $(C', S', F', E', t')$, either in biology itself or in some other discipline which bears some resemblance to those aspects of the system under study and for which a tested and "worked out" mathematical model $\mathfrak{M}'$ has already been constructed. Such analogues for biosystems have been found, for example, in physics, chemistry and, more recently, in the engineering sciences, and even in economics and the social sciences—wherever "systems" are studied from a mathematical point of view. In fact, it would appear that the majority of mathematical models in biology may have arisen in this way.

The use of such analogues, however, requires great caution, for there is the danger that the systems which they represent may become misleadingly identified with the biological systems, without of course being completely equivalent to them. In other words, while the analogous model $\mathfrak{M}'$ may suggest a model $\mathfrak{M}$ for the biosystem $\mathfrak{S}$, it must be kept in mind that $\mathfrak{M}'$ itself represents a measure on a system whose composition, structure, and function are not in one-to-one correspondence with the system under study. If, on the other hand, the system can be redefined legitimately in a manner which establishes such a correspondence, then $\mathfrak{M}'$ may be used with greater validity. However, this would depend on defining new concepts or redefining old biological concepts and would thus carry the danger of imposing an artificial pattern on the biological system. Dangers of this type possibly may be avoided by using the analogue and its associated model $\mathfrak{M}'$ simply as an indicator of a discipline of mathematics which is useful in quantifying the type of system under study, and then proceeding to construct $\mathfrak{M}$ independently of $\mathfrak{M}'$ (as though it were the first step in the direct approach just discussed).

It has been said with some justification that virtually every branch of mathematics will turn out to be of use in mathematical biology. And indeed, one can delineate classes of topological or "relational" or set theoretic models, geometric models, symbolic logic models, statistical (including information- and decision-theoretic) models, probabilistic models, algebraic models, analytical or deterministic models, stochastic models, etc., and mixtures of these. In the time remaining, the latter two classes of models will be discussed and their interrelationships stressed. In some cases, these models can lead to closely related results; in other cases to differing predictions. In general, they are unfortunately regarded as competing classes, whereas it would be more scientific to regard them as coextensive.

## IV. Deterministic or Analytic Models

A deterministic mathematical model is characterized by the representation of certain quantifiable aspects of a biosystem in the form of ordinary algebraic variables, or analytic functions of time. The actual construction of such a model generally involves the application of a direct association procedure $\mu$ between the biosystem and the mathematical domain of ordinary algebra: algebraic variables $x$ are identified with the pertinent components of $\mathfrak{S}$, and then known biological or other scientific principles are applied to these algebraic variables to obtain, usually, algebraic or differential equations that constitute the model.

Implied by such a representation is the expectation that, given the initial value $x_0 = f(0)$ of such a variable $x = f(t)$, the entire future course is determinable from the mathematical expression for $f(t)$; i.e. once $f(t)$ is obtained, the value of $x$ at any subsequent time $t = t_j$ is obtained by substituting $t_j$ into $f(t)$.

Most often such a model involves an approximation of a discrete quantity such as the cardinality of a compositional set by a continuous variable— it is therefore an idealization to this extent, at least. If one uses the model in a purely theoretical sense, i.e. to make theoretical predictions about the states of the system on the basis of formal mathematical deductions, then some interesting und useful new predictions about the behavior may be obtained as analytical hypotheses. On the other hand, as soon as one begins to test such hypotheses by *in vitro* or *in vivo* experiments, say or to use constants appearing in the algebraic or analytic expressions of the $x$'s as convenient indices for describing the behavior of the system, then problems of estimation using pertinent experimental data are encountered. And whether one turns to simple curve-fitting-by-eye procedures or statistical estimation procedures, one requires a statistical or probabilistic context. This does not violate the original deterministic assumption—which may be feasible in certain circumstances—but it does point to the inadequacy of the analytical model as a framework for data analysis.

In recognition of the enlargement of the original biosystem by the addition of an experimental environmental system or component, one must enlarge the deterministic model or representation accordingly. In a sense this amounts to obtaining from it a very simple kind of stochastic model. But it might be more appropriate to refer to the result as a "statistically-adjusted-deterministic model". These points may be demonstrated by the following simple example, following which this section will be concluded with a general mathematical formulation of the concept of a statistically-adjusted-deterministic model.

*Example: A Deterministic Model of the Kinetics of the Unimolecular Reaction Process.*

Perhaps a "lower bound" to biological systems particularly amenable to this kind of mathematical treatment is a simple biological or chemical transformation, $T: A \to B$, in which a set of objects $a \in A$ of some biological or chemical species $A$ are transformed directly into objects $b \in B$. Here the system $\mathfrak{S}$ consists of $C$-components, $A$ and $B$, and the function $T \in F$. If this is a chemical transformation, then the rest of the description of $\mathfrak{S}$ would consist of noting that the transformation may be considered as a homogeneous one, spacewise, and carried out *in vitro* in an environment $E$ consisting of a pressure and temperature controlled reaction vessel with associated measuring devices to follow the progress of the reaction, etc. If it is a biological transformation of some kind such as, say, the destruction of a group $A$ of organisms, then a complete $\mathfrak{S}$-like description would include the pertinent structures of the organisms, the nature of the destructive process $T$, temperature conditions of the environment $E$ and a description of other systems in the environment of significance to this $\mathfrak{S}$.

Suppose further, that it is simply the kinetics of this process in which we are interested, and simply at an empirical level, say. Then, in the chemical case, for the simplest possible formulation, the *Law of Mass Action* is invoked as a fundamental principle. In the biological case an equivalent principle of the *Law of Malthus* states that the rate at which population $A$ diminishes in a process of *simple destruction*, again, is proportional to the number of organisms present. And so, by identifying an algebraic variable $x$, in the first case with the cardinality of the set $A$ divided by volume $V$ (i.e., in terms of concentration units) and in the second case with the number of organisms in the experimental medium—and by making the convenient mathematical assumption in both cases that $x$ is a continuous, in fact, differentiable function of time, we bring the problem into the domains of algebra and the Calculus. Thus, the simple differential equation

$$\frac{dx}{dt} = -kx, \tag{5}$$

where $k$ is a positive constant (fixed for a given set of environmental parameters, say), together with an initial condition $x(0) = x_0 > 0$, constitutes a mathematical model of the "kinetics" of the process.

Note that in this procedure we have, however, represented a *discrete* quantity, the number of molecules or the number of organisms, by a *continuous* mathematical quantity. In very large systems no error may result, though in small systems a significant error could result. Such a simplification does result in making the process of the Calculus directly available to the study of the kinetics of the system, so that we are in a position to trace out implications of this representation of the underlying natural process by using a well-known mathematical methodology. Thus, in this case, the more direct relationship of $x$ to $t$ is obtained in a formal way by solving or integrating the differential equation of that model to obtain:

$$x = x_0 e^{-kt} \tag{6}$$

and, by applying the formal rules of algebra, this equation can be put in a convenient "linear form":

$$\log x = \log x_0 - kt, \tag{7}$$

showing $\log x$ as a linear function of $t$ and $k$ as the slope of the straight line. And so we would expect that, if all we have assumed about the transformation $T$ is true, then a semilogarithmic plot of the data should approximate a straight line.

As pointed out in the preceding, however, the representation given by equations (6) or (7) should be enlarged to accommodate statistical or curve-fitting procedures aimed e.g. at estimating from experimental data the so-called rate constant $k$, usually taken as the index of the strength of such a transformation. In this case this amounts to replacing (6) e.g. by the more realistic (stochastic) equation:

$$x = x_0 e^{-kt} + \varepsilon(t) \tag{8}$$

where $\varepsilon$ is in theory a random function of time such as a Gaussian random process. In dealing with experimental fluctuations such an extension seems to be quite generally and tacitly subsumed.

Speaking more generally, the utilization of such models thus involves the compounding of the original model-construction procedure $\mu$ with some "statistical" procedure $\bar{\mu}$:

$$\mathfrak{S} \xrightarrow{\mu} \mathfrak{M} \xrightarrow{\bar{\mu}} \overline{\mathfrak{M}} \tag{9}$$

which results in imbedding $\mathfrak{M}$ in a practical statistical framework. In this, $\overline{\mathfrak{M}} = \mathfrak{M} \times \mathfrak{P}$ refers to the probabilistically, or statistically, enriched form of the original model $\mathfrak{M}$. Once this is done, then the methodology of statistics becomes available to the problem of relating the mathematical interpretation implied to the biology itself through the medium of experiment and observation. The total $\mu$-procedure may be diagrammed as follows:

$$\begin{array}{ccc} \mathfrak{S} & \xrightarrow{\;\mu\;} & \mathfrak{M} \\ \downarrow{\scriptstyle \mu_e} & & \downarrow{\scriptstyle \bar{\mu}} \\ \mathfrak{S}\times\mathfrak{E} & \xrightarrow{\;\bar{\mu}_e\;} & \overline{\mathfrak{M}}=\mathfrak{M}\times\mathfrak{P} \end{array} \tag{10}$$

The left-hand portion of this diagram refers to the imbedding $\mu_e$ of the original system $\mathfrak{S}$ in an experimental context resulting in the augmentation of $\mathfrak{S}$ by the Experimental system $\mathfrak{E}$, represented by the "Cartesian product" notation $\mathfrak{S}\times\mathfrak{E}$. And the lower line refers to the correspondence $\bar{\mu}_e$ set up between the experimentally studied system $\mathfrak{S}\times\mathfrak{E}$ and an appropriately, statistically enlarged mathematical model $\overline{\mathfrak{M}}=\mathfrak{M}\times\mathfrak{P}$.

A further use and modification of such models (and indeed of other classes of models as well), is required for rather complex situations such as the mathematical representation of the kinetics of a metabolic process, in which the mathematical model may consist of a system of non-linear differential equations. Because it is not possible usually to obtain the solution of such systems, numerical methods must be tried. But numerical approximations yield only special solutions and are tedious to apply. For such purposes one turns to an electronic digital or analog computer. And so, the mathematical model $\mathfrak{M}$ (or $\overline{\mathfrak{M}}$ now may be understood) must undergo further transformations $\mu$. The first of these may consist of introducing scaling factors into the model $\mathfrak{M}$, transforming the variables and using the methods of numerical analysis to put the mathematical expressions into a form that may be implemented on a computer—call this entire procedure $\mu_1$, and the result $\mathfrak{M}_1$. Then for later convenience the entire numerical procedure may be flow-diagrammed (procedure $\mu_2$), resulting in a third form $\mathfrak{M}_2$; viz., the computer-orientated form of the model $\mathfrak{M}_2$. Then $\mathfrak{M}_2$ is programmed (e.g., using FORTRAN) into a symbolic language (procedure $\mu_3$). The resulting program in effect is a fourth form, call it $\mathfrak{M}_3$, of the model. Implementation of the program on a computer results in the translation of the symbolism (procedure $\mu_4$) into machine language resulting in an electronic realization of the model, $\mathfrak{M}_4$. And finally the computer generates (procedure $\mu_5$) outputs such as sets of curves $\mathfrak{M}_5$ which allows us to visualize, say, graphically the predictions of the representation. These steps are summarized in the following diagram:

$$\mathfrak{S} \xrightarrow{\;\mu\;} \mathfrak{M} \xrightarrow{\;\mu_1\;} \mathfrak{M}_1 \xrightarrow{\;\mu_2\;} \mathfrak{M}_2 \xrightarrow{\;\mu_3\;} \mathfrak{M}_3 \xrightarrow{\;\mu_4\;} \mathfrak{M}_4 \xrightarrow{\;\mu_5\;} \mathfrak{M}_5 \tag{11}$$

In this sense, then, such a mathematical model is seen to be a prerequisite to the programming of such biological problems on a computer. This combination of mathematical and digital computer methodologies creates for very complex biological systems a new kind of biological experimentation, which the author has referred to as *"in numero" studies* [7, 9, 11], in contrast to *"in vitro"* and *"in vivo"* studies. This methodology of course applies to other classes of models as well; most notably to stochastic models wherein numerical methods combined with Monte Carlo procedures allow the generation of artificial curves referred to also as "simulations". More details on this aspect will be given in the next section. A number of such studies have already been undertaken in our laboratory and are referred to in the list of references; e.g. a theoretical study of glycogen metabolism [18], a theoretical study of hepatocyte proliferation processes [21, 22], studies in the simulation of first order kinetic systems, [7, 11] and the computer simulation of electrocardiographic analysis [10, 17]. Time does not permit us to discuss all of these studies here.

Mathematical models used in this way and such combinations of experimental and theoretical methodologies, then, allow us to experiment on the system, numerically or theoretically, without disrupting the actual system, in order to discover the conditions required within the total complex of the hypothetical biosystem to account for what is observed *in vivo*. Such a technique is essential, for example, in studying reversible reactions involved in a whole chain of reactions, each single step of which could not be varied directly, or separated from the rest of the chain. These same *in numero* techniques become a means also of simulating abnormalities of medical significance, providing some extra insight in situations that are not amenable to more direct experimental study. The previously quoted references discuss some theoretical studies of diabetes and toxic cirrhosis and their correlations with *in vivo* data, all from this point of view.

## V. Stochastic Models

An empirical first test for the necessity of a stochastic formulation would involve repeating a set of experimental, timed determinations on the system $\mathfrak{S}$ as a means of detecting the presence of significant, unpredictable random variations. The superposition of all curves onto a common pair of axes in such a case might reval a strong central tendency, but with a detectable spread, or range, of curves to either side, say. It is in fact the purpose of a stochastic model to provide a mathematical rationale for such random ensembles.

The most primitive underlying principle of construction in the case of deterministic models is stated in terms of large collections of the basic units of composition or structure. But in the stochastic approach, one begins with the pertinent individual discrete unit of composition $C$ or structure $S$, say, introducing, in effect, a probability space for representing the effects of the associated transformation $T$ of each such unit resulting from possible random aspects of the mechanism in the system, and then translates the unitary probabilistic assumptions into probabilistic statements about the number of such units, i.e., into suitably defined "random" variables.

For example, when we consider the kinetic aspects of a biological system, the usual stochastic $\mu$-procedure consists of imposing on each of the discrete events of the pertinent transformation $T$ in $F$ of $\mathfrak{S}$, a homogeneous probability measure or distribution relative to which the cardinalities of the sets of $C$- or $S$-components of $F$ are conceived as random variables. In this way the kinetic aspects of the underlying system become mathematically expressed in terms of a unified collection of random variables $\{x(t)\}$ (as opposed to the ordinary algebraic variables of the deterministic model), one corresponding to each possible value of time $t$ (either discrete or continuous); or in a multivariate case, of collections $\{x_1(t)\}$, $\{x_2(t)\}$, ... of such families. This formulation makes available to the theoretical study of $\mathfrak{S}$, the entire mathematical machinery of the theory of probabilities and stochastic processes, and also a statistical basis for a complete discussion ultimately of random fluctuations in related experimental data. In practice one attempts first of all, to match up the properties of the theorized underlying random biological process with a known stochastic process containing an established mathematical formalism from which further properties of the system may be deduced.

Very extensive use of the so-called "Markov Process" has been made in studies of biological as well as chemical kinetics [1, 12, 15, 16]. A whole mathematical theory has grown up around this particular kind of Process, which makes it especially useful. And in some of these cases (see [2, 3, 5, 6]) very close relations between the alternative deterministic model and the stochastic model have been demonstrated.

*Example: A Stochastic Model of the Kinetics of the Unimolecular Reaction Process.*

In this case an immediate relation with the earlier deterministic model is established by incorporating the Mass Action principle in the probability space corresponding to the transformation of each molecule. Thus, following the preceding general suggestions, the first step here is the assignment to each molecular element $a \in A$ of a basic probability:

$$p(\varDelta t) = \mu \varDelta t + 0(\varDelta t), \tag{12}$$

the same for each molecule and for any time $t$ after initiation of the reaction, that in time $(t, t + \varDelta t)$ this molecule becomes transformed into $b = T(a)$, an element of the new component $B$ of the evolving system $\mathfrak{S}$. This very simple probability space independently associated with each molecule of $A$ then leads to the construction of a probability space for the whole process by deducing from a statistical assumption of independence that

$$\mu n \varDelta t + 0(\varDelta t) \tag{13}$$

gives the probability that a single transformation will take place in $(t, t + \Delta t)$ among the $n$-molecules of $A$ assumed to be present at time $t$.

The relation of this result to the deterministic Mass Action Principle has been conceived [3] as follows. Substituting $n$ for $x$ in equation (5), and approximating $dn$ by the infinitesimal $\Delta n$:

$$|\Delta n| = k n \Delta t + 0 (\Delta t) \tag{14}$$

where $0 (\Delta t)$ is a higher order infinitesimal, vanishing in the limit as $\Delta t \to 0$. This may be interpreted as the approximate number of $A$-molecules transformed in the interval $(t, t + \Delta t)$, so that, where $n$ is the cardinality of $A$ at time $t$, the ratio

$$\frac{|\Delta n|}{n} = k \Delta t + 0 (\Delta t) \tag{15}$$

may be interpreted in the context of the frequency theory of probabilities as the probability of a simple transformation in this time interval. Then by identifying the deterministic rate constant $k$ with the stochastic parameter $\mu$, equation 15 is seen to correspond exactly to the basic probability measure (eq. 12), of the current representation.

The discrete, integral-valued time developing random variable $x$ is then defined over this space, where $x = 0, 1, 2, \ldots$, $x_0$ gives the number of molecules of $A$ per fixed arbitrary volume of the homogeneous reaction space. Previous studies of stochastic models [2] suggested the reasonableness of identifying the random variable $x$ with a continuous-time Markov process with finitely many states $i = 0, 1, 2, \ldots$, $x_0$ and stationary transition probabilities (see references [1, 12, 13, 15, 16]). Using this as the basic mathematical model $\mathfrak{M}$, the so-called "transition probabilities" $\{p_{ik}(\tau)\}$ (where $p_{ik}(\tau)$ is the probability of a transition in time $\tau$ from any state $i$ occurring at arbitrary time $t$ to a state $k \leq i$), which contain the information for defining the whole process, were deduced to be:

$$p_{ik}(\tau) = \binom{i}{k}(e^{-\mu\tau})(1 - e^{-\mu\tau})^{i-k} = b(k; i, e^{-\mu\tau}) \tag{16}$$

And from this we obtained the *prior* probability:

$$p(x; t) = \binom{x_0}{x}(e^{-\mu\tau})^x (1 - e^{-\mu\tau})^{x_0 - x} \tag{17}$$

that, given an initial cardinality $x_0$, at subsequent time $t > 0$ the cardinality will have been reduced to the value $x$. This distribution considered as a function of $t$ was shown to have a *mean value*

$$\mathfrak{E}(x; t) = x_0 e^{-\mu\tau} \tag{18}$$

and *variance function*

$$\sigma^2(x; t) = x_0 e^{-\mu\tau} (1 - e^{-\mu\tau}) \tag{19}$$

The fact that the mean value function of the stochastic model, on identification of the parameter $\mu$ with the deterministic rate constant $k$, is identical with the negative exponential smooth time course (see equation 6) demonstrates consistency of this particular stochastic model with the deterministic model. This theoretical property would seem to guarantee that the sample curves of the stochastic process should appear at least qualitatively very similar to experimental runs on such reactions which had already been classified as satisfying the accepted deterministic criterion of unimolecularity. For further comparison, one could therefore begin with the ordinary variable $x$ of the earlier deterministic model and transform it into a "stochasticized" variable $x'$ which reflects the hypothesized intrinsic stochastic component of the present model by writing

$$x' = x_0 \, e^{-kt} + \varepsilon_2 \, (t) \tag{20}$$

where now $\varepsilon_2 \, (t)$ is the subsumed unimolecular stochastic process "normalized" to the mean value $x_0 \, e^{-kt}$. Then, a further statistical enlargement of this expression to accommodate an experimental error process, call it $\varepsilon_1 \, (t)$ would lead to:

$$x' = x_0 \, e^{-kt} + \varepsilon_1 \, (t) + \varepsilon_2 \, (t) \tag{21}$$

Comparison of (21) with (20) and (8) shows that it represents a combination of the two models, in the sense that if $\varepsilon_1$ is negligible then (21) reduces to (20); and if $\varepsilon_2$ is negligible, then (21) reduces to (8) (where $\varepsilon_1$ corresponds to $\varepsilon$ used in eq. 8).

This raises the practical question of assigning the actual origin to fluctuations that may be observed in experimental determinations of first order kinetics. Accordingly, if one wishes to verify experimentally and directly the stochastic hypothesis of the present model, it would be necessary to perform error-free *in vitro* experiments. Since this is not possible, *in numero* studies have been initiated for studying each random component independently; i.e. for obtaining simulations of both kinds of random processes amenable to statistical analyses. Such studies have been carried out using the IBM 1620 and have recently been reported briefly [7, 11]. In the case of the stochastic model itself (the $\varepsilon_2$ process), the method whereby the simulations were effected is a *Monte Carlo Procedure* based on the transition probability functions (eqs. 16 and 17) of the stochastic model adjusted to the conditions of two classical well-established unimolecular reactions: the *sucrose inversion reaction* and the *decomposition of di-t-butyl peroxide*. The data published by Pennycuick [19] on the first of these and the data published by Raley, Rust and Vaughan [20] on the second had been analyzed quite intensively (see [3]) and established as classical examples of unimolecular reactions.

Time does not permit our going further into the details and results of this study, some of which are in the references previously cited. Suffice it to say, that agreements between experiment and the theoretical, *in numero* procedures have been obtained which satisfy both qualitative and statistical criteria of significance. This work is still continuing in an effort to point up the necessity of considering that, in equations such as (8), the random component $\varepsilon(t)$ must at least be considered to have two components: $\varepsilon_1(t)$ due to experimental and hence extraneous, controllable fluctuations; and $\varepsilon_2(t)$, inherently and uncontrollably related to the mechanism of transformation. Most recently attention has also been given to the study of the characteristics of the experimental error component $\varepsilon_1(t)$, so that eventually we may obtain a deeper knowledge of the interrelationships of these sources of error and their experimental significance.

Those interested in stochastic models feel that the neglect of the probabilistic aspect of biological systems is not justifiable even in the face of mathematical difficulties, some of which are newer than those encountered in ordinary analytical work and hence may appear larger than equivalent analytical difficulties with deterministic models for which direct or indirect treatments already exist. Hearsay to the contrary, it is not true that those interested in the stochastic approach do this to the exclusion of the deterministic approach. In cases where no real discrepancies may result, it is readily agreed that an easier deterministic treatment commends itself—in fact, is usually the place where one begins. On the other hand, a doctrine of complete determinism leads one to ignore or minimize the importance of well-established evidence of inherent biological variability in biosystems; and at the substratum of molecular biology such neglect can at times lead to costly misinterpretations of kinetic data.

## VI. Conclusion

It is hoped that these remarks and examples may be of some use in obtaining an overview of the application of mathematical methodology and rationale to the study of biological systems generally. It is also hoped that the biomathematical formulations and conceptualizations may be productive in suggesting at least a possible approach to the construction of a whole theory, or axiomatization, of mathematical models. What has so far been presented is of course merely an indication that such a theory is conceivable. Whether or not the particular approach suggested turns out to be fruitful in this direction is really less important at this time than the principal aim of the present paper which is to provide hopefully a little insight into what constitutes a mathematical model and to the differences between the important classes of deterministic and stochastic models.

## References

1. Bailey, N. T. J.: The Elements of Stochastic Processes, with Applications to the Natural Sciences. New York: John Wiley 1964.
2. Bartholomay, A. F.: Bull. Math. Biophys. **20**, 97 (1958).
3. — Bull. Math. Biophys. **20**, 175 (1958); **21**, 363 (1959).
4. — Bull. Math. Biophys. **22**, 285 (1960).
5. — Biochemistry **1**, 223 (1962).
6. — In: N. Rashevsky (Ed.), Physico-mathematical Aspects of Biology, p. 1. New York: Academic Press 1962.
7. — IBM 6th Medical Symposium, cosponsored by Brookhaven National Laboratory, Oct. 5—8, p. 325. New York: Poughkeepsie 1964.
8. — Bull. Math. Biophys. Special Issue **27**, 235 (1965).
9. — (Abstract) in Proc. of the Third Ann. Symp. on Biomath. and Computer Sc. in the Life Sciences. Houston, Tex.: 1, 1965.
10. — (Abstract) SIAM Review **8**, 558 (1966).
11. —, Y. Li and E. Fuglister: (Abstract) — Abstracts of the Biophysical Society 10th Annual Meeting, Feb. 23—25, 1966 p. 27,. Boston Massachusetts, USA: 1966.
12. Bharucha-Reid, A. T.: Elements of the Theory of Markov Processes and Their Applications. New York: McGraw-Hill 1960.
13. Doob, J. L.: Trans. Amer. Math. Soc. **58**, 455 (1945).
14. Goodwin, B. C.: Temporal Organization in Cells. New York: Academic Press 1963.
15. Gurland, J. (Ed.): Stochastic Models in Medicine and Biology. Madison: Univ. of Wisconsin Press 1964.
16. Harris, T. E.: The Theory of Branching Processes. New Jersey: Prentice-Hall 1963.
17. Levine, H. D.: Proc. 9th International Congress of Internal Medicine, Amsterdam, Holland, Sept. 7—10, 1966. In: "Excerpta Medica International Congress Series", No. 137, p. 429 (1966).
18. London, W.: J. Biol. Chem. **241**, 3008 (1966).
19. Pennycuick, S. W.: J. A. C. S. **48**, 6 (1926).
20. Raley, J. H., F. F. Rust and W. E. Vaughan: J. A. C. S. **70**, 88 (1948).
21. Shea, S. M.: Bull. Math. Biophys. **265** (1964).
22. — and A. F. Bartholomay: J. Theoret. Biol. **9**, 389 (1965).
23. Waddington, C. H.: The Strategy of the Genes. London: George Allen and Unwin Ltd. 1957.

*Discussion*

Walter:

In relation to the suggestions that $\varepsilon_1(t)$ and $\varepsilon_2(t)$ are not separable:
1. In 1957 (or so) Dr. Schnoll (USSR) found large fluctuations of inorganic phosphate liberated after a constant interval of hydrolysis of ATP catalyzed by myosin. The fluctuations were reported to be much larger than the expected experimental error. 2. Dr. J. A. Christiansen suggested a few years ago in "Advances in Enzymology" that enzyme catalyzed reactions might be synchronized. This would mean that enzyme reactions would not be strictly continuous. *If* this suggestion is correct it would be important in relation to the calculation of $\varepsilon_1(t)$ and $\varepsilon_2(t)$.

ARLEY:

Of course, I do not know what answer Dr. BARTHOLOMAY himself would give to your comment. However, personally I would like to stress most emphatically here—as I also do in my textbook (ARLEY, N., and K. R. BUCH: "Introduction to the theory of probability and statistics", Science Edition, Wiley, New York, 1966)—that the experimental errors—Dr. BARTHOLOMAY's term $\varepsilon_1(t)$ here—can never be got rid of in any actual experimental observation. Personally, I therefore wish to protest very much against the expression "experimentally *controllable* fluctuations" in Dr. BARTHOLOMAY's paper at this point.

BERGNER:

Dr. BARTHOLOMAY has earlier—in his analysis of the invertase reaction—tried to determine separately the random fluctuations implied by the stochastic character of the system itself. If I recall it correctly he calculated, on this basis, the confidence interval (95%?) for the concentration curve and obtained an interval that satisfactorily covered the experimental data. Hence, his analysis implies that the pure experimental error can be neglected. But we know, I believe, that this is not correct and, consequently, we here have an inconsistency.—One may also raise the question whether BARTHOLOMAY's result suggests that the molecular events are not independent of each other. From a physical viewpoint, however, this is not a very attractive conclusion.

ARLEY:

As a physicist I would also say on this point: certainly it is not since it is one of the most well established facts of atomic physics that the behaviour of the single atoms and molecules are independent stochastic events, in any case when we are neglecting the coupling between them which occurs when they are bound together in crystal lattices. We therefore need more knowledge about the state of matter considered in Dr. BARTHOLOMAY's paper Dr. BERGNER quotes.

KRÜGER:

Does the stochastic approach also include growth phenomena? Is it possible to relate a mathematical description of macroscopic phenomena—which is inevitably the starting point for biomathematics—to microscopic events, as has been done in the transition from classical to statistical mechanics and as could be exemplified by the progress of genetics from classical terms to molecular genetics?

ARLEY:

Although I do not know the details of Dr. BARTHOLOMAY's papers, I would guess that the answer is yes, if $x$, as here, denotes a concentration or a number of molecules in a certain volume, then the equations are valid for any value of the variables, large or small, i.e. they also cover the macroscopic behaviour of the phenomenon considered. It is well known even for biologists that any actual measurements can only give us a *finite* number of *points* and that, when these $n$ points are plotted in a coordinate system $(x, y)$, we always can, in an infinite number of ways, construct a continuous curve passing exactly through each of these points (e.g. by a polynomial in the independent variable $x$ of degree at least equal to $n-1$). However, anyone of such interpolation curves— the growth functions being a special class of them—does not give us any new information about nature and her basic mechanisms unless they have been deduced mathematically from basic principles, which are, for governing the behaviour of atoms and moleculres, the statistical or stochastic laws of quantum theory. In biology there is neither place nor need for extensions of Bohr's complementarity philosophy, originally proposed to overbridge the gap between the two physical pictures of matter, viz. wave and particle, because in biology no such gap between different pictures of living matter exists. I am therefore convinced that in *all* branches of science, including biology and psychology, no mathematical model of any phenomenon can be considered as being a theory of that phenomenon in the scientific sense of the word unless the model is based upon and directly derived from quantum theory. Curves expressing macroscopic phenomena can only be considered as having heuristic value.

Bartholomay (final comment given by correspondence):

To Walter: My idea in delineating two main components of random error was not to suggest either that they are independent (or separable) or the opposite, but that in dealing with experimental data we should consider the possibility that in certain cases unusual or unexpected random fluctuations may not signify necessarily a poorly executed experimental run; the explanation may be deeper than this.—To Arley (first comment): The fact that the standard validation of any method of chemical determination rests on experiments designed to estimate statistically the repeatability, reproducibility and precision indicates that the observer is in a position to exert some measure of control over random fluctuations. Certainly human error, coupled with instrumental imprecision, are important factors contributing to the random features, apart from inherent random aspects of the chemical mechanism itself. It is in this sense that I refer to the former as "controllable". I do not mean to imply that I consider that they are in any case controllable to the extent of being eradicable in any practicable sense. If this were so, then one could examine the residual inherent components in the ordinary experimental *in vitro* sense. The fact that they appear not to be is the reason that I have introduced the *in numero* approach, using the computer to simulate various patterns of random fluctuation as a means of perhaps gaining a little additional insight into the whole complicated question.—To Bergner and Arley: In my paper referred to by Bergner (Bull. Math. Biophys. **21**, (363 1959)) I derived a very simple statistical formula for estimating the stochastic parameter $\mu$ which corresponds to the classical first-order rate constant $k$. In applying such a formula directly to data, one is guilty of ignoring the presence of extraneous or "controllable" random fluctuations. On the other hand the usual regression methods for estimating $k$ directly from kinetic data ignore the presence of inherent fluctuations. In the various cases treated in that paper it was of some interest to note that the estimates of $k$ and $\mu$ by these different rationales turned out to be indistinguishable from a statistical point of view. I therefore drew an heuristic inference from such observations that the formula in the paper might be utilized as a convenient, simple, numerical way (indirectly) of avoiding the more cumbersome classical curve-fitting and regression methods of estimating $k$.—To Krüger: I am afraid that I disagree with his comment that the starting point for biomathematics is a mathematical description inevitably of macroscopic phenomena; the body of my text should make this abundantly clear.—To Arley: With all of its progress over the past two decades, biology has not yet advanced to that same point as physics where a transition from classical deterministic to statistical physics (and stochastic approaches) was seen as essential to further progress. We may conjecture that current findings point strongly to such a biological day. In such a case we should be turning our minds and work toward that eventuality. In the meanwhile, like all theories and models in science, including physics, validated simpler models are of value until proved otherwise. But there appears to be no such thing as a complete or unmodifiable or completely accurate or precise formulation of any natural or physical phenomenon. Thus, when Dr. ARLEY cites quantum theory as his standard for testing the ultimate validity of a mathematical model, I wonder if this suggestion itself is not more heuristic or even artificial than scientifically rigorous.

# Turnover Compartmentalization.
# An Approach to Analysis of Whole-Body Retention Data

P.-E. E. Bergner

With 2 Figures

*Abstract*

In the early days of biologic tracer analysis, whole-body retention curves were frequently interpreted in terms of so-called rapidly and slowly exchanging compartments. Classic compartment analysis soon proved this approach incorrect; but, in the following, a modified form of it is presented that can be given a precise physicomathematical justification. The purely deductive approach is made possible by the introduction of a specific stochastic concept: the turnover compartment. The theory is based on previously published ergodic relations for open, heterogeneous systems.

Consider an open system: Every particle entering the system will, with probability one, eventually leave it without loss of identity. To fix ideas let the system be a cow, and the particles stable calcium ions. I have shown [1] that for this kind of system a precise microphysical representation can be constructed in terms of a stochastic language: The movement of each individual $Ca^{2+}$ through the cow is visualized as a time-homogeneous Markovian process. That is, the system (the beast) is considered as a set of "states" on which is defined a transition probability function with appropriate mathematical behavior (e.g. the function does not depend on the distribution of particles, i.e. the particles are statistically independent of each other).

By the steady-state distribution we mean the particle distribution that results after there has been, for an infinitely long time, a constant input of particles into the system. I have shown [2] that the steady-state distribution is, for this type of system, a logically natural reference distribution, similar to thermal equilibrium in statistical mechanics.

The retention probability $P_S(t)$ is the probability of a particle that enters the system at the time zero still being in the system at the time $t$. Problem: How is the transient time process $\{P_S(t); t\varepsilon [0, \infty)\}$ related to the steady-state distribution? It has been possible to demonstrate [3] that $\int_0^\infty P_S(t)\, dt$ can be interpreted as the inverse of the "mean exit-probability", with the mean taken with respect to the steady-state distribution. But, as explained below, from a practical viewpoint, this result is not enough nor satisfactory [4].

Experimentally one tries to realize the steady-state distribution by keeping the daily uptake of stable calcium equal to the output. $P_S(t)$ is estimated by injecting a small amount of tracer (e.g. $^{45}Ca^{2+}$) and observing the retention of radioactivity as a function of time. But the observations must be stopped at some time $t = T \ll \infty$, and the previously mentioned integral cannot be estimated. Problem: Can any quantities be constructed that characterize the steady-state distribution of mother substance (here, stable calcium) and that can be estimated from this kind of tracer data? I have recently presented a set of such quantities [4], but to determine their practical usefulness they must be applied to actual experiments, and must therefore be introduced to the consumer of theory—the experimenter. Usually, in this field, the experimenters have minimal mathematical training, and the following is an attempt to give a non-mathematical presentation of the new concepts; the approach is inspired by the success statistics has had in presenting itself in terms of white and black balls in an urn.

The upper part of Figure 1 is supposed to represent the cow, with its input and output of calcium. Steady-state is assumed. Obviously the different calcium ions stay varying lenghts of time in the system, i.e., the ions have different time of sojourn and in these terms we classify the steady-state distribution: The black $Ca^{2+}$ with a time of sojourn between 0 and $t_1$, the white ions with time of sojourn between $t_1$ and $t_2$; the square ions, with time of sojourn longer than $t_2$, are not considered. Steady-state implies that the numbers of black and white $Ca^{2+}$ ($m_1$ and $m_2$ respectively) are time independent; and so also are the turnovers, $r_1$ and $r_2$, i.e. the number of black and white particles, respectively, that leave the system per unit of time.

The procedure implies a subdivision of the steady-state population of particles into two mutually exclusive *turnover compartments*, $K_1$ and $K_2$. The compartments have well-defined masses or sizes ($m_1$, $m_2$) and turnovers ($r_1$, $r_2$), but are in general undefined geometrically and physicochemically. $K_2$ can be subdivided into two parts: an initial part $K_{21}$, and a final part $K_{22}$, with the masses $m_{21}$ and $m_{22}$ ($m_2 = m_{21} + m_{22}$). $K_{21}$ consists of those white particles that, at the moment of observation, have stayed in the system less than $t_1$, whereas the particles in $K_{22}$ have stayed longer than $t_1$ and are therefore just about to leave the system. Only $K_{22}$ contributes to $r_2$; i.e., the particles in $K_{21}$ have zero exit probability. Note that $K_1$ consists only of a final part (instead of $K_1$ it would therefore be more logical to write $K_{12}$).

As shown in Figure 1, the different steady-state quantities ($m_1$, $m_{21}$, $m_{22}$, $r_1$, $r_2$) can be represented by a *turnover diagram*: The heights of the columns give the masses, and the lenghts of the arrows give the turnovers. In addition we have other quantities, the turnover factors (rate constants) $\lambda_{E_1}$ and $\lambda_{E_2}$ ($m_1 \lambda_{E_2} = r_1$; $m_{22} \lambda_{E_2} = r_2$); they can be interpreted as conditional mean exit-probabilities with respect to the steady-state distribution of particles in $K_1$ und $K_{22}$ respectively.

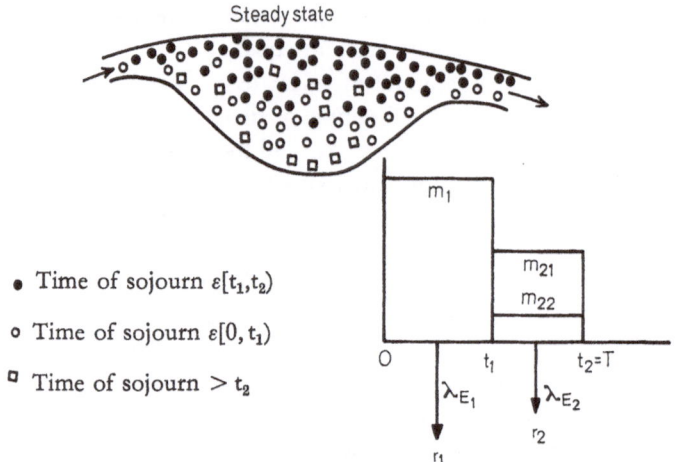

• Time of sojourn $\varepsilon[t_1, t_2)$

○ Time of sojourn $\varepsilon[0, t_1)$

□ Time of sojourn $> t_2$

Fig. 1. The principles of turnover compartmentalization as explained in the text. The figure shows how the steady-state masses or compartment-sizes ($m_1$, $m_{21}$, $m_{22}$), the turnovers ($r_1$, $r_2$), and the turnover factors ($\lambda_{E_1}$, $\lambda_{E_2}$) can be diagrammatically represented; the heights of the columns denote the compartment sizes. Note that the lengths of the arrows represent $r_1$ and $r_2$ respectively, not $\lambda_{E_1}$, $\lambda_{E_2}$

As I have shown elsewhere [4], in addition to $\lambda_{E_1}$ and $\lambda_{E_2}$, one can by simple calculations determine, from the retention data alone, the *relative* masses and turnovers; these quantities become, after multiplication with the total daily uptake (input) of mother substance, equal to $m_1$, $m_{21}$, $m_{22}$, $r_1$ and $r_2$ respectively. The calculation of these relative values is possible provided the total time period $T$, during which the retention $P_S(t)$ is observed, is not less than $t_2$; i.e., considering all the retention data obtained, the turnover diagram ends at $T$, as indicated in Figure 1.

The upper part of Figure 2 shows such a turnover diagram of the relative compartment-sizes and turnovers, but instead of only two turnover compartments, we now consider a subdivision into six compartments (each interval is 1.5 days; i.e., $T = 9$ days). The diagram is obtained from a $^{45}Ca^{2+}$-retention study on a parathyroidectomized cow (G. P. MAYER et al., School of Veterinary Medicine, University of Pennsylvania; personal communication). The lower diagram is from a similar study in the same animal, after receiving parathyroid extract. It seems reasonable to assume that the final parts of the turnover compartments are dominated by the plasma and fecal calcium. Hence, the diagrams indicate that parathyroid extract increases the plasma and fecal calcium. Also the sum of the relative turnovers increases from 0.544 to 0.628 and, because this sum for the complete process (i.e. from $t = 0$ to $t = \infty$) equals unity, it may be concluded that the turnover of the slowly exchangeable bone calcium is decreased by the parathyroid extract; it seems reasonable to identify the later turnover compartments, which are here not observed, with the slowly exchangeable bone calcium.

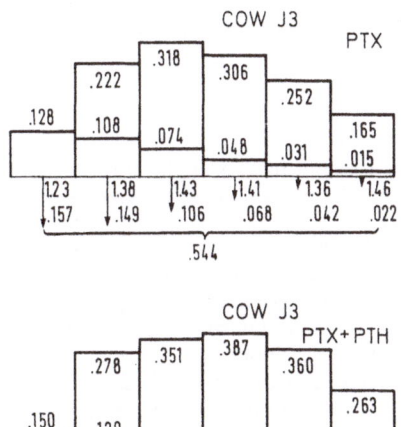

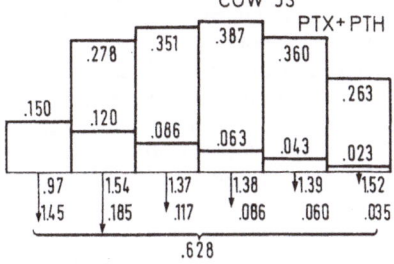

Fig. 2. Turnover diagrams obtained from $^{45}Ca^{2+}$-retention studies in a parathyroidectomized (PTX) cow, without and with supply of parathyroid extract (PTH). The diagrams show the relative values of $m_{ij}$ and $r_i$ according to the principles of Fig. 1; the numbers at the base of the arrows give the turnover factors ($\lambda_{E_i}$). Each interval is 1.5 days, i.e. $T = 9$ days

This example illustrates how the parameters introduced here might possibly be used for a physiologic interpretation of whole-body retention data. But it is an open question whether, in this regard, the present form of the turnover diagram is the most convenient one, and the main purpose of this investigation has instead been to show that the problem formulated at the beginning has at least one formal solution: For a broad class of transient Markovian processes there exist definite relations between the time process and the underlying time-independent probabilistic structure, when this structure is defined in terms of a steady-state distribution. It is practical significance that one can construct steady-state parameters that can be estimated from the first, finite part of the time process (here the incomplete tracer retention curve).

From the viewpoint of general methodology the results show that, at least occasionally, a microphysically deductive approach is possible also in truly macroscopic biology. That is, observable quantities can be constructed, which have precise physical significance in advance of the observation; this is similar to entities in physics like temperature and energy, which have their meaning independent of any actual experiment. In my opinion it is only under such conditions that a theory can serve as the basis for truly efficient design of experiment: analysis of errors, calibration, standardization and, eventually, comparison between different observations.

## References

1. BERGNER, P.-E. E.: J. Theor. Biol. **1**, 120 (1961).
2. — USAEC Report, ORAU-103 (1967).
3. — J. Theor. Biol. **6**, 137 (1964).
4. — USAEC Report, ORAU-104 (1967).

*Discussion*

SMITH:

I would like an explanation of the compartments $m_1$, $m_{21}$ and $m_{22}$ in relation to (a) physiological compartments and (b) output compartments.

BERGNER:

$m_1$ is the steady state amount of particles (e. g. in grams) which stay in the system no longer than $t_1$. $m_{21}$ is the corresponding number of particles that have stayed in the system less than $t_1$, but are known to leave before $t_2$ but not before $t_1$ (note, $t_1 < t_2$). $m_{22}$ is the steady state amount of particles which have stayed in the system longer than $t_1$ but are known to leave before $t_2$.—The identification of all the turnover compartments with different physiological compartments is not given by the present theory, but must be done separately. That is, such an identification is by necessity hypothetical and must be tested by separate experiments. The specific identification I made of the "final" compartments in the cow diagrams is in agreement with separate observations of calcium content in blood and feces.—The identification I made of the change in relative turnover cannot, so far as I know, be verified by any other method. However, as this identification is rather obvious, it suggests that we here have a method for studying the change of the bone calcium turnover.

# Functioning of a Cell Model

Z. Simon, D. Farcaș, and A. Cristea

*Abstract*

The functioning of a cell model is studied in which DNA synthesis starts at moment $\tau$ when the concentration of a DNA precursor $C$ reaches a threshold value. The synthesis rate of $C$ is considered to be proportional to the quantity of a mitotic enzyme and the cell volume to the total protein quantity. A model of this type with only four proteins tends towards a stable, periodic behaviour. Relations between gene activation degree, cell volume and cell cycle length are discussed.

In the present model we suppose that the DNA-synthesis is triggered at moment $\tau$ when the concentration $c = C/V$ of a DNA precursor $C$ reaches a threshold value $c_\pi$; in reality, a periodic DNA synthesis in synchronized bacterial cultures becomes continuous if all four deoxyribonucleoside-triphosphates are present in concentrations of $10^{-4}$ M [2]. The other periods of the cell cycle (DNA synthesis S, antephase $G_2$ and mitosis M) are supposed to set in automatically and to be completed during a constant time length $\tau_1$, once the above condition is satisfied. The cell volume $V$ is considered as proportional to the total protein amount. These assumptions were used in a preceding note [3] in which, for the sake of simplicity, $V$ and the protein concentrations $P_i/V$ were considered constant. Our aim here is to study the behaviour of such a model in which the volume variation during the cell cycle is also taken into account.

The following equations describe the model:

$$\frac{dP_i}{dt} = \psi A_i - \lambda_i P_i \qquad i = 1, 2, \ldots \quad (1)$$

$$\frac{dC}{dt} = \varphi P_1 - \eta C \qquad (2)$$

$$V = K \sum_i P_i \qquad (3)$$

$$c(\tau) = c_\pi \qquad (4)$$

with the initial conditions:

$$P_i^{(n+1)}(0) = \frac{1}{2} P_i^{(n)}(\tau + \tau_1) \qquad (5)$$

$$C^{(n+1)}(0) = \frac{1}{2}\left[\int_0^{\tau+\tau_1}(\varphi\, P_1 - \eta\, C)\,dt - \Theta\right] \tag{6}$$

$\lambda_i$ and $\eta$ are decay rates; $P_1$ the enzyme which catalyses the synthesis of $C$ at a rate $\varphi$; $\tau$ the length of $G_1$ (early antephase), $\tau_1$ that of $S + G_2 + M$. $\Theta$ is the amount of $C$ used up during DNA-synthesis. The upper index enumerates the cell cycle. $A_i$ is the gene activation degree (fraction of time during which gene $i$ is de-repressed). The gene activation degrees $A_i$ are assumed to be fixed for a given model. The synthesis rate $\psi$ can be expressed in terms of DNA-synthesis rate and protein on template synthesis rate, on the basis of the relevant equations given in [4]. The decay times of RNA and templates are assumed to be short relative to the cell cycle $\tau + \tau_1$, while the decay times of proteins are assumed to be long.

There can exist different situations, depending on the decay processes and on the competition of macromolecular synthesis processes for micromolecular precursors. One situation will be discussed here chosen so as not to conflict with results given in [3]. We assume that there are two classes of proteins: for "class I", which contains also $P_2$, the synthesis processes are saturated in micromolecular precursors. For "class II", which contains also $P_1$, the synthesis processes compete for precursor $X$ (for example, an aminoacid available in small amounts, which is not contained in "class I") synthesized by protein $P_2$ at the rate $\nu\, P_2$. Thus:

$$\psi_{\mathrm{I}} = \text{const}, \qquad \psi_{\mathrm{II}} = \nu P_2 / \sum_{\mathrm{II}} A_j \tag{7}$$

If such a model is to describe cells unchanged by division cycles (stable cells), the daughter cells resulting from division should be identical with the parent cell at the beginning of the previous cycle:

$$\begin{aligned} P_i^{(n+1)}(0) &= P_i^{(n)}(0) \\ C^{(n+1)}(0) &= C^{(n)}(0) \end{aligned} \tag{8}$$

The behaviour of such a model was studied on a digital computer [1]. The model consisted of only four proteins: $P_2$ and $P_3$ of "class I", $P_1$ and $P_4$ of "class II". The decay was considered to be catalysed by $P_3$ and all four proteins were considered to be competing, in the decay process, for $P_3$:

$$\lambda_i = \frac{\lambda_i^0\, P_3}{P_1 + P_2 + P_3 + P_4} \tag{9}$$

The synthesis and decay rates, $c_\pi$ and $\tau_1$ were equated to: $\psi = \varphi = 1$, $\nu A_1/(A_1 + A_4) = \nu A_4/(A_1 + A_4) = 1$, $\lambda_i^0 = 4$, $K = 1$, $c_\pi = 0.025$, $c_1 = 0.03$. Instead of (6), the simplified initial condition:

$$C^{(n+1)}(0) = 0 \tag{10}$$

was chosen. For a great variety of initial protein quantities $P_i^{(1)}$ (o) (all 0.1000; all 0.0050; $P_1 = 0.0060$, $P_2 = 0.0400$, $P_3 = 0.0200$; $P_4 = 0.0000$), the $P_i^{(n)}$ (o) values tend towards $P_1 = P_4 = 0.0061$, $P_2 = P_3 = 0.0195$ and the cell cycle length towards $\tau + \tau_1 = 0.37$. Condition (8) is thus fulfilled for this model. With stabilized behaviour the protein concentrations remain unchanged, during the whole cell cycle and from one cell cycle to the next.

If certain linearizing assumptions are made, i.e. neglecting the decay processess ($\tau + \tau_1 \ll 1/\lambda_i$, $1/\eta$) and substituting the simplified initial condition (10) for (6), $\tau_1 \ll \tau$, equations (1)—(10) describing the model can be integrated and relations between gene activation degrees, duration of cell cycle and cell volume can be obtained. The results are (with $V = K \sum_{i \in I} P_i$; i.e. for class I proteins only);

for class I: $\qquad P_i = \psi\, A_i\, (t + \tau + \tau_1); \quad 0 < t < \tau + \tau_1$ $\qquad$ (11)

for class II:

$$P_j = \frac{\nu\,\psi\, A_2\, A_j}{\sum_{II} A_j} \left[ \frac{3}{2}\, (\tau + \tau_1)^2 + (\tau + \tau_1)\, t + \frac{t^2}{2} \right]; \quad 0 < t < \tau + \tau_1 \qquad (12)$$

$$V = K\,\psi\,(t + \tau + \tau_1) \sum_{I} A_i; \quad 0 < t < \tau + \tau_1 \qquad (13)$$

$$\tau \approx \left( \frac{12}{13}\, \frac{k\, c_\pi \sum_{I} A_i \sum_{II} A_j}{\nu\, \varphi\, A_2\, A_1} \right)^{1/2} \qquad (14)$$

The model studied in this paper really tends towards a stable, periodic behaviour, and thus seems to be appropriate to the description of stable cell types. The extension of $\tau + \tau_1$, produced via competition by enhancing activities ($A_i$ — s) of non-mitotic genes (eq. 14) is consistent with the fact that high mitotic activity and high histospecific activity are mutually exclusive.

## References

1. FARCAS, D., and Z. SIMON: Studia Biophysica **2**, 339 (1967).
2. LARK, K. G.: Biochem. Biophys. Acta **45**, 121 (1960).
3. SIMON, Z.: J. Theoret. Biol. **16**, 294 (1967).
4. TSANEV, R., and BL. SENDOV: J. Theoret. Biol. **12**, 327 (1966).

*Discussion*

KLAMERTH:

What kind of cells did you use? Are the results based on synchronized cells? If not, would not this interfere with your proposed constant concentrations of proteins? What precursor for DNA synthesis did you measure? The thymidine kinase should be measured rather than a nucleotidetriphosphate. How does the formation of mRNA fit in your hypothesis?

Simon:

> Our considerations refer to cells in culture of not too high a density. In principle they would also be valid for tissues, although perhaps here other factors, e.g. the cell volume, could intervene.—Concerning the thymidine kinase, measurements have indeed been done with the extracted enzyme, to which labelled substrates have been added in well-known and high concentrations. However, we would need information on those enzymes in the cells, whose substrate concentrations (for TMP, TTP) are not known.

Klamerth:

> Such activities would indeed be difficult to obtain since in the cell precursors of TMP (for example) could be used also as precursors for RNA-synthesis.— How does your model account for the experimental fact that exogenously given thymidine blocks the progression from $G_1$ to S?

Simon:

> As you know, even TTP inhibits some earlier steps in its synthesis. Our considerations would nevertheless hold good if $C$ is a precursor of TTP, rapidly transformable to it, and if a proportionality between the concentrations of C and TTP exists.

Kiefer:

> Cell cycle time is not constant. It would be desirable to introduce stochastic considerations to take account of the variation of the parameters.—E. coli does not have a $G_1$-S-$G_2$ cell cycle, but rather a continuous S-phase. One has to be careful to use this bacterium for experimental verification.

Simon:

> The test on E. coli is not valid for the whole model. We supposed only linear volume increase $dV/dt = v$ and that $C$ synthesis is catalyzed by some growth-stable protein $dC/dt = \varphi V$, decay processes being neglected. Condition (6) becomes (without implication on eq. 4):
>
> $$C_0 = \frac{2}{3} v \varphi \tau_c^2 - \Theta$$
>
> or, as the initial cell volume $V_0 = v \tau_c$, if the initial precursor quantity is neglected ($C_0 = 0$), one obtains:
>
> $$\tau_c V_0 = \frac{2\Theta}{3v} \qquad \begin{array}{l} \tau_c \text{: cell cycle} \\ \Theta \text{: precursor quantity used} \\ \quad \text{up in DNA synthesis} \end{array}$$

According to some data on E. coli at various division rates ($\tau_c$ between 82 and 25 min) (when E. coli double 2 chromosomes per cycle: K. G. Lack, Bact. Rev. **30**, 3, 1966), the product $\tau_c$ dry mass is approx. constant.—Other tests were carried out with rapidly dividing mammalian cells (in culture). $P_1$ and its RNA-template were assumed to be unstable (In regenerating liver, thymidine kinase and thymidylate synthetase really have half-lives of approx. 3 hrs—their templates also.)—By means of a relation between $V_0$, $\tau_c$, $\tau$ and $c_\pi$, and $\varphi$ the first variables are known, we may calculate $c_\pi$. One obtains values lying between $1 \cdot 10^{-4}$ and $4 \cdot 10^{-3}$ mol/l, i.e. not a constant value.

# A Digital Computer System
# for the Construction and Analysis of Steady State Models
# of Enzyme Catalysed Networks

J. A. Burns

With 1 Figure

*Abstract*

A fairly general method for constructing computer models of enzyme systems in solutions is described. The method allows for high order systems and for metabolic inhibition but does not cater for enzyme repression. Both the structure of the enzyme system and the values of the parameters already in the system can be readily adjusted. The problem of finding the steady state and of discovering the rate controlling effect of an enzyme are discussed.—The methods are applied to some simple systems of order 10 enzymes by way of illustration.

## Introduction

A model of a metabolic system will, most naturally, appear as an "interconnection" diagram which displays the important transformations and influences thought to be present in a real system. The "structure" of the system is contained in this diagram. Attached to the diagram will be information of a quantitative nature concerning the values of parameters in the system.

However the problem of "modelling" is not that of writing a single fixed program and running it with variable data. The model is not a fixed thing and will, almost certainly, change frequently in the course of an investigation. For example extra enzymes may be added or perhaps an extra inhibitory effect, "just to see what happens", which may be immediately removed if it seems of no interest. Furthermore the investigator does not wish to be tied to a fixed format of operations but needs to be free to adjust parameters and ask "questions" about the enzyme system with complete flexibility. A further complication from the point of view of "programming strategy" is that the questions which the investigator wishes to ask cannot all be clearly defined in advance, indeed some new "questions" almost always arise as an investigation proceeds.

The above remarks apply particularly to a situation, such as our own, where we have a theoretical interest in the properties of enzyme networks, in addition to our more specific interest in interpreting the arginine pathway in *Neurospora*. A final requirement of "modelling" is that by its nature it demands a turn round of at least twice per day, which in turn requiers, on a shared machine, that the answers to a significant group of questions can be computed in a time of order 15 min.

A general method for modelling must therefore make it easy for the investigator to move from his "annotated diagram" to the computer and also facilitate his consideration of the effects in the model of changes in structure or parametric values. The method must also allow the investigator to ask "pre-programmed" questions about the system in a flexible way, i.e. in any order, and to carry out calculations, as required, on variables and parameters in the model using their normal names. The ability to extend the list of pre-programmed questions and operations as new ideas occur to an investigator is most important and the overall programming technique should make it easy to do this.

## Programming Methods

The machine, an E.E.L.M. KDF9 had Core: 16K of 48 bit words Cycle time: 8 micro-secs; Floating point multiply: 15 micro-secs. The KDF9 is available to all university users and provides a turn round of the order twice per 24 hrs for programs running less than 15 min. For longer runs turn round is unpredictable.

The single pass "load and go" "Atlas Autocode" compiler commonly used in Edinburgh [1] compiles very rapidly from an "Algol like" source language taking of the order 4 min for the largest programs possible with the 16K of core. In view of the availability of a fast Algol type compiler, the requirements mentioned in the introduction can be met most simply using a program structure as follows:

## Start of Program

A. Algebraic description of system under investigation, preceded by list of all names used.

B. Routines which enable analysis to be carried out on system in A and which facilitate setting variables and parameters; (for examples see later).

C. List of commands for the analysis of A subject to what is "pre-programmed" in B, simple calculations etc.

## End of Program

The above layout is a program which does not require any external data but has merely to be compiled and executed. Clearly sections A and C can be very simple, amounting, respectively, to an algebraic description of the system and a sequence of commands taken from a permitted list. The more sophisticated user is still quite free to exploit the full power of a high level language.

If it is desired to alter the structure of the system under study this will mean making adjustments to A or even, if a new system is to be considered, replacing A entirely. Altering parameters and asking "questions" about the system is done in C and C will be completely replaced at each simulation run. Finally, the provision of new "questions" for use in C will mean adding routines to section B.

At no time therefore is there a fixed program which merely accepts data. This leaves us with the considerable task of altering an extensive program quite frequently in sections A, B, and C.

To solve this problem a "pre-compiler" was written in collaboration with Dr. J. G. Burns[1]. This is essentially a program for merging, in a very flexible way, text received from paper tape with text already held on magnetic tape in various files. At present the "pre-compiler" maintains seven program files and a library and has a set of commands enabling the user to modify the contents of one or more files, to build up programs from parts held in different files, and to present the programs thus generated, sequentially to the compiler. All these operations being on a "load and go" basis, the complete set of operations requires only one paper tape as input and thereafter proceeds automatically.

A person without "computer know-how" can easily combine a system A with a set of analytic operators contained in B, about which he need know very little, and then ask "questions" in a flexible way in part C. Furthermore, systems previously studied can be left on file and again placed in A when required, similarly, one of several alternatives could be chosen for B.

The cost of the pre-compiler phase is about 30 seconds overhead and 15 seconds for each simulation assembled and presented to the compiler, which is trivial compared with time spent in compilation: 3 to 4 min and in actually running the simulation: 12 min or more.

## Type of Metabolic System

The computer methods so far discussed impose no limitation on the model but in what follows the model is limited to be that of enzymes and their substrates in solution in a well stirred space.

Since our primary interest is in the properties of the steady state, it is advantageous to make no mention of enzyme-substrate complexes but to use "net-flux" expressions similar to those described by Cleland [2]. A

---

[1] Dr. J. G. Burns, Dept. of Natural Philosophy, Univ. of Edinb.

slight extension enables us to consider enzyme quantity as a variable of the system thus including such effects as repression and induction. The use of "net flux" expressions has several advantages:

1) The number of differential equations needed to describe the system is now only the number of free metabolic pools.

2) High frequency terms, associated mainly with enzyme substrate complexes, have been largely removed from the system which results in a very large gain in computer efficiency.

3) The steady state and the properties of the steady state will not be affected and even the "dynamics" should approximate to the true dynamics, provided enzyme concentrations are several orders of magnitude smaller than their substrates, which is often the case.

4) Most of the parameters involved in describing an enzyme now have an intuitive and operational meaning, e.g. Michaelis constant.

## Analysis

Routines to carry out the necessary operations of analysis are included in part B of the program layout among these are:

1) Steady state $(x)$

This will cause the system to move from its initial position to an exact steady state, such that all the enzymes are within x per cent of the values desired for them. For many purposes a steady state close to a given point in enzyme space is all that is required.

The routine works by integrating the differential equations of the system, the time required on the computer depending critically on the ratio of the highest and lowest frequency in the system i.e. the bandwith. Systems with a large excess of one or more enzymes will have a high bandwith. If the system has no steady state then the routine will "give up" after a preset time.

2) Coefficients

This is a routine for calculating the rate controlling effect of any enzyme on any flux in the steady state system.

Let $f_i$ be a flux at some point in a steady state system and $e_j$ the amount of any enzyme then if $e_j$ is increased by $de_j$ and the system assumes a neighbouring steady state there will be a change $df_i$ in $f_i$. Let $C_{(i,j)}$ = Limit as $de_j$ 0 of $((df_i/f_i)/(de_j/e_j))$ or sensitivity coefficient of flux $f_i$ with respect to enzyme $e_j$ is fractional change in $f_i$ divided by fractional change in $e_j$ which produced it. The routine will calculate and print out the coefficients over all $i, j$ thus enabling the investigator to study the rate-controlling effects of enzymes in his model system.

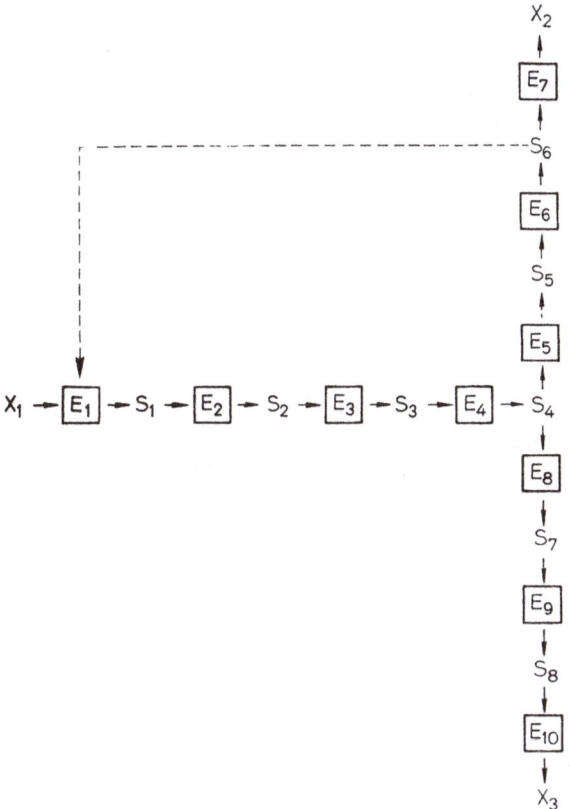

Fig. 1. Test System

The system consists of 10 reversible enzymes all having identical parameters except for enzyme $E_1$ which is subject to non-competitive inhibition from $S6$. External substance $X_1$ is converted via the system into either $X_2$ or $X_3$, the branch point occurring at metabolite $S4$.

With the given parameters, which have no particular significance, the system settles to a steady state with the flux divided equally between the two branches, as would be expected, and with $E_1$ inhibited by 50%.

## Discussion

At present the methods described have been applied only to test systems of order 10 enzymes; one such system is shown in figure 1. The result of asking for "coefficients" on this simple system is shown in Table I, where it can be seen, for example, that $E_7$ has the greatest degree of control, 46%, over fluxes in the main pathway.

The sum of the coefficients in any one column can be seen to be approximately 100 and this has turned out to be a rather general property of steady state enzyme systems (in preparation).

Table 1. *Sensitivity coefficient of fluxes with respect to enzymes.*

Coefficient of $f_i$ with respect to $e_j = 100 \times (df_i/f_i)/(de_j/e_j)$

|  | F1 | F2 | F3 | F4 | F5 | F6 | F7 | F8 | F9 | F10 |
|---|---|---|---|---|---|---|---|---|---|---|
| E1 | 24 | 24 | 24 | 24 | 25 | 25 | 25 | 25 | 25 | 25 |
| E2 | 15 | 15 | 15 | 15 | 15 | 15 | 15 | 15 | 15 | 15 |
| E3 | 4 | 4 | 4 | 4 | 4 | 4 | 4 | 4 | 4 | 4 |
| E4 | 1 | 1 | 1 | 1 | 1 | 1 | 1 | 1 | 1 | 1 |
| E5 | −17 | −17 | −17 | −17 | 17 | 17 | 17 | −49 | −49 | −49 |
| E6 | −6 | −6 | −6 | −6 | 6 | 6 | 6 | −16 | −16 | −16 |
| E7 | 46 | 46 | 46 | 46 | 51 | 51 | 51 | 43 | 43 | 43 |
| E8 | 17 | 17 | 17 | 17 | −16 | −16 | −16 | 51 | 51 | 51 |
| E9 | 5 | 5 | 5 | 5 | −5 | −5 | −5 | 17 | 17 | 17 |
| E10 | 1 | 1 | 1 | 1 | −2 | −2 | −2 | 6 | 6 | 6 |
| Sum | 90 | 90 | 90 | 90 | 95 | 95 | 95 | 96 | 96 | 96 |

# References

1. Edinburgh University Computer Unit. Report No. 4.
2. Cleland, W. W.: Biochem. Biophysica Acta **67**, 104 (1963).

## Discussion

Walter:

Do you assume all steps are far from equilibrium?

Burns:

We make no approximations other than those implicit in our use of "net flux" expressions.

# Computation of Equilibria and Kinetics of Chemical Systems with Many Species

H. J. Bremermann

*Abstract*

An example is given to show that the solutions of a system of differential equations, describing chemical reactions of the second order, are generally of a non-elementary nature. Thus, the solutions cannot merely be analytically "written down"; at best approximate solutions of limited accuracy can be obtained. The number of variables required for metabolic studies might turn out to be quite considerable. The numerical solution is a non-trivial problem. It should be advantageous to start by calculating the equilibrium concentrations which satisfy a system of non-linear equations. Theory and practice of the numerical solutions of such equation systems are underdeveloped—especially if the number of variables is considerable. The author's experimental calculatory methods, still in the process of development, are outlined briefly.

Consider a second order chemical system: Let $X_1, \ldots X_n$ denote the concentrations of the interacting chemical species, $\alpha_i, \beta_{ij}, \gamma_{ijk}$ constants, $\dot{X}_i$ the time derivatives. The equations for the reaction rates have the form:

$$\dot{X}_i = \alpha_i + \sum_{j=1}^{n} \beta_{ij} X_j + \sum_{j,k=1}^{n} \gamma_{ijk} X_j X_k, \quad i = 1, \ldots n \qquad (1)$$

In addition the $X_j$ have to satisfy mass balance equations of the form

$$\sum_{j=1}^{m} c_{ij} X_j + c_{i0} = 0, \quad i = 1, \ldots m, \qquad (2)$$

where the $c_{ij}$ and $c_{i0}$ are constants. Also, the conditions $X_i \geq 0$ must be satisfied.

For some metabolic systems a set of equations (1) and (2) gives an adequate description while others must take into account a distribution of $X_i$ between several compartments and transfer mechanisms between compartments [7a; 16a, b]. Equations (1) also arise in ecology (VOLTERRA theory) (comp. GARFINKEL [16]).

If the equations (1) reduce to linear equations, then their general solution can be written as the sum of at most $n$ different exponential functions. The world, however, does not seem to be linear, while mathematics ceases to be elementary when we have to deal with non-linearities in many variables. M. ROSENLICHT (Berkeley) has communicated to the author the following example which demonstrates this fact: Consider:

$$\dot{x} = x, \quad \dot{y} = xy, \quad \dot{z} = y. \qquad (3)$$

As a general solution we obtain successively through integration: $x(t) = e^{t+a}$, $y(t) = Ce e^{t+a}$, and $z(t) = C \int e e^{t+a} dt$, where $a$ and $C$ are constants. Substituting $u = e^{t+a}$ the integral becomes $\int (e^u/u) du$. ROSENLICHT [17] has given a simple proof that $\int (e^u/u) du$ is *not an elementary function*: it cannot be obtained through finitely many algebraic operations on rational functions, exponentials, logarithms, the trigonometric functions and their inverses. This example shows that the equations (1) in general (except for $n = 1$ and possibly $n = 2$) do not have elementary solutions. No amount of ingenuity will produce formulas for the $X_i(t)$ that are algebraic expressions in $t$ and the elementary transcendentals (as is the case when the equations are linear). In order to obtain a numerical approximation of a solution to (1) one must use computers.

The metabolism of *E. coli* involves, according to WATSON [18], some $2000-3000$ different species of molecules. Even if all of the species and their reactions were known, the numerical solution would transcend the present state of the art of computation. Analog computers are limited to a small number of variables by hardware problems. Analog work is being done by HEINMETS [12] and by J. J. HIGGINS of the Johnson Foundation, Univ. Pennsylvania, Philadelphia, Penna. At the same institution GARFINKEL is doing digital work (Comp. [15] and the bibliography there); he uses a straightforward stepwise integration method (EULER's method). D. F. DeTAR [8], working independently on similar problems with similar integration methods, reports: "It turns out in practice that these quite primitive methods of numerical integration give very good results and are actually more efficient for kinetics problems than many of the more elaborate methods".

Difficulties arise when steady state concentrations of some species are very small while their reaction rates are large. In this case it can be of advantage to compute steady state concentrations by different numerical methods. (Comp. [10].) For a bibliography of computer methods see BARD [2], p. 178. In addition there are methods reported in [6, 7, and 9]; the methods of [7] and [6] are based on minimizing the free energy of the system and the application of non-linear programming techniques.

Knowledge of the steady state concentrations (which we will denote by $X_i^{(0)}$) permits transformation of the equations (1) into a more convenient form. We substitute $X_i = Y_i + X_i^{(0)}$. The new variables $Y_i$ are the deviations from equilibrium. For sufficiently small $Y_i$ the linear terms dominate and the solutions can be approximated by the solutions of a linear system which can be written as a sum of exponentials (a fact much used and abused in metabolism studies). Perturbation methods can also be used.

Computation of the steady state means finding the solution of equations (1) and (2) with $\dot{X}_i = 0$. Alternatively one may use the equilibrium equations of the system which are of the form

$$\sum_{j=1}^{n} X_j^{p_{ij}} - K_i, \quad i = 1, \ldots N; \tag{4}$$

where the $p_{ij}$ are integers between $-2$ and $2$, $K_i$ the equilibrium constants. (In some systems an accurate description is more complicated.) In both cases we have a system of simultaneous non-linear equations of many variables. Linear equations pose a challenge to computers when the number of variables is large ($>100$) and the upper limit of today's technology seems to be of the order of 1000. Non-linear equations are more difficult. Methods found in the literature are sometimes impractical except for small $n$. H. ZASSENHAUS pointed out in his lecture at the Annual Meeting of the German Mathematical Society at Karlsruhe, Sept. 14, 67, that the most efficient computational methods (for many tasks) must be determined by experimentation and through competitive comparison. At present little such comparison of methods developed by different groups takes place due to lack of communication.

In cooperation with J. and N. GOGUEN this author has recently begun experimentation with a variety of computational procedures for equilibrium computation. Let the left hand terms of the equations (4) and (2) be denoted by $\Phi_i$, let $M = m + N$ and let $f = \sum_{i=1}^{M} \Phi_i^2 (X_1, \ldots X_n)$. The point at which $f = 0$ is a minimum of $f$ and is a solution of (4) with balance equations (2). The solution of equations (4) and (2) is thus reduced to optimizing the function $f(X_1, \ldots X_n)$. ANTHONY and HIMMELBLAU [1] have followed a similar approach except that they used a different method (by HOOKE and JEEVES [14]) to minimize $f$. To optimize $f$ we tried to solve $f_{X_i} = 0$, $i = 1, \ldots M$, where $f_{X_i}$ is the partial derivative of $f$ with respect to $X_i$. NEWTON's method for several variables may be used for this purpose (Comp. [11] Sect. 25, and [13]). The method is computationally rather cumbersome for large $M$ since it involves at each iteration the inversion of an $M$ by $M$ matrix of the second partials of $f$. Instead we have tried to use NEWTON's method in each variable separately. In the proximity of the solution good convergence was obtained while with a "bad initial guess" divergence occurred in some cases. Previous experimentation with optimization procedures by the author (BREMERMANN, ROGSON, and SALAFF [4], [5], [3]) has led to a method of "optimization along random rays" (unpublished) which can be used alone, or in order to provide an "initial guess" close enough for the modified Newton's method to converge. The latter turned out to be very effective in a case of linear $\Phi_i$-s, derived from a system of linear equations in 64 variables with condition number 2. A solution accurate to 13 decimal places was obtained in 10 sec of CDC 6400 time. The program was written in FORTRAN IV. In some experiments the balance equations (2) were not included in the $\Phi_i$ but were solved to reduce the number of independent variables. This approach, however, was not very successful. Convergence difficulties occurred and the expressions for the derivatives of $f$ became cumbersome.

Experimentation has been in progress only for a short time. It is hoped that it will lead to efficient programs that can solve equations (4) as well as (1) up to $M = 100$ and greater. A different approach based on minimizing a free energy function rather than $f$ has been pursued by Shapiro [17a]. Our optimization method could be applied to the free energy function as well.

## References

1. Anthony, R. G., and D. M. Himmelblau: J. Phys. Chem. **67**, 1080 (1963).
2. Bard, A. J.: Chemical Equilibrium. New York: Harper and Row 1966.
3. Bremermann, H. J., and H. de Grasse: Logistics Review **3**, 15 (1967).
4. —, M. Rogson and S. Salaff: In: E. A. Edelsack, L. Fein, A. B. Pattee and A. B. Callahan (Eds.), Fundamental Biological Models. Washington: Spartan 1965.
5. — In: H. L. Oestreicher and D. R. Moore (Eds.), Cybernetic Problems in Bionics. New York: Gordon and Breach 1968.
6. Clasen, R. J.: The numerical solution of the chemical equilibrium problem, RAND Corporation (Santa Monica, Calif.) Report RM-4345-RP Jan. 1965.
7. Dantzig, G. B, J. Folkman and N. Shapiro: J. Math. Analysis and Appl. **17**, 519 (1967).
7a. —, J. C. de Haven, I. Cooper, S. M. Johnson, E. C. De Land, H. E. Kanter and C. F. Sams: Persp. Biol. Med. **4**, 324 (1961).
8. De Tar, D. F.: J. Chem. Educ. **44**, 191 (1967).
9. — J. Chem. Educ. **44**, 193 (1967).
10. — and D. E. De Tar: J. Phys. Chem. **70**, 3842 (1966).
11. Forsythe, G., and C. B. Moler: Computer Solutions of Linear Algebraic Systems. New Jersey: Prentice-Hall, Englewood Cliffs 1967.
12. Heinmets, F.: Analysis of Normal and Abnormal Cell Growth. New York: Plenum Press 1966.
13. Henrici, P.: Elements of Numerical Analysis. New York: Wiley 1964.
14. Hooke, R., and T. A. Jeeves: J. Assn. Comp. Mach. **8**, 212 (1961).
15. Garfinkel, D.: J. Biol. Chem. **241**, 3918 (1966).
16. — J. Theoret. Biol. **14**, 46 (1967).
16a. Ličko, V.: Bull. Math. Biophys. **25**, 141 (1963).
16b. — Bull. Math. Biophys. **28**, 379 (1966).
17. Rosenlicht, M.: Pac. J. Math. early **24**, 153 (1968).
17a. Shapiro, N. Z.: A generalized technique for eliminating species in complex chemical equilibrium calculations. RAND Corporation (Santa Monica, Calif.) Report RM-4205-PR, Sept. 1964.
18. Watson, J.: Molecular Biology of the Gene. New York: Benjamin 1966.

## Discussion

WALTER:

With regard to the definition of the steady state and the assertion that some $\dot{X}_i = 0$: (1) for closed systems $\dot{X}_i(t) = 0$ $(0 < t < \infty)$ only at extrema (instantaneously). However, for closed systems there exist mass constraints of the form: $\Sigma X_i = $ constant. In this case the conditions for an approximate steady state derive from $\Sigma \dot{X}_i = 0$. Whenever some $|\dot{X}_i|$ are small compared to other $|\dot{X}_i|$ the small $\dot{X}_i$ are considered to be approximately zero. Thus the steady state approximation is never exact (except at extrema or equilibrium) in closed systems. For this reason one refers to "quasi-steady-states" for closed systems (see for example: C. WALTER, J. Theoret. Biol. **11**, 181 (1966)). (2) For open systems $\dot{X}_i(t) = 0$ at extrema or during the stable state period of minimum entropy production (the steady state). Thus one can refer to "steady states" for certain open systems.

BREMERMANN:

The equations can describe both an open system and a closed system depending upon the coefficients. With $a_i$-s non-zero, and appropriate mass balance equations, there can be a net flow. "Steady state" means $\dot{X}_i = 0$; if the system happens to be closed $\dot{X}_i = 0$ would correspond to the equilibrium.

BURNS:

We are interested in a rather similar problem, that of finding a stationary state for the equations of motions of an enzyme catalyzed network. For this problem we have found that "netflux" expressions, similar to those described by W. W. CLELAND (Biochim. Biophys. Acta **67**, 104 (1963)), are advantageous. These reduce substantially the number of differential equations for the motion of the system by eliminating equations concerned with the movement of enzyme substrate complexes. Furthermore the equations so eliminated are often those containing the "fast" movements in the system; the remaining differential equations can then be integrated, using a much larger step length, until a stationary state is achieved.

BREMERMANN:

Using the system itself for adjusting the $X_i$ seems like a good idea. We plan, however, to experiment with a variety of methods to determine the most useful ones. There is a lack of theory about the efficiency of numerical methods (theoretical error estimates are usually above practically obtainable values). Different groups of workers have developed their own methods, but there is a lack of communication and publication. We have just started and plan to experiment with a variety of methods. We have a few ideas that look promising.

# The Relation between Heat Production, Oxygen Consumption and Temperature in Some Poikilotherms

J. N. R. GRAINGER

With 2 Figures

*Abstract*

A calorimeter is described which is suitable for measuring the heat production of aquatic poikilotherms. The relation between the rates of oxygen consumption and heat production is given for bakers yeast between 20° and 35°. 1 ml oxygen was equivalent to 2.6 to 4.9 calories with glucose as substrate. Similar equivalences are given for *Asellus*, *Nucella*, *Limnaea*, *Lumbriculus* and *Rana* tadpoles.

LAVOISIER in 1780 was the first to measure the amount of heat produced by an animal; RUBNER [5] was the first to show that the heat calculated from metabolism was very close to that which could be measured directly in a calorimeter. Much direct and indirect calorimetry [1, 2] has since been carried out on man and domestic animals. Little has however been done on poikilotherms [3]. A programme has been started in this Department on the energetic efficiency of poikilotherms at different temperatures. This paper is concerned with one aspect of this, namely the relation between the rate of oxygen consumption and the rate of heat production.

*Methods:* The calorimeter used in the heat measurements consisted of a Dewar flask of 230 ml capacity closed by a rubber bung. Three glass tubes projected through the bung. One of these carried the leads to a nichrome heating coil, another carried leads to two glass-covered thermistors (F 23, Stantel) and the third was connected to a small glass pipette for bubbling oxygen or nitrogen through the flask contents. The thermistors were connected to the circuit shown in Fig. 1. To ensure adequate electrical insulation the thermistor probes projected from a small rubber bung attached to a glass tube, inside which lay the junctions between the thermistors and their leads, and which was joined by rubber tubing to one of the three glass tubes projecting through the flask bung. This tube contained some silica gel.

In most experiments 50 ml of medium was used in the flask. The flask was totally immersed in a constant temperature tank controlled to $\pm 0.02°$ by a mercury toluene regulator with proportioning head. It was found necessary to immerse the fixed resistor (C) in a tube in the constant temperature tank. At temperatures of 30° and higher (C) was replaced by a 1 kohm resistor. Heat flow between the flask contents and the bath was measured in a series of calibration experiments by observing the rise (or fall) of temperature in the flask for various temperature

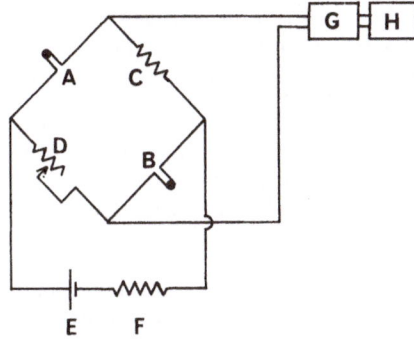

Fig. 1. Calorimeter circuit. A, B, thermistors (Stantel F 23); C, 2 k. ohm wire wound resistor; D, decade resistance box reading to 0.1 ohm; E, 2v. accumulator; F, 15 k.ohm resistor; G, amplifier (Pye nanoammeter); H, galvanometer (Pye Scalamp)

differences between flask contents and bath. Known amounts of heat were added in a further series of calibration experiments by means of the nichrome heating coil and from the rise in temperature the heat capacity of the flask and contents was calculated. With this apparatus heat outputs of 3 calories per hour could readily be measured. It was found necessary when bubbling gas through the flask contents to ensure that this was completely saturated with water vapour and at the same temperature as the bath. This was done by bubbling it through four flasks containing water, which were immersed in the constant temperature tank. A typical experiment is shown in Fig. 2.

The bakers yeast was obtained daily from a local shop. The suspension medium was 0.066 M $KH_2PO_4$—$Na_2HPO_4$ buffer, pH 5.2 containing glucose to a concentration of 1%. Owing to the absence of a N source there was no significant

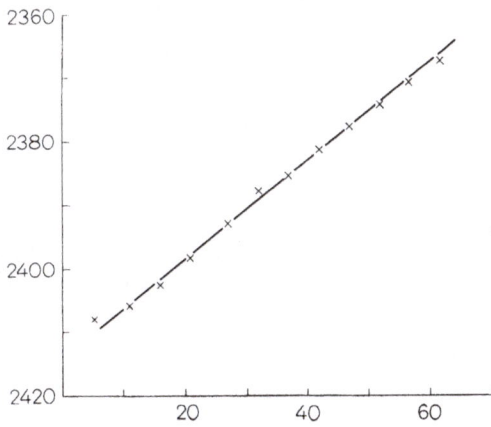

Fig. 2. A typical experiment with bakers yeast. Vertical axis: reading of D (Fig. 1) in ohms; Horizontal axis: min. The slope corresponds to a heat production of 15.5 cal./hr. When corrected for heat flow (1.36 cal./hr in this experiment) and N content the final result is 4.79 cal./mg N/hr at 20.4°

growth of cells in these experiments. Oxygen consumption of the same yeast suspension was measured with a Warburg apparatus. Nitrogen was estimated using Kjeldahl's method followed by steam distillation of aliquots of the neutralised diluted digest into boric acid which was then backtitrated with 0.01 N. HCl using mixed indicator. The same methods were used with the other organisms. With *Nucella* the suspension medium was sea water. Tap water which had been passed through activated charcoal was used for the remaining organisms.

The bridge was balanced by altering the variable resistance D (Fig. 1) until the galvanometer H showed no deflection. The reading of D was then noted. Readings were taken every five minutes for a period of at least 1 hr (after an initial equilibriation period of 1—2 hrs). Oxygen was slowly bubbled through the flask for $^1/_2$ min every 5 min but in some of the yeast experiments bubbling was continuous. Tests showed that adequate oxygen was present throughout each experiment. In most experiments the flask contents were slowly stirred by a magnetic stirrer.

*Results*: The results are summarised in the Table. Also included are some of the findings of other authors. These have been recalculated to convert them to the same units as are used in the present work. The dog data is that obtained by LUSK as quoted in [2]. At the bottom of the Table are the calculated oxygen/heat equivalences for four substances. The values for these are based on their complete oxidation. The $\Delta H$ values were taken from [1]. In the case of alanine and glutamic acid the oxidation

Relation between heat production and $O_2$ consumption

| | No. of Exps. | Temp. °C | $O_2$ consumption $\mu l/mg\ N/$ hr | Heat production cal/mg N/ hr | Equivalent of 1 ml $O_2$ (calories) |
|---|---|---|---|---|---|
| Bakers yeast | 9 | 20.4 | 1170 | 5.76 | 4.92 |
| Bakers yeast | 5 | 25.8 | 2305 | 6.16 | 2.67 |
| Bakers yeast | 5 | 30.0 | 2435 | 9.47 | 3.89 |
| Bakers yeast | 6 | 34.9 | 3359 | 12.12 | 3.61 |
| Lumbriculus sp. | 2 | 20.4 | 22.1 | 0.25 | 11.26 |
| Asellus aquaticus | 2 | 20.4 | 27.5 | 0.12 | 4.47 |
| Nucella lapillus | 1 | 20.4 | 16.8 | 0.08 | 4.69 |
| Limnaea pereger | 1 | 20.4 | 20.3 | 0.05 | 2.60 |
| Rana temporaria tadpoles | 1 | 20.4 | 43.2 | 0.30 | 6.95 |
| Bird erythrocytes (5) | | | | | 4.3—5.0 |
| Vibrio metschnikoff (5) | | | | | 5.6—6.4 |
| Strongylocentrous larvae (4) | | | | | 3.9 |
| Dog (2) | | | | | 4.85 |
| Glucose | | | | | 5.01 |
| Palmitic acid | | | | | 4.65 |
| Alanine | | | | | 4.61 |
| Glutamic acid | | | | | 5.21 |

is to $CO_2$, $H_2O$ and $N_2$. Some other amino acids have higher oxygen/heat equivalences but almost all dietary substances lie between 4.5 and 6.0 cal.

The rate of oxygen consumption is a measure of the activity of the oxidation systems in the cells of an organism. The rate of heat production on the other hand is the sum of the enthalpy changes ($\Delta H$) for all the reactions going on in the cells. Each measures a somewhat different aspect of metabolism. One of the weaknesses of the present experiments is that they were not measured simultaneously. Respiration in the Warburg flask may not have been quite the same as in the calorimeter. This, together with such factors as the possible intermittent use of oxygen or the repayment of temporary oxygen debts may account for the low value obtained with *Limnaea*. The high value for *Lumbriculus* is almost certainly due to some anaerobic breakdown of substrates.

In the yeast experiments the values tended to be lower with increasing temperature. Here the main energy source was glucose and should have given a value of 5. This was only attained in the 20° experiments. The explanation of the lower values at the other temperatures is not known but could possibly be ascribed to side reactions. There are grounds for thinking that the very low results at 25.8° were due to a slight settlement of cells which may have occurred in the experiments at this temperature. Other authors [7] have found that when the pattern of metabolism of an organism is accurately known agreement is very good between expected and observed heat productions.

## References

1. BLAXTER, K. L.: The energy metabolism of ruminants. London: Hutchinson 1962.
2. BRODY, S.: Bioenergetics and growth. New York: Reinhold 1945.
3. DAVIES, P. M. C.: Ph. D thesis University of Dublin 1965.
4. MEYERHOF, O.: Biochem. Z. **35**, 280 (1911).
5. — Pflügers Arch. ges. Physiol. **146**, 159 (1912).
6. RUBNER, M.: Z. Biol. **30**, 73 (1894).
7. SENEZ, J. G., and J. P. BELAÏCH: Colloques Internationaux du Centre National de la Recherche Scientifique. Marseille 1963. No. 124, 357.

*Discussion*

WOODSTOCK:

I wish to ask you two questions: (1) Was respiration also measured as $CO_2$ production? Oxygen uptake might also have occurred by non-respiratory reactions. (2) Were P/O ratios determined to estimate the efficiency of respiration in producing ATP? If ATP production were uncoupled from respiration, how would this affect the conversion factor between respiration and heat production?

GRAINGER:

(1) $CO_2$ production was measured in the yeast experiments but not in the others. Some $O_2$ may have been used by non-respiratory reactions but the amount is not likely to be large. — (2) This ratio was not determined. I do not know what effect uncoupling ATP would have on the heat production/$O_2$ consumption ratio.

SMITH:

In comparisons of direct calorimetry with oxidative exchange it is most important to establish steady states and states of material balances.

GRAINGER:

I quite agree. Tests showed that the steady state was approached in these experiments. It must also be remembered that conditions inside the calorimeter are rather unnatural and different from the conditions in the Warburg flask.

# On Structure and Metabolic Rate of the Skeletal Muscle of the Eel (*Anguilla vulgaris* L.)

H.-D. Jankowsky

With 1 Figure

*Abstract*

The metabolism of the intact eel and of its excised muscles show the same pattern of temperature acclimation; but under certain conditions opposite responses can be obtained. Thus, the oxygen consumption of the fish does not completely follow the respiratory activity of muscle tissue. In the eel the oxidative capacity of the muscle mainly depends on the small amount of red fibres; there are great differences in the mitochondria content between white and red fibres. These histological results correlate quite well with the reported measurements of in vitro metabolism.

The temperature acclimation of the eel has been the subject of several investigations [5], which have established some similarities between the metabolism of the fish and of its isolated muscles; both have the ability to partially compensate: when measured at the same test temperature the metabolic rates of eels kept at a colder acclimation temperature are higher than the rates of fish kept for the same time in warmer water. But if an eel was maintained in an artificial "two-temperature condition" (the anterior end and the main body mass adapted to different temperatures), only the oxygen uptake of the muscles reflected the different acclimation conditions. The oxygen consumption of the whole fish was similar to that of animals which were kept completely at that temperature to which in this fish only the head was exposed. Attempts to assess quantitatively the relative contributions of the body musculature to changes in the oxygen consumption of the entire eel were complicated by the fact that the respiratory activity of the muscles in vitro is not equal in all parts of the body. The anterior body muscles show only 74% of the respiration of similar tissue from the tail. The different oxygen consumption could not be explained by changes in water, protein or fat content.

Even in a macroscopic investigation, however, there are to be seen two types of muscle in cross-sections through the eel body, especially in the posterior end of the fish. Some distinct bundles of red fibres are found in the peripheral parts, and the main muscle mass consists of white fibres.

In fish the red and white skeletal muscles have different structures and functions [1, 3]. As the cytochrome oxidase activity gives a measure of the oxidative metabolic capacity of a tissue, we compared manometrically the activity of this enzyme in homogenates from different parts of the musculature of the silver eel. The cytochrome oxidase activity of the deeper white muscles is approximately the same from the front and the tail of the fish ($O_2$-uptake 160 $\mu l$/100 mg wet weight/hr). When one takes a peripheral muscle sample from the tail containing red fibres the enzyme activity is much higher ($O_2$-uptake 350 $\mu l$/100 mg wet weight/hr). Thus the difference in the oxygen uptake of muscle from the anterior and the posterior parts of the body is due only to the red fibres, although the amount of these fibres is very small. Frozen sections across body musculature of adult eels and 10 cm long young eels enable us to estimate, by measuring the areas, that about 5 % of the muscle from the tail region consists of red fibres; in the head region the value is still lower.

From the histological point of view the two types of muscle fibres are markedly different. In the young eel the red musculature is 100—120 $\mu$ thick; the sarcoplasm-rich fibres have a diameter of about 37 $\mu$. They contain a lot of fat (sudan red test) and show a high succinic dehydrogenase activity (nitro-BT test). The diameter of the white fibres is 77—110 $\mu$ and the succinic dehydrogenase activity is much weaker. Electron microscopic studies show that in longitudinal sections the proportion of the total fibre area occupied by mitochondria is less than 1 % in white fibres, but in the red fibre approximately 15 %. The white fibres contain only a few small mitochondria in the sarcoplasmic surface layer; between the fibrils they are absent (Fig. 1a). In the red muscle fibre the sarcoplasmic coat is 1.5—2 $\mu$ thick and contains many mitochondria, especially in the neighbourhood of capillaries. Only in the red fibres are elongated mitochondria found between the fibrils (Fig. 1b): (1 $\mu$ broad, up to 8.5 $\mu$ long); attached to and between the mitochondria lie many fat droplets.

Some preliminary investigations indicate that there are variations of mitochondrial ultrastructure in the red muscles of eels adapted to different temperatures; mitochondria of cold acclimated fish seem to have more tightly packed cristae. As the oxidative capacity of certain tissues has been correlated with the number and structure of their mitochondria, such a change would agree with the metabolic data of eel muscle. In the Cyprinid fish, Idus idus, an increase in the mitochondria content of muscles takes place during the process of cold acclimation as was previously shown by conventional staining techniques [2].

Little is known about the physiological properties of the red muscle in fish. PROSSER [4] has measured the tonic electrical activity in the body wall muscles of eels. There was a striking correspondence between tonic action potentials and the tissue oxygen consumption with regard to the two ends

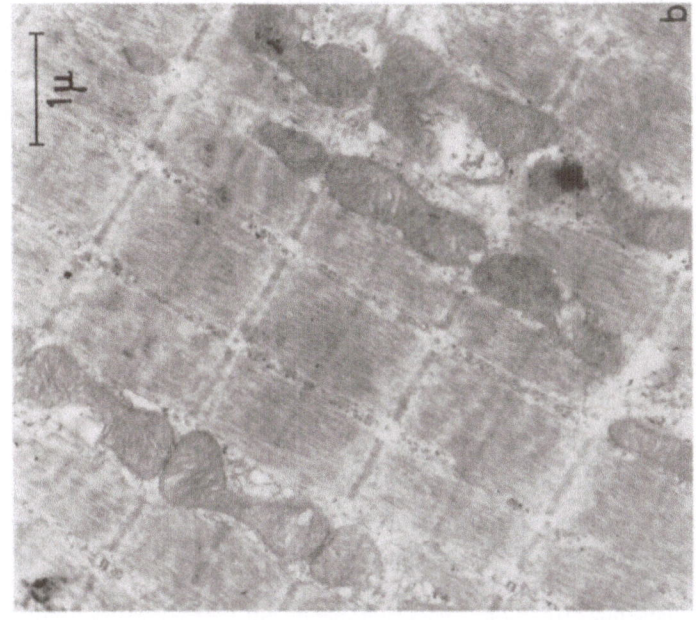

Fig. 1. Longitudinal sections of muscle fibres in the dorsal skeletal muscle of the eel; *a*: white fibre, *b*: red fibre. 16500 × magnification

of the fish. These observations together with the present results indicate that the red fibres have a tonic function. WITTENBERGER's [6] investigations suggest that in the red muscles substances are metabolized which are produced in the white muscle as a result of anaerobic metabolism. The high mitochondria content and the good vascularisation of these fibres agree with this concept.

## References

1. BARETS, A.: Arch. Anat. Micr. Morph. Exp. **50**, 91 (1961).
2. JANKOWSKY, D., u. H. KORN: Naturwissenschaften **52**, 642 (1965).
3. KROMPECHER, S., et al.: Acta Biol. Hung. **16**, 389 (1966).
4. PROSSER, C., et al.: Naturwissenschaften **52**, 168 (1965).
5. SCHULTZE, D.: Z. wiss. Zool. **172**, 104 (1965).
6. WITTENBERGER, C.: Rev. Roum. Biol. (Zool.) **12**, 139 (1967).

*Discussion*

GRAINGER:

Have you done any experiment on isolated red muscle kept in organ culture at different temperatures?

JANKOWSKY:

No, we have not carried out such experiments, but we soon plan to start with experiments on cultured leucocytes of fish.

KRÜGER:

Did you measure the $O_2$-consumption of red and white muscles?

JANKOWSKY:

The red muscle can in vitro consume more than nine times more oxygen than the white muscle.

# Analysis of Heart Rate Adaptation to Temperature

E. Zerbst

With 3 Figures

*Abstract*

The capacity adaptation of the heart rate to increasing or decreasing temperature is determined by ion concentrations of the extracellular space and by the ion permeabilities of the pacemaker membrane. Parameters (driving forces, velocity coefficients and compartment dimensions) are qualitatively determined by experimental arrangements. The parameter-relations are identical in frog, rat and dog heart. The causal mechanisms of adaptation are derived by coordinating the experimental data with the requirements satisfying flux-equilibrium and irreversible thermodynamics.

A sudden increase of temperature at the pacemaker of the heart results in a steep frequency acceleration; however, in spite of the constancy of the newly achieved temperature level the pacemaker impulses subsequently decrease to a medium steady state rate. The phenomenon of this overshooting frequency in short term adaptation to temperature reminds us of influences expected by feedback regulations. The question may be raised, whether another possibility for understanding this regulatory mechanism can be considered [6].

L. v. Bertalanffy described the characteristics of first-order-regulations in thermodynamically open reactions [1]. The kinetics of these systems exhibit overshooting and undershooting processes as a consequence of a parameter variation; it is exactly the same phenomenon which is observable in capacity adaptation to temperature.—In our previous studies we tried to clear up the problem whether an equivalent mechanism for the adaptation of heart rate to temperature exists. These experiments were done on hearts of frogs, rats and cats; the question was: What components of the ion transport kinetics at the membrane of heart pacemakers (i.e. driving forces, velocity coefficients and fluxes) are parameters of adaptation? The experimental results obtained have been correlated with the kinetics of simple steady state systems. After having coordinated the experimental results with the properties of thermodynamically open systems, a hypothesis on short term capacity-adaptation was proposed and examined carefully by specially designed experiments.

The simplest flux equilibrium model of metabolism is the monomolecular reaction with concentrations $A$ and $B$. In order to maintain the conditions for an open system, $A$ must be replenished from a source $S$ and $B$ must be removed to some sink $Z$. One way to establish this is by considering diffusion; by adding in diffusion processes the system's scheme will be like figure 1. Here $k_1$ and $k_3$ are diffusion constants or permeabilities; the

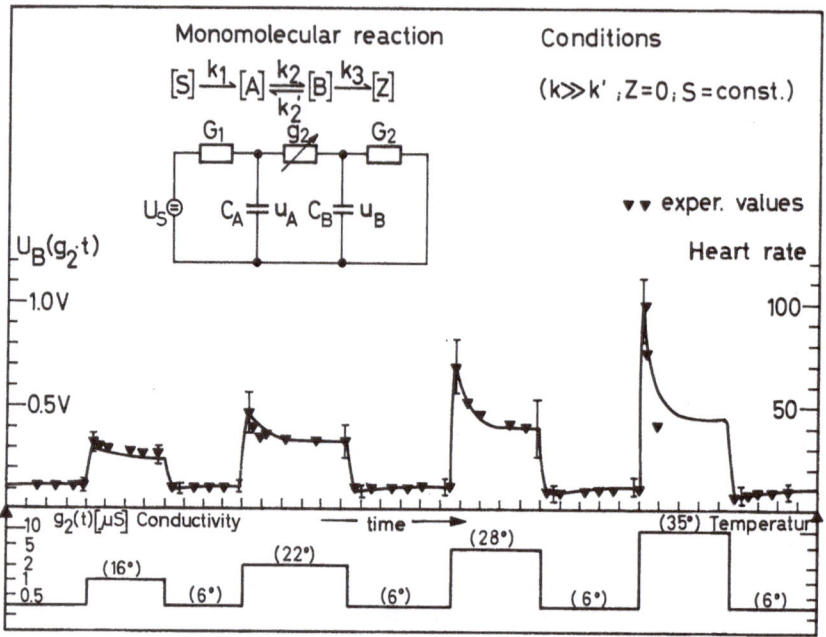

Fig. 1. Flux equilibrium model of metabolism and its circuit equivalent. Simulation of adaptation curves by the model and parameters of the model

processes may be either free or membrane diffusion. Probably the latter is common in the pacemaker systems discussed here. The chemical velocity coefficient $k_2$ depends on temperature to a higher degree than diffusion. After having raised $k_2$ suddenly by temperature, the concentration of $B$ is elevated as an overshoot and then decreases in dependence on the decrease of the concentration $A$. Thus, the driving force of the reaction from $A$ to $B$ will be diminished as a function of time and $k_2$. The detailed description is given according to the transfer of MICHAELIS-MENTEN's reaction to thermodynamically open systems [2].

Fig. 1 demonstrates the experimental results on frog heart and on an electrical equivalent circuit of the monomolecular reaction in open systems. To simulate the effect of temperature on $k_2$ the conductivity $G_2$ was varied exponentially; jumps in conductivity resulted in an overshooting timefunction of the electrical potential on the capacitance $C_2$. This potential was recorded by an automatic pen. In the figure the uninterrupted line shows the response of concentration $B$ to rectangular temperature jumps; the parameters of the model were chosen in such a manner as to simulate the results of frog heart capacity adaptation.

In this way, the parameters of the pacemaker metabolism were summarized by the parameters of the steady state model and the biological results examined by variation of the experimental conditions. A close analogy was

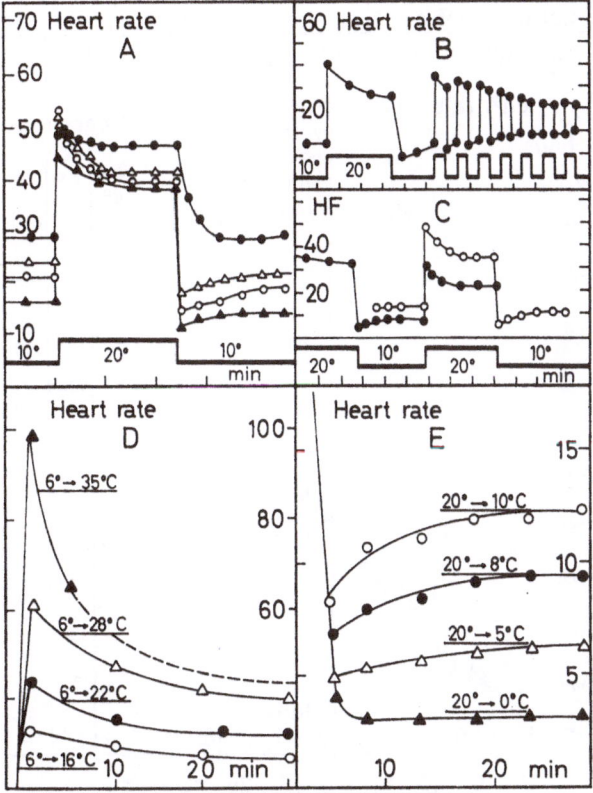

Fig. 2. Results of various experiments on the frog's heart capacity adaptation to temperature

maintained in every variation, for example after a slow temperature rise, with stepwise jumps or with very frequent temperature variations (Figure 2). In the diagram of fig. 3 the dynamic (overshoot) and static (steady state) values of heart frequency are plotted against temperature together with the resulting value of $k_2$ of the model system. A sigmoid plot of the temperature response curve results; the equivalent of the model reactions were calculated with the aid of a nomogram [4]. The values of the concentration $B$—now associated with the energetic source of "temperature response"—are plotted against those of $k_2$ in a logarithmic manner ($k_2$ depends on temperature exponentially). Furthermore the figure demonstrates the equivalence of the pacemaker systems in all three species, viz. frog, rat and cat, with respect to temperature reactions. If an equivalent scale is chosen for the specific temperature and frequency ranges, then the stimulus-response curves in the three species become identical. The inflection of the curves lie in ranges either of physiological body temperature or of optimal ambient

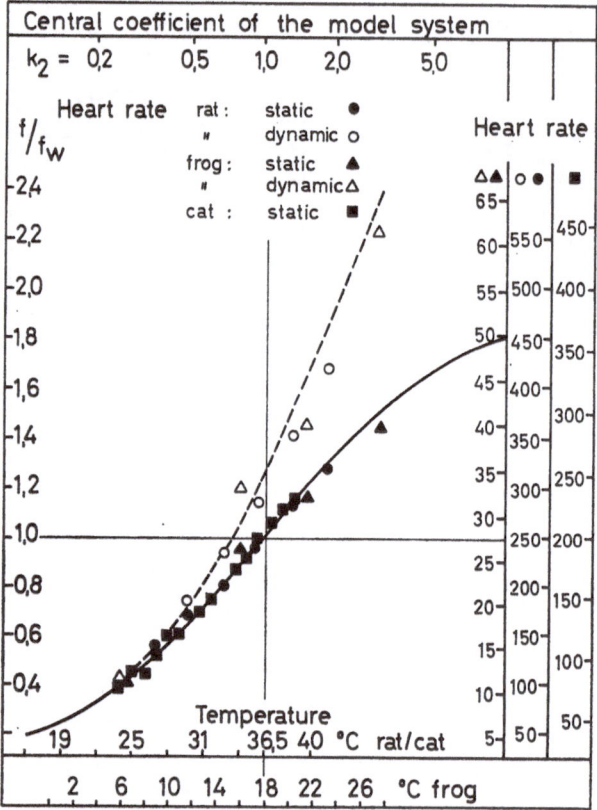

Fig. 3. Dynamic and static values of heart frequency plotted against temperature. Equivalent temperature—response curves of the model

temperature. The sigmoid shape of the temperature-response curve results from the limiting character of diffusion or other more or less temperature-independent processes. This fact may also be responsible for the well known decrease of the temperature quotient $Q_{10}$ with rising temperature range. We have dealt with this fact in a special paper [4].

Having discussed the equivalence of model and pacemaker reactions, the following question remains to be answered: What are the *real* metabolic factors (driving forces, coefficients and fluxes) in the pacemaker system that influence the steady state transition of heart rate in capacity adaptation to temperature? It is well known that heart frequency is a function of the rapidity of diastolic depolarisation of the pacemaker tissues, the rapidity being a function of the ion diffusion potentials and the specific permeabilities. According to NERNST's law, diffusion potentials are dependent on the actual ion concentration differences in- and outside the cell; they are real steady state values. Of course, they are dependent on the active and passive ion transport across the membrane. If the unidirectional passive sodium influx is greater than the active transport from inside the cell to the intercellular space, the sodium quotient $Na_0/Na_i$ is lowered and a decrease of the heart frequency results; reciprocal conditions hold true for the potassium potential.

The inactivation of acetylcholine at the membrane is increased by temperature more markedly than its rate of replenishment; permeability of potassium depends on the actual concentration of acetylcholine. The lowered permeability as a consequence of temperature rise increases the rapidity of diastolic membrane-depolarization. In the time course of heart frequency the balance of potassium transport will be shifted toward a higher rate of active transport; whereas, because of frequent depolarizations, the balance of sodium transport is shifted towards a higher passive influx rate. Thus, the higher temperature leads to a gradually increasing potassium potential and a gradually decreasing sodium potential. Following a rise in temperature, therefore, the heart rate will be lowered starting from the first maximum-value; it will reach a steady state according to the new ion transport-balance.

This briefly outlined hypothesis has been tested by experiments with capacity adaptation after certain variations of permeability and ionic potentials. This was done by adding to the perfusion medium of the heart various substances, such as ACh, atropine, eserine, epinephrine and β-receptor blocking agents. The ionic potentials have been varied by diminishing or increasing the concentration of extracellular sodium or potassium (or chloride). Under these special circumstances, the steady state transitions behaved exactly in the mode predicted by our hypothesis.

Finally we carried out another test in order to support the validity of the hypothesis. At constant temperature the heart rate was doubled for a certain time by means of electric pacemakers; thus we varied the permeabilities and ionic fluxes electrically. When the driving was stopped, the heart rate exhibited an undershoot before attaining its spontaneous frequency. The curves of heart rate were equal to those which reached its new steady state following a sudden drop of temperature. In the last-mentioned experiments the permeabilities and ionic concentrations have been simultaneously varied by means of humoral and ionic factors, as was done in the experi-

7*

ments on capacity adaptation to temperature. The results obtained are in principle well consistent with predictions made on the basis of the hypothesis on flux equilibrium in the metabolism of the pacemaker system. The special set-up of these experiments was derived from the theories of open systems and irreversible thermodynamics. The application of these theories to physiological experiments appears promising.

## References

1. Bertalanffy, L. v.: Biophysik des Fließgleichgewichts. Braunschweig: Vieweg 1953.
2. Burton, A. C.: J. cell. comp. Physiol. **14**, 327 (1939).
3. William, E., u. E. Zerbst: Z. wirtsch. Fertigg. (to appear 1968).
4. — — Intern. elektron. Rdsch. **4**, 100 (1967).
5. Zerbst, E.: Helgoländer wiss. Meeresunters. **14**, 51 (1966).
6. —, G. Hennersdorf u. H. v. Bramann: II. Intern. Kongr. Biophysik, Wien, Sept. 1966.

*Discussion*

Grainger:

Denbigh, Hicks and Page have shown that it is possible to get "false start" transients in your theoretical system. Have you found these?

Zerbst:

We never found "false start" transitions in the object under discussion.

Smith:

The model would appear to require consideration of the possible role of divalent cations.

Zerbst:

We plan to investigate the role of $Ca^{++}$ and of other divalent cations in further experiments. So far, we only know that $Ca^{++}$ influences the permeability, i.e. the coefficients for passive fluxes of sodium and potassium in our more complex model.

Christophersen:

You have described the changes of metabolic rates induced by change in the experimental temperature as temperature adaptation. I would like to draw your attention to the concept of temperature adaptation after Precht, Prosser et al., who agree in the opinion that 3 phases are to be distinguished in the time course of temperature-induced changes of performance activities: (1) a step change of the rate, frequently demonstrating an "overshoot" in activity. (2) a readjustment of a new steady state and (3) a slow compensation of the rate which in the ideal case may lead back to the original activity. The last mentioned compensation is accompanied by changes in the parameters, e.g. enzyme concentrations, etc. Only such a compensation should correctly be termed "temperature adaptation" in the proper sense, whereas a readjustment of the steady state occurring immediately after the temperature shift should not be understood as adaptation.

ZERBST:

I agree with you that the term "adaptation" should be used only in clearly specified cases (In human physiology we have found 38 different definitions!). I prefer, therefore, to use the expression "short-term capacity adaptation". Of course, one cannot observe isolated tissues or organs long enough to justify the use of the term "adaptation to temperature" in the sense of PRECHT's definition.

# The Heat Production of Goldfinches and Canaries in Summer and Winter

S. Gelineo

*Abstract*

Investigations of heat production were carried out in July and January on 6 male goldfinches *Carduelis carduelis* and 6 male canaries *Serinus canaria* in a resting 16 hrs unfed state. By a number of previous experiments the birds were familiarized with the investigator and the experimental technique. The experiments have shown that both species have a lower heat production in summer compared with that in winter at the neutral temperature and at 1 °C.

Whereas the results of several authors indicate that the basal metabolic rate (BMR) does not depend on the season, there are many other papers recording either an increase of BMR in winter and correspondingly a decrease in summer, or the opposite [6, 8]. This means that the problem of seasonal heat production in homoiotherms is still open.

Our present interest is focussed on the following point: how do BMR and heat production (at about 1 °C) behave in two kinds of birds, whose geographical distribution and, hence, average environmental temperatures are different? One of them, the goldfinch *Carduelis carduelis*, is distributed throughout Europe and even reaches the polar region [9], whereas the other, the canary *Serinus canaria*, perfers warmer regions.

## Methods

Both the summer and winter experiments on male *C. canduelis* and *S. cararia* were done on the same individuals caged in open air and placed several times in the respiratory apparatus prior to the experiment in order to adjust them to the experimental conditions. Measurements of gas exchange (60—80 min) were done according to [5] between 7 and 10 a.m. (relative humidity 52—70%) both in July (2—3 months after nesting) and in January. The environmental temperature at which the canaries lived was 23—28° C in summer 4 weeks prior to the experiment, for the goldfinches 22—28 °C; in winter the temperature was—6 to +8 °C for canaries and 0—10 °C for goldfinches. — Metabolism was calculated as calories/m²/24 hr; the body surface was calculated by Meeh's formula (K = 10).

## Results

### 1. General Remarks

Birds brought in summer to winter temperature (1 °C) soon fell into hypothermia of 3.3 °C (goldfinch) or 3.8 °C (canary). In *C. carduelis* panting occurred at ambient temperature of 29.2 °C in winter and of 33.6 °C in summer. In winter canaries at ambient temperature of 36.1 °C panting occurs, whilest in summer canaries at 37 °C. These results confirm earlier observations on birds [4, 5].

## 2. Basal Metabolism

By analyzing BMR obtained at neutral temperatures, we find that BMR appears in winter at 30 °C environmental temperature, after a stay of appr. 3—4 weeks at —6 to +8 °C; in summer it occurs at 33 °C, if the birds are living at temperatures between 23 and 28 °C. The neutral temperature of the goldfinch is in summer (for environmental temperatures varying from 22 to 26 °C) appr. 32 °C and in winter (for environmental temperatures from 0—7 °C) appr. 28.3 °C. This phenomenon, thoroughly examined in several kinds of birds, depends directly on adaptation temperatures. The results on heat production of the two species of birds, measured at the respective neutral temperature, are given in Table 1. By comparing BMR of the birds, obtained in July and in January, with earlier findings at different adaptation temperatures during one season [4, 5], it is seen, that a close correspondence exists: in cold season, high BMR; in warm season, low BMR occurs (the differences are statistically highly significant).

## 3. Heat Production at about +1 °C

The problem of thermoregulation in different seasons is an age old one, known since EDWARDS [3] hinted at the seasonal variation in the capability of birds to produce heat. In confirming the early observations made by EDWARDS, our results (Table 1) show that also at low experimental temperature of about 1 °C birds have in winter a higher metabolism than in summer.

## Discussion

In the range of temperature allowing BMR in both warm and cold seasons, the heat production of goldfinches is higher than that of canaries, which could be interpreted as a condition for the wide distribution of the goldfinches in Central Europe. However, the observed result also seems to demonstrate that in goldfinches the morphological type of seasonal adaptation has not yet been achieved as a consequence of the improved insulation by pelt or plumage [8], this type of adaptation allows animals of the northern regions to produce less heat at the same external temperatures than animals of southern regions.

Table 1

| Investigation | Species | N | Summer | | | N | Winter | | |
|---|---|---|---|---|---|---|---|---|---|
| | | | Weight (g) | Expt. Temp. (°C) | Cal/m²/24 hr | | Weight (g) | Expt. Temp. (°C) | Cal/m²/24 hr |
| BMR | S. canaria | 6 | 15.62±0.70 | 33.1 | 1183±106 | 6 | 15.80±0.66 | 30.4 | 1694±167 |
| | C. carduelis | 6 | 16.67±0.64 | 32.1 | 1209±158 | 6 | 16.60±0.75 | 28.4 | 1532±82 |
| Metabolism ~1°C | S. canaria | 6 | 15.70±0.73 | 0.8 | 3249±178 | 6 | 15.80±0.71 | 0.9 | 4146±216 |
| | C. carduelis | 6 | 16.70±0.56 | 0.5 | 3361±305 | 6 | 16.63±0.72 | 0.9 | 4747±173 |

# References

1. BENEDICT, F. G.: Vital Energetics. Carnegie Inst. Wash. Publ. No. 503, 1938.
2. —, and E. L. Fox: Biocentenary Nr. Amer. Philos. Soc. Proceedings **66**, 511 (1927).
3. EDWARDS, W. F.: De l'influence des agents physiques sur la vie. Paris: Chez Crochard, Libraire 1824.
4. GELINEO, S.: Glas Serbian acad. of sci. **192**, 115 (1949).
5. — Arch. Sci. Physiol. **9**, 225 (1955).
6. — Handbook of Physiology. Sec. **4**. Adaptation to environment. Amer. Physiol. Soc. Washington, D. C. 259 (1964).
7. HART, J. S.: Handbook of Physiology. Sec. **4**. Adaptation to environment. Amer. Physiol. Soc. Washington, D. C. 295 (1964).
8. IRVING, L., H. KROG and M. MONSON: Physiol. Zool. **28**, 173 (1955).
9. NEUNZIG, K.: Die Einheimischen Stubenvögel. Magdeburg: Creutz'sche Verlagsbuchhandlung 1922.

# Quantitative Thermogenesis of Brown Fat in Hibernation and Cold Adaptation[1]

R. E. SMITH, BARBARA A. HORWITZ[2], and Y. IMAI[3]

With 3 Figures

*Abstract*

Brown fat thermogenesis, relative to that of the intact animal, was estimated quantitatively in cold-adapted rats and arousing ground squirrels. The caloric output of the rat interscapular pad was calculated from temperature differences between arterial and venous flows through the pad. Extrapolating from these data, the total brown fat accounted for 8.2% of the rat's heat production at 4 °C. In the arousing *C. lateralis*, brown fat thermogenesis was estimated by approximating the *in vivo* $qO_2$ with rates measured *in vivo*. Estimates of the heat from this tissue ranged from 10% initially to 5% as the squirrels approached normothermic body temperatures. In view of the assumptions involved in these calculations, the 7.5% average is considered a minimal estimate of the tissue's contribution in support of arousal from deep hibernation.

Thermal evolution from the brown fat has been well demonstrated in arousing hibernators [5, 6, 12, 14], in cold-exposed neonates [1, 8] and in cold-exposed adult non-hibernators [3, 4]. To allay assertions [10, 11] that in rats this heat is insignificant because the brown fat/body weight is small, the brown fat thermogenesis relative to total heat output was re-examined in cold-exposed rats and in arousing ground squirrels (*C. lateralis*).

Rats exposed to 5 °C for 3—4 weeks and returned to 26 °C for 2 weeks were anesthetized (Na Pentabarbitol, 80 mg/kg body wt.). Copper-constantan thermocouples (TC) were placed in the interscapular pad and over the fat pad on the outer and inner skin surfaces. TC were also fixed upon the thoraco-dorsal artery and the deep central venous outflow, and one was inserted 6 cm into the colon. In the hibernating ground squirrels, TC were acutely inserted into the right axillary brown fat and colon. These animals were placed in a closed-chamber volume meter and the oxygen consumption continuously recorded at a constant chamber temperature.

[1] Supported in part by NASA Research Grant NGR-05-004-035 and USPHS Research Grant HD-03268-01.
[2] Postdoctoral fellow, U.S.P.H.S. (1-F2-GM-13,445-01).
[3] Present address: Department of Physiology, Kyoto Prefectural University of Medicine, Kyoto, Japan.

The total oxygen consumption of the rat and the temperatures of the 6 loci monitored are shown during cold exposure and 25 °C (Fig. 1). At 25 °C, temperatures of the interscapular pad ($T_f$) and the venous outflow ($T_v$) were

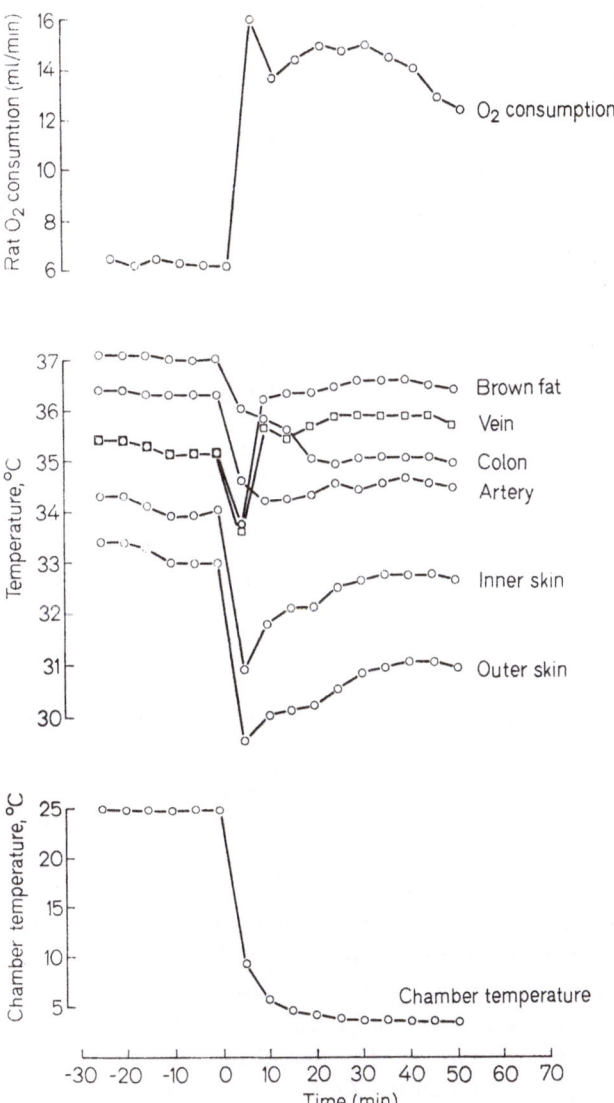

Fig. 1. Effect of cold exposure on the oxygen consumption of the rat and the temperature of the 6 sites described in text

quite similar; the arterial temperature ($T_a$) however, was generally 0.5 to 1.0 °C higher than $T_v$ and 0—1.0 °C higher than $T_f$. The colonic temperature ($T_b$) averaged 1.35 ± .17 °C higher than $T_f$. Upon exposure of the rat to cold, $T_a$, $T_v$, $T_f$, and $T_b$ usually fell initially, while the rate of total oxygen consumption (MR) rose (Figure 1). Within 10 min, however, $T_f$ and $T_v$ had begun to increase, followed thereafter (18 min after cold exposure) by a rise of $T_a$ and sometimes $T_b$.

After the chamber temperature reached 4 °C (within 10—15 min.), $T_v-T_a$ was relatively constant for each rat, averaging 1 °C. During the same time interval, $T_f$ averaged 2.9 °C warmer than the inner skin surface and 5.9 °C warmer than the outer skin surface.

From the observed temperature differences [$T_v-T_a$; $T_f-T_{outer\ skin}$; $T_{f(final)}-T_{f(initial)}$], the heat output of the interscapular pad was calculated

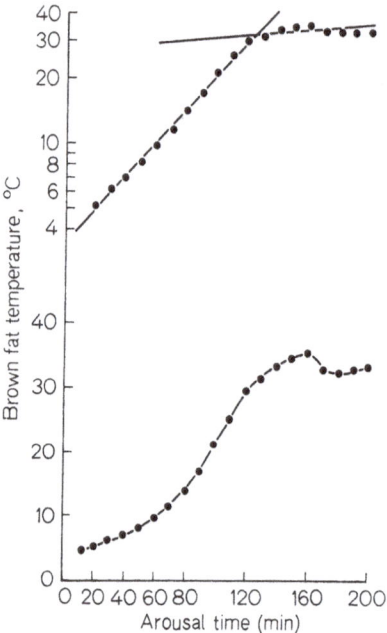

Fig. 2. Brown fat temperature as arithmetic and semilogarithmic functions of time during the arousal of *C. lateralis* at 2 °C ambient temperature. Substituting the values of *a*, the intercept, and *b*, the slope, into the equation (log *y* = log *a* + *bt*), the time *t* at which *y*, the brown fat temperature, reached 37 °C was found. Using *t*, and *b*, and setting *y* equal to the rate of oxygen consumption measured *in vitro* at 37 °C, the equation was solved for *a* in terms of the metabolic rate of the tissue. Thus, "*y*" becomes an expression of the rate of oxygen consumption of the brown fat as a function of time

as the sum of the convective and conductive heat transfers from the tissue plus the heat attributable to the rise in $T_f$ [9]. On the premise that the metabolism of the interscapular pad was like that of the other brown fat areas [15], the rate of heat production calculated for this pad (.076 ± .004 kcal/hr/gm tissue) was also assigned to that of the total brown fat. Hence, the interscapular pad accounted for 2.4% of the total heat production of the rat while the whole of the brown fat contributed 8.2%.

However, since shivering was observed in almost all animals, the calculated thermal contributions of the brown fat are relatively lower than would have been obtained in lieu of shivering.

In the ground squirrel, brown fat thermogenesis during arousal at 2 °C was estimated by assuming that (1) the metabolism *in vitro* reflected that *in vivo* and (2) the metabolic rate of the tissue was linear with brown fat temperature. Accordingly, oxygen consumption *in vivo* at 37°C was obtained from the maximum $qO_2$ of brown fat slices incubated *in vitro* with catecholamines at 37°C. Using these data, equations describing the rate of oxygen consumption of the tissue as a function of arousal time were derived from the "warming" curves (cf. Figure 2). By integration, the amount of oxygen utilized by the brown fat over given time intervals was calculated and compared to the total body heat production.

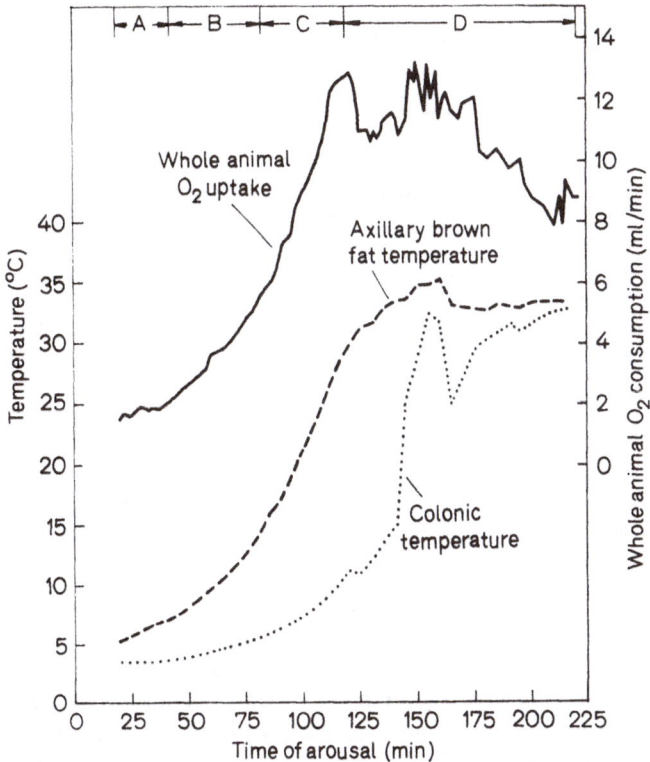

Fig. 3. Total body oxygen consumption, brown fat temperature, and colonic temperature of *C. lateralis* aroused from deep hibernation at 2 °C. A, B, C and D refer to the various phases of the total body oxygen consumption as indicated in text

In all squirrels with colonic temperatures below 6°C, the rate of total oxygen consumption showed an initial lag phase ("A" in Fig. 3) which was followed by an interval ("B") where the MR slowly rose. This second phase extended between brown fat temperatures of 7.3 ± 0.9 °C and 12.0 ± 1.0 °C, preceding the period of rapid rise of MR ("C"), which culminated in a maximum. The period following the peak MR ("D") was characterized by an apparent "hunting" phenomenon wherein the rates were erratic but tended to decrease.

Notably, the maximum MR of the squirrel was attained at an average $T_f$ of $30.7 \pm 1.9\,°C$ and well before the colonic temperature reached normal levels. That the peak rate actually occurred near the time that the posterior vascular area opened was indicated by the rapid rise in colonic temperature, and suggests that the tissues in the anterior portion of the body were primarily responsible for the total heat production during the greater part of arousal.

Concordantly, the estimated thermogenesis of brown fat (Table 1, Estimate I) was greatest during the initial stages of arousal (10.4%) and became reduced as the peak MR was approached (5.1%). These values,

Table 1. *Average brown fat thermogenic contribution (% of intact animal heat production \*)*

| Arousal Phase | n | Estimate I | Estimate II |
|---|---|---|---|
| A | (84) | $10.43 \pm 0.37$ | $41.46 \pm 1.46$ |
| B | (46) | $6.78 \pm 0.30$ | $26.99 \pm 1.16$ |
| C | (65) | $5.11 \pm 0.16$ | $20.29 \pm 0.63$ |
| D | (40) | $6.13 \pm 0.26$ | $24.37 \pm 1.02$ |
| A—C | (195) | $7.79 \pm 0.25$ | $30.97 \pm 0.99$ |

\* Means $\pm$ S. E. Estimate I calculated from *in vitro* $qO_2 = 804\ \mu l\ O_2/100$ mg tissue/hr; Estimate II from a $qO_2 = 3402\ \mu l\ O_2/100$ mg tissue/hr as projected from the differences between the effect of norepinephrine on rabbit brown fat *in vitro* and *in vivo* [2, 7].

however, are dependent upon the *in vitro* $qO_2$ used to approximate the *in vivo* metabolism. Since arousal undoubtedly constitutes a systemic stress, a $qO_2$ obtained *in vitro* in the presence of catecholamines was considered a better approximation than measurements without these agents. However, norepinephrine is reported to raise the metabolism of rabbit brown fat *in vivo* 3—4 times more than it does *in vitro* [2, 7]; hence these differences, if applying also to the ground squirrel, would increase the calculated thermogenesis from brown fat to 41.5% initially and 20.3% at the peak metabolic rate (Table 1, Estimate II).

Thus, the values ranging from 10.4% initially to 5.1% appear to represent a minimum estimate of the thermal contribution of brown fat in support of arousal at low temperature.

In perspective, however, the thermogenic importance of brown fat, whether in hibernators or non-hibernators, depends not only upon the actual amount of heat produced, but its bodily distribution as well. As previously emphasized [13, 15], the vascularity and special localization of the brown fat serves the requirement for localized application of heat to the spinal cord, aortic arch, the sympathetic chain, and the vital organs of the thorax.

## References

1. Brück, K., u. B. Wünnenberg: Pflügers Arch. **283**, 1 (1965).
2. Dawkins, M. J. R., and D. Hull: J. Physiol. **172**, 216 (1964).
3. Donhoffer, Sz., F. Sardy, and Gy. Szegvari: Nature **203**, 765 (1964).
4. —, and Z. Szelenyi: Acta Physiol. Hung. **28**, 349 (1965).
5. Hayward, J. S., and E. G. Ball: Biol. Bull. **131**, 94 (1966).
6. —, C. P. Lyman, and C. R. Taylor: Ann. N. Y. Acad. Sci. **131**, 441 (1965).
7. Heim, T., and D. Hull: J. Physiol. **186**, 42 (1966).
8. Hull, D., and M. M. Segall: J. Physiol. **181**, 449 (1965).
9. Imai, Y., B. A. Horwitz, and R. E. Smith: Proc. Soc. Exp. Biol. Med. **127**, 717 (1968).
10. Jansky, L.: Fed. Proc. **25**, 1297 (1966).
11. Joel, C. D.: In: Renold, A. E., and G. F. Cahill Jr. (Ed.), Handbook of Physiology, Section 5, Adipose Tissue, p. 59. Washington, D. C.: Am. Physiol. Soc. 1965.
12. Smalley, R. L., and R. Dryer: Science **140**, 1333 (1963).
13. Smith, R. E.: Fed. Proc. **21**, 221 (1962).
14. —, and R. J. Hock: Science **140**, 199 (1963).
15. —, and J. C. Roberts: Am. J. Physiol. **206**, 143 (1964).

### *Discussion*

Woodstock:

(1) Why is the brown fat brown, i.e. is the brown color associated with differences in the respiratory system? (2) Upon chilling in rats, do marked changes in $QO_2$ occur in non-brown fat similar to those in brown fat?

Smith:

(1) The color of the brown fat is ascribed largely to the cellular heme pigments e.g. cytochromes, and hemoglobin in the rich vascular supply to the tissue. Additionally, there is evidence that flavoproteins and lipoproteins may also contribute to the color. The apparent color also changes with seasons in some hibernators, and in the course of cold exposure the light brown color of the tissue becomes more reddish in appearance. (2) So far as I am aware, no data are available on the $QO_2$ of white adipose tissue during cold acclimation.

Langer:

How large is the amount of total mass of brown adipose tissue in the ground squirrel as a percentage of total body weight and how did you get this figure? How much weight of brown fat is lost during the course of arousal from hibernation?

Horwitz:

(1) In the ground squirrels employed in this study, the mass of the brown fat relative to that of the total body weight averaged near 4.8%. This value was obtained by weighing the brown fat excised respectively from individually

weighed animals. Since this study was done in the latter part of the hibernating season (February and March), it is likely that the relative amounts of brown fat in the squirrels were considerably less at this time than would be seen at the beginning of the hibernating season (cf. A. T. Rasmussen, J. Morphol. **38**, 147, (1924)). (2) We did not measure the loss in brown fat weight during the course of arousal. Joel, however (Handbook of Physiol., Sect. 5, Adipose Tissue, Amer. Physiol. Soc., Washington, D. C., 1965, p. 59) has reported that during the arousal of the 13-lined ground squirrel (*C. tridecemlineatus*), the brown fat loses about 0.95 gm of lipid.

Arley:

Can you measure the phenomenon of shivering in a quantitative way? And if so, could it be used to observe the metabolic processes of your hibernating animals visually and thus without interfering with their normal life processes? It seems to me that avoiding such interference would be one of the most important problems encountered in biological experimentation.

Smith:

In answer to your question, let me try to clarify an apparent misunderstanding. Since we sought to evaluate the brown fat thermogenesis relative to that of the whole animal, there was no necessity to determine the quantity of heat resulting solely from shivering. Mention was made, however, of the fact that shivering usually occurred during some stage of the experimental procedure. This observation was reported to emphasize that the total caloric output of both the rats and squirrels included the thermal contributions from both shivering and non-shivering activities. Therefore, the percentages calculated for the brown fat thermal contributions are lower than would be obtained if the heat production of the tissue had been compared only to the non-shivering thermogenesis of the animal. On the other hand, I entirely agree with you that the optimal conditions for any experimentation are those which interfere least with the normal physiological status of the animal.

# Contributions to the Energetics of Animal Growth

F. KRÜGER

*Abstract*

It is shown that the mathematical formulation of PÜTTER-BERTALANFFY of the anabolism-catabolism concept of growth has to be accomplished by a third term connecting the energetic output of anabolism with the production of living matter. This third term leads to the concept of the "Struktur-Energie" and opens methods for its numerical evaluation. The "Struktur-Energie" represents a positive entropy. Some examples are given for the application of the concept of the "Struktur-Energie".

## Introduction

It is a well known fact that growth not only consists of the addition of new matter to the existing organism, but that there are always processes which destroy a part of the living structure. Growth represents an equilibrium between these building processes which we call anabolism and the destroying processes which we call catabolism.

The matter for the organization of new body substances is delivered by the food, but the organization of this new living matter not only represents chemical problems but also important energetic problems, which have not been dealt with thoroughly until now.

Energy is a physical concept. If we wish to understand and to describe the part played by energy in the course of growth we need a mathematical model for the description of the quantitative relations during growth. There are many proposals for the mathematical description of growth, but most of them are applicable only to special cases.

Nearly half a century ago PÜTTER [11] introduced into the mathematical description of growth the concept of the equilibrium between anabolism and catabolism. He assumed that the processes which build up new matter approximately follow the surface rule and that the catabolic processes are proportional to weight. This assumption can be supported by many experimental facts.

In 1934 Von BERTALANFFY [1] — without knowledge of PÜTTER's paper — started with the same concept and arrived principally at the same formulation of growth processes as Pütter found. Considering the following derivations I have changed somewhat the symbols of the Bertalanffy-equation:

$$\frac{dw}{d\tau} = a \cdot w^\alpha - k \cdot w \tag{1}$$

$$\text{(anab.) (catab.)}$$

($w$ = weight; $\tau$ = age; $a$, $k$, $\alpha$ are parameters)

Since that time the Bertalanffy-equation has been frequently cited in literature. At the last Symposium I derived [8] that the formulation of growth, which I proposed some years ago, also contains this Bertalanffy-equation. Therefore, these two growth functions are based on the same concept of Pütter.

The Bertalanffy-equation uses the allometric formula for both the anabolic and catabolic processes:

$$y = b \cdot x^a \tag{2}$$

This function was found to be adequate to describe relative growth in a very wide range. The power $\alpha$ in the expression for anabolism indicates the surface rule and the power 1, which is not written, expresses proportionality to weight for catabolism.

The very simple equation of Pütter-Bertalanffy seems to me to be the best starting point for a first approach to the problem of balance in energy in the growing animal.

### Critique of the Pütter-Bertalanffy-Equation

Using the Pütter-Bertalanffy-equation and assuming a very short time the differential quotient $\frac{dw}{d\tau}$ approaches zero. The same would be found in a starving animal. In this cases the form of (1) would be:

$$a \cdot w^a \approx k \cdot w \tag{3}$$

From a mathematical point of view this equation is not possible because by no value of $\alpha$ can we reach equality on both sides of the equation for a wider range of $w$.

There exists also a physical objection against equation (3). The expression for the anabolism is based on measurements of heat production or oxygen consumption. It therefore represents an energetic value. The expression for catabolism is derived from experiments on loss in weight or loss in nitrogen caused by the destruction of body substance. The expression for catabolism therefore represents a mass value. It is not possible to equalize mass values and energetic values or to subtract them from another. On both sides of the equation the same physical dimensions must be used.

One may reach this claim by introducing a third term into the equation describing the living mass produced by the energetic output of anabolism. In this case, we would have mass values on both sides of the equation. From a mathematical point of view this third term would have the power $(1 - \alpha)$. Introducing this third term into the PÜTTER-BERTALANFFY-equation we would get in the non-growing animal.

$$c \cdot w^{(1-a)} \quad \cdot \quad a \cdot w^a \approx k \cdot w \qquad (4)$$
$$\text{(conversion)} \quad \text{(anab.)} \quad \text{(catab.)}$$

This equation is only valuable and applicable for a nongrowing animal.-

My theoretical considerations had reached this point. Somewhat later and much by chance, the results of Dr. PANDIAN from the University of Madras (India) came to my hands, giving me the possibility to test my theory.

### The Results of Dr. PANDIAN

PANDIAN [10] measured the food intake, the food consumption and the food conversion in two Indian fishes. The regular growth of fishes, which mathematically shows no breaks in the growth curves, seems to be a very appropriate example for the study of growth phenomena. PANDIAN determined all values in calories and therefore delivered comparable values for all measurements. Measuring the metabolism of small and large fishes, his results allow an allometric evaluation. It is also important that he related all parameters of the allometric formula to body weight. The factors and exponents of the allometric formula are given in Table 1.

Table 1. *Allometric evaluation of the results of* PANDIAN [10] *on food-consumption, food-absorption, consumption and conversion in two Indian fishes*

|  | Megalops | | Ophiocephalus | |
| --- | --- | --- | --- | --- |
|  | Exponent | Factor | Exponent | Factor |
| Food-consumption | 0.71 | 0.085 | 0.76 | 0.057 |
| Food-absorption | 0.70 | 0.081 | 0.77 | 0.050 |
| Consumption | 0.77 ($\alpha$) | 0.045 | 0.88 ($\alpha$) | 0.025 |
| Conversion | 0.54 ($\gamma$) | 0.040 | 0.55 ($\gamma$) | 0.028 |

In this table "food consumption" means the total food the fish has eaten. The "food absorption" is the food consumed minus the faeces cast and therefore represents the amount of food really taken up. "Conversion" represents the gain in weight measured in calories. "Food consumption" represents the difference between the food absorbed and the food converted. It therefore exhibits the matter which has not been used for the gain in weight and hence must have been catabolized by oxydation.

At this time we only are interested in the values of the different exponents. A value of about 0.7 has been found for the food consumption. The same value is found for the absorption. This signifies that in relation to body weight the uptake of food decreases with increasing size and a constant percentage of the food taken up is not absorbed.

The exponent for the food consumption representing the abolished matter is found to be about 0.8. This exponent agrees with the exponent found in respiration measurements in fishes.

The exponent of 0.54 for the conversion is astonishingly low. Other authors stated its value to about 0.7, but they related their results to the food consumption. Under these conditions the experiments of PANDIAN would give the same value. For that reason the low value of the exponent found by PANDIAN is verified by other results. The allometric formula describes the relative growth and we may compare parameters only if all our measurements are related to the same standard. This is, in our case, represented by the weight of the fishes.

To understand the low value of the exponent of conversion, we assume that conversion is related to anabolism by the allometric formula:

$$\log \text{conv} = \log c + \delta \log \text{anab} \qquad (5)$$

Anabolism is described by the function:

$$\log \text{anab} = \log a + \alpha \log w \qquad (6)$$

If we set (6) into (5) we get:

$$\log \text{conv} = \log c + \delta \log a + \alpha \delta \log w$$
$$= C + \alpha \delta \log w \qquad (7)$$

The product of the two exponents less than one is less than each of them. This seems to be a simple explanation for the low value of the power of conversion. The low value of the exponent for conversion means that the production of body substance decreases to a higher degree than anabolism.

The value of the product of the two exponents was found in the experiments as $\gamma$ and so we may calculate: $\delta \equiv \frac{\gamma}{\alpha}$. This exponent for the relation between anabolism and conversion would be 0.63 for *Ophiocephalus* and 0.75 for *Megalops*. It seems to be interesting that these two exponents come near to the power of the surface rule.

You may be at first astonished — and I was also — that the theoretical value of $(1 - \alpha)$ was not found. It would be in the neighbourhood of 0.2, but we have to remember that this value was derived for the non-growing animal. In the experiments we have to deal with growing animals acquiring weight. Under these circumstances we cannot expect the derived value for the power, it must be higher. I suppose that we find the theoretical value if we relate the allometric expressions for conversion to anabolism:

$$\frac{a \cdot w^{0.8}}{c \cdot w^{0.54}} = \frac{a}{c} \cdot w^{0.26} \tag{8}$$

This would be the expected low value for the exponent of conversion.

## The Energetic Relations between Anabolism and Catabolism

1) If we divide the BERTALANFFY-equation by $w$ we get:

$$\frac{dw}{w \cdot d\tau} = a \cdot w^{(a-1)} - k \tag{9}$$

The exponent of the catabolism disappears. That means that in our respiration experiments we measure the exponent of the anabolism. This conclusion is easy to understand. The catabolism related to weight in a graphical plot is represented by a straight line running parallel to the abscissa. Therefore the deviation of the curve from the parallel is caused by the process differing from proportionality, which is the anabolism.

2) The energy delivered by catabolism is proportional to the catabolized mass.

3) The energy delivered by oxydation is proportional to the amount of oxidized mass.

Now we may understand the meaning of the exponent in the allometric formulations of growth processes, if we make use of the growth formulation I proposed some years ago [5].

The logarithmic form of this function is:

$$\log y_\chi = \log Y_{max} - \frac{1}{\chi + \xi} \cdot \log N \tag{10}$$

($y_\chi$ = dimension at the age $\chi$; $Y_{max}$ = maximum size; $N$ = constant of velocity; $\xi$ = mathematical prenatal age).

This mathematical model of growth processes has the advantage that it stands in a narrow mathematical relation to the allometric formula [6, 7]. The velocity of growth in this formula (10) is represented by two parameters: the constance of velocity $\log N$ and the variable reciprocal value of the mathematical age. The allometric exponent is given by the quotient of the two constants of velocity of the compared growth processes:

$$\alpha = \frac{\log N_1}{\log N_2} \tag{11}$$

$\log N_1$ would be the parameter for anabolism in our respiration experiments as I explained before, $\log N_2$ would be the constance of velocity for the weight. But the value of $\log N_2$ is the same for all processes proportional to weight and therefore also applies to catabolism. Hence the exponent $\alpha$ in the exponents on metabolism represents the energetic relation between anabolism and catabolism. The same would apply for the exponents of other energetic processes.

### The "Struktur-Energie"

After having explained the meaning of the exponent of the allometric function as an expression of the energetic relation between anabolism and catabolism we may go a step further in the evaluation of the results of Dr. Pandian. Anabolism is described by:

$$\text{an} = a \cdot w^{0.8}$$

and conversion by:

$$\text{conv} = c \cdot w^{0.54}$$

From these two equations we may derive the energy needed for the organization of the unity of weight and find:

$$a \cdot w^{0.8} : c \cdot w^{0.54} = x : w$$

$$x = \frac{a}{c} \cdot w^{(1 + 0.8 - 0.54)} = \frac{a}{c} \cdot w^{1.26} \tag{12}$$

The exponent being higher than unity means that with growing size more energy is needed for the organization of the body substance. Unfortunately Pandian has not given numbers from which it should be possible to estimate the value of catabolism, but we may be sure that the quotient $\frac{a}{c}$ has a higher value than $a$ or $c$. Therefore the amount of the energy balance (4) is higher than on the right side. If the matter produced by anabolism is catabolized only energy proportional to weight is disengaged. The organization of the living body substance needs more energy than the chemical composition assigns.

This mathematical deduction is easy to understand because there is energy needed for the transport of the nutritious matter from the gut to the organs, for concentration energy etc. From this point of view it is also easy to understand that the energy needed for the organization of body substance increases with increasing size.

Bertalanffy (1942, 1951) already perceived this energy and used for it the denotation "Zellarbeit". I would propose the denotation "Struktur-Energie". The progress we reached by our mathematical derivations consists of an approach to its numerical value. We find it, if we evaluate the energetic relation between anabolism and conversion in the non-growing animal. The experiments of PANDIAN confirm the theoretical conclusion that the structure energy is distinguished by a very low value of the exponent.

## The Character of the "Struktur-Energie"

Here the question arises as to the physical character of this "Struktur-Energie". Physicists have tried to give an answer to this problem. The living matter represents a state of highly improbable order. For physicists such states of improbable order are defined by negative entropy. They therefore see in the structure of the living matter a state of negative entropy. This physical definition of entropy seems to me not to be applicable to the order in the living matter. The living matter does not represent a stationary state, but rather a dynamic state, and the energy consumed for the organization of the living structure is not recovered, if it is destroyed by catabolism.

The living structure is continuously destroyed and rebuilt and there are no signs that the energy expended for the organization of the living structure may be regained. From the modified BERTALANFFY-equation (4) it may be also seen that the "Struktur-Energie" does not reappear in the catabolism. Therefore we have to see a positive entropy in the "Struktur-Energie".

We may understand the situation easily with an example from our daily life: The energy wasted for winning and transporting coal from the mine to our homes is not regained for if we heat our homes with this coal, it has disappeared as positive entropy.

In connection with the problem of the role of entropy in the living matter the question arises whether there exists a principal difference between the energy balance in the organism and in inanimate nature. Are there processes violating the second principle of thermodynamics? I think there are now more signs for such an assumption. The living structure does not represent a state of negative entropy; it is a dynamic state sustained by a continuous supply of energy which is degraded to positive entropy. The living state represents a continuous struggle of the organism against the physical trend toward disorder.

The concept of the "Struktur-Energie" seems to me very important for the evaluation of experiments on energetics in organisms. As a first example we may regard the kinetics of an experiment on the utilization of food for mechanical labor. We find a certain relation between the energy consumed in the food and the work performed. It is a well known fact that we find a rather high exploitation of the energy content of the food; but the utilization of the energy by the contractile mechanisms itself is higher. The muscle oxidizes glycogen for its activity, but this glycogen has first to be built into the structure of the muscle; this is a part of the "Struktur-Energie" which escapes without being utilized as entropy. These derivations seem to me rather obvious and also very important for the evaluation of our experiments.

Another example are the experiments cited by VON BERTALANFFY [4] at the first symposium four years ago. He showed graphs of the respiration of rats of different size under basal and active conditions. Both curves show a break at a weight of about 100 g. I think that this break coincides with the puberty where the rate of growth decreases, as VON BERTALANFFY described 20 years ago [2]. I confirmed this break at the last Symposium [8], using my own growth formula. This break in the growth curve shows the decrease of anabolism closely allied with the rate of growth as we have seen. It is also found in other growth phenomena in the rat [13].

The curves for the respiration of the active rats are more highly situated, which may easily be understood. They also show the break at puberty. An astonishing fact is that the curves for the active animals show less of a slope. LOCKER [9] was the first to observe this phenomenon.

I think that the concept of the "Struktur-Energie" gives a simple explanation for this phenomenon. If we introduce the concept of the "Struktur-Energie", we have to distinguish three forms of energy metabolism:

1) the metabolism which supplies the need for the "Struktur-Energie" for the existing substance,

2) the need in energy for the organization of new substance,

3) the need for the activity of the animal.

Among these the "Struktur-Energie" is distinguished by a low allometric exponent, as we have seen. The active animal consumes the glycogen in its muscles. This glycogen in the muscles represents a part of the body mass which must be restored. This restauration is not connected with addition of new matter and therefore represents an unalloyed "Struktur-Energie". Hence, in the active animal "Struktur-Energie" with its lower exponent represents a larger share of the total metabolism and reduces the value of the exponent. Introducing the concept of the "Struktur-Energie" we are able to explain the LOCKER-phenomenon.

### Value of Mathematical Models of Growth

Growth represents a fundamental quality of the living matter. It is not possible to describe growth exactly without mathematics. Thence the search for suitable mathematical formulations of growth is a very important biological problem.

The process of growth may be described mathematically only by exponential functions. The same is the case in other vital processes. Therefore, the more intimate analysis of growth processes requires the insight into the interaction between rather complicated exponential functions. Our intellect is not able to understand such interactions of exponential functions. All these problems require the use of mathematics.

We started with the allometric formula as an approach to the relative growth in animals. It has been shown to be applicable in numerous examples for the description of growth processes in morphology and physiology. Thence it may be regarded as a highly secured biomathematical expression. PÜTTER [11] and BERTALANFFY [1] applied this allometric formulation to the basic processes of growth: anabolism and catabolism. In this manner they derived a very good formulation of the growth of animals as a function of time. I could show that the PÜTTER-BERTALANFFY-equation has to be accomplished by a third term connecting the energetic output of anabolism with the formation of living substance. This third term allowed the derivation of the "Struktur-Energie", which seems to me a very important concept for the understanding of growth processes, as could be shown by some examples.

By its mathematical relationship to the allometric formula, the growth function which I proposed, was a further important tool; because of this property it is possible to give an explanation for the parameters of the allometry formula, so that they do no more remain pure numerical values. In this manner it was possible to explain the exponent calculated from respiration measurements as an energetic relation, and we gained further insights into the energetics of growth. Here we see that it is very useful, if we are able to establish mathematical connections between our formulations.

The new growth formula also includes the anabolism-catabolism concept of the PÜTTER-BERTALANFFY-equation. According to this hypothesis a part of the living matter continuously decays. The loss in mass caused by catabolism disturbs the equilibrium between anabolism and catabolism and is compensated by increased anabolism. This increase in the rate of anabolism continues until the "Soll-Wert" is reached. This compensatory growth is a well known fact in animals whose growth has been retarded by illness or starvation. Here we have to assume regulatory mechanisms between anabolism and catabolism linked together by metabolites. This assumption would describe the pathological growth as a disengagement of the anabolic and catabolic processes—each running without control. In the course of normal growth anabolism only repairs the loss in mass caused by catabolism. If the connection between anabolism and catabolism is disturbed, living matter will be organized without limits. This seems to me a very interesting aspect of the much discussed problem of pathological growth, derived from the anabolism-catabolism concept of growth.

The allometric formula, the new growth formula and the completed PÜTTER-BERTALANFFY-equation seem to me useful tools for a further theoretical and experimental approach to the quantitative relations in the process of growth. It may be possible that corrections will be necessary or that better formulations will be found, but I think that simple models will be more useful than complicated ones. The latter perhaps offer better approaches to special cases, but they do not allow further derivations.

It was an essential feature that the models which I used resulted in solutions which are intellectually perspicuous and therefore offer the possibility to control experimentally the mathematical conclusions. Here seems to me given the heuristic value of the mathematical formulations going beyond the reproduction of series of numbers.

## References

1. Bertalanffy, L. v.: Entw. Mech. Org. **131**, 613 (1934).
2. — Hum. Biol. **10**, 181 (1938).
3. — Theoretische Biologie, Bd. II, Stoffwechsel, Wachstum.
   1. Aufl. Berlin: Borntraeger 1942. 2. Aufl. Bern: Francke (1951).
4. — Helgol. Wiss. Meeresunters. **9**, 5 (1964).
5. Krüger, F.: Naturwissenschaften **49**, 54 (1962).
6. — Zool. Anz. (Suppl. Bd.) **27**, 249 (1964).
7. — Helgol. Wiss. Meeresunters. **12**, 78 (1965).
8. — Helgol. Wiss. Meeresunters. **14**, 302 (1966).
9. Locker, A.: Naturwissenschaften **48**, 445 (1961).
10. Pandian, T. J.: Marine Biol. **1**, 16: (1967).
11. Pütter, A.: Pflügers Arch. **180**, 298 (1920).
12. Schrödinger, E.: Was ist Leben? Bern: Francke (1946).
13. Zucker, L. and Zucker, Th. F.: J. gen. Physiol. **25**, 445: (1941).

*Discussion*

Klamerth:

If your formula is quite a general one, being applicable not only to growing organisms but also to multiplying entities, e.g. a culture of *E. coli*, it seems to me that the term $\chi$ should be replaced by generation time.

Krüger:

As I have shown the formula may be applied to the growth of yeast cells as well as of *Paramecium*; however, I did not attempt to apply it also to the growth of cells in culture. I suppose that in this instance $\chi$ would represent the time elapsed since the starting of the culture.

Meyer-Döring:

What is the significance of the terms $\chi$ and $\xi$?

Krüger:

In my formula $\chi$ means age starting from birth, whereas $\xi$ represents mathematically prenatal age. Zucker (1941) was the first to propose this formula, but he was not able to calculate the value of $\xi$ and therefore used arbitrary values, e.g. age from conception onwards.

MEYER-DÖRING:

It seems to me that the significance of the word "Strukturenergie" you introduce is somewhat different from that familiar in molecular biology.

KRÜGER

Indeed the concept of "Strukturenergie" is used here not in the same way as in biochemistry, what would probably be the reason to search for a better suited and less ambiguous expression.

WALTER:

Could you give a precise definition of what you mean by "anabolism" and "catabolism"? There is no way to distinguish many biochemical reactions as reactions that "form" materials and reactions that "decompose" materials, because often the decomposition of one material is necessary for the formation of another material.

KRÜGER:

It is certainly difficult to distinguish between anabolic and catabolic processes at the molecular level, but in the overall balance of an organism we are nonetheless able to observe processes consisting of building up new material and processes destroying at least a part of it.

BREMERMANN:

Since energy requirements for creation of structure have come up, I would like to mention that I have just published a paper (Progr. Theor. Biol. 1, 59 (1967)) in which I argue: E. coli, in the process of growth, must compensate the structural entropy that is created through giving off heat. The latter amounts to 600 watts/kg under exponential growth conditions. A substantially higher figure would most likely boil E. coli to death. The heat production, however, limits the time in which E. coli can reproduce itself—16—20 min. It turns out that this time is about as short as is possible under the laws of physics.

SIMON:

The $X = \dfrac{a}{C} w^{1.26}$ relation, i.e. $\dfrac{dX}{dw} \sim w^{0.26}$ could perhaps be interpreted in terms of statistical thermodynamics as being due to constraints imposed on the individual cells by complexity interrelations in the organism. The bigger the organism the more order imposed on individual cells, which means the less the per cell entropy. Perhaps some measure of the intercellular dependence existing in a multicellular organism could be obtained, if $w$ is made equal to cell number and cell size is approximately constant. Then we have

$$\frac{dX}{dw} = \frac{d\,(H - T\,S_1 - T\,S_{ord})}{dw} = C\,w^{0.26}$$

wherein $H - TS_1$ means free energy necessary for the creation of one cell which we assume constant:

$$\frac{d\,(H - T\,S_1)}{dw} = C_1$$

If we define the number $W(w)$ of "microstates" — irrespective of the possibility of such a definition — where one cell is allowed to take on in a $w$-cellular organism, we have:

$$S_{ord} = k \ln \frac{W(w)}{W(1)} \;;\; -\,T\,\frac{d\,S_{ord}}{dw} = C\,w^{0.26} - C_1$$

By integration one would obtain:

$$\ln \frac{W(w)}{W(1)} = \frac{W(1)}{kT}\left[C_1(w-1) - \frac{C}{1.26}(w^{1.26}-1)\right]$$

as an empirical relation for cellular interdependence vs. size of the organism.

LOCKER:

I have to raise the following objections against the assumptions made in Prof. KRÜGER's paper: (1) Provided one accepts the obsolete anabolism-catabolism-concept then it seems to be by no means justified to ascribe to anabolism energetic values (e.g., cal/g body weight/hr) and to catabolism mass values (e.g., g/g body weight/hr), since a converse procedure would equally well be possible, namely to define anabolism as the body mass formed by the organism and catabolism as the heat produced by the organism. Since both together are expressions for metabolizable energy, i.e. the difference between the chemical energy in the food and the chemical energies in feces, urine, etc., it would be most sensible to ascribe to both forms energy dimensions, viz. calories. If so, the point of critique of the BERTALANFFY-equation based on the anabolism-catabolism-concept would become untenable. (2) The statement that in weight-specific respiration measurements only anabolism would be recorded is erraneous and at least not generally applicable since the supposed-weight proportionality of catabolism contradicts experimental facts (cf. ADOLPH, E. F., Science **109**, 579 (1949); MUNRO, H.N. and DOWNIE, E.D. Nature **203**, 603 (1964))—by the way also KRÜGER's own deductions drawn from PANDIAN's results. (3) It seems to be absolutely inappropriate to lay claim to an identity of "structural energy"—which, in KRÜGER's opinion would be the same as "cellular work"—with entropy. Whereas from the viewpoint of a statistically interpreted thermodynamics of closed systems— which is, however, inapplicable to the organism — the final state of such systems is of higher probability than the initial state, we have to consider in the open system, i.e. the organism, the occurrence of continuously structure forming processes the energy of which is supplied as freely convertible energy from exergonic reactions in metabolism. It should be beyond any doubt that building up the ordered structures of living organism is associated with a decrease of entropy which is made possible by means of coupling with processes of increasing entropy. The former may occur within a compartment or within the organism *in toto*, the latter in another compartment or in the organism's environment. This is somewhat a spatial correspondence to the well-known split-up of the term of entropy-change in the irreversible thermodynamics. During growth entropy decreases permanently as long as structure forming processes occur. Thus, growth is characterized by a tendency toward a minimum entropy production which is achieved at steady state. This minimum condition allows formation of structures with minimum energetic requirements. Free energy (enthalpy) cannot only be converted to work in general but also to structure forming processes. In the thermodynamically determined structural metabolism (OPITZ, E., and D. LÜBBERS, Hb. allg. Path. Bd. **4** (I) 2, p. 395 (1957)) the part of free energy serving as "structural energy" is used in different ways, depending on the rapidity with which entropy is produced as a consequence of destruction. The greater part of "structural energy" is needed for the formation of stable structures, the lesser part for labile structures, whereas for the maintenance of the structures formed the amount of energy required is just in an opposite sequence. This means that only the destruction of cellular structures raises the level of entropy in the organism.

KRÜGER:

(1) My derivations are based on the PÜTTER-BERTALANFFY equation. It is unwarranted to call it "obsolete", since no other useful concept has yet been proposed. Both authors have derived the mathematical expression for catabolism from the loss in weight on starving animals and the expression for anabolism from experiments on heat production. It is therefore a fact that the BERTALANFFY equation uses for both terms physically different dimensions. If we use caloric values for both terms, we find a higher value for the anabolism than for the catabolism, and meet the same difficulty. If Dr. LOCKER defines "catabolism as the heat produced by the organism", catabolism would stand in an exponential relation to body weight, and this would mean that the BERTALANFFY equation is no longer valid. (2) I have never said that we measure the anabolism in respiration measurements. I derived only from the BERTALANFFY equation the fact that we measure the *exponent* of the allometric formula in those experiments. The literature cited by Dr. LOCKER is not at my disposal at the moment. Therefore I do not know in what manner catabolism was estimated by these authors. I am however convinced that the earlier observations of PÜTTER and other authors are reliable. The reputed mistake in my derivations from the results of PANDIAN would need an accurate exposition to give an answer. (3) Dr. LOCKER's objection against my hypothesis of the structural energy as an entropy is not convincing. I derived this concept from the BERTALANFFY equation completed by the conversion term. I shall try to clarify my concept: The animal needs much more food than is formed into body substance. What is the fate of energy consumed with the food? Naturally it is partly used for physical and chemical work, but it is my opinion that the construction of any order also needs energy and that the energy expended for this aim is no longer freely convertible. Not suppositions, only experiments may confirm or disprove my ideas.

LOCKER:

In Prof. KRÜGER's paper has been neglected the distinction between physical and informational entropy; only the former bears physical dimensions, viz. erg/degr., while the latter does not. An equivalence, however, between the two can be accomplished in the form of $H = -aS$, by concerning physical entropy with the total number of ways of which a structure can be assembled. Then $a = 0.73$. For the production of information in an organism a conservation law holds, namely that it is equal to the rate of information leaving the system (by performing functions) plus the rate of storage (e. g. for growth) and for maintenance and repair. Since in the organism $\Delta S = S_{waste} - S_{food}$, the net entropy in the organism can be transformed to information (or $H_{food} - H_{waste} = H_{org}$). During growth the increase of free energy in the organism associated with the conversion of nutrient material into protoplasm involves a heat and an entropy term. In case that no other work than protoplasm synthesis is done

$$\Delta S_{int.} = \frac{1}{T} (\Delta F_{ext.} + \Delta H_{int.})$$

($\Delta H$: heat absorbed by reactants from the environment; $-\Delta H/T$: entropy change of environment; $\Delta S$: entropy change of reactants). $\Delta S_{int.}$ may thus be obtained by experimental assessment of $\Delta F_{ext.}$ (free energy of food ingested and released as waste) and the addition of $\Delta H_{int.}$ (difference of heat content between cell structure and a stoichiometrically equivalent amount of food;

all reactions being referred to unit time). This results in a negative value for entropy change during growth. The energy balance principally consists of 2 processes, linked together by a parameter expressing gram of waste products cast per gram cell produced and having a minimum value if $\Delta F_{ext.}$ is negative: (1) Concentration of food in the cell and conversion to protoplasm, (2) conversion of food to waste. We distinguish the following entropy changes: (a) decrease associated with condensing nutrient molecules into the cell, (b) decrease by ordering food molecules into cellular structure, (c) increase through converting food to waste products and (a) change involved in the heat transfer from and to the environment. The thermodynamic function of an organism which consists of selecting molecules with a lot of free energy can be seen already from the example of the reversible denaturation of globulin ($T\Delta S = +54.0$ kcal/mole, 27 °C), which in its reversed form can be comprehended as the last step in building up protein from constituent amino acids during growth: $\Delta F = -111,3$ kcal/mole, a great part of which ($T\Delta S = +54.0$ kcal/mole) dissipates as heat, such that it cannot be used further, whereas ($\Delta H$) 57.3 kcal/mole remains in the protein thus illustrating the free energy cost for constructing highly ordered molecules (LINSCHITZ 1953, MOROWITZ 1960, WILKIE 1960). — Conclusively it should be emphasized—at the same time by repealing unqualified assertions like those that "there are signs for assuming a violation of the second law"—that an experimental proof can only be expected from formulations that have been presented in precise quantitative terms.

# Models of the Growth of Organisms under Nutrient Limiting Conditions

P. R. Payne, and Erica F. Wheeler

With 3 Figures

*Abstract*

Mathematical models of the growth of animals have been based on consideration of possible internal factors which regulate and limit metabolism so that
the resultant growth equations contain only so-called genetic growth constants.
In most practical situations, however, the growth of animals can be influenced by
changing the relative proportions of essential nutrients in the diet. Two growth
equations are described: (1) one which applies to the growth when the diet fed
has a correct balance of essential nutrients and growth is limited only by the total
intake of metabolizable energy; (2) the other, when the ratio of utilizable protein
to energy in the diet is the factor deciding growth rate and maximum body size.

Since the beginning of this century, numerous authors have attempted
to express their ideas about the nature of animal growth in the form of
mathematical models. Most of these have been based on a consideration of
the growth process as the interaction between an early self-accelerating
phase, and a later self-inhibiting phase, both of which are regarded as being
under the control of genetically determined factors. Thus Brody [2] used
the two equations:

$$\text{Weight } (W) = a\, e^{k_1 t}$$
$$\text{and } W = W_{\max} - b\, e^{-k_2 t}$$

to describe the two phases of growth, where $t$ stands for time and $k_1$, $k_2$
and $W_{\max}$ were regarded as genetic growth constants. These constants are
of great importance in the study of the comparative physiology of growth.

However, under most practical conditions both phases of growth are
influenced not only by genetic factors but also by the quality of the food
supply; this includes both the proportions of essential nutrients in the diet
and the total quantity of food consumed. Thus no mathematical description
of growth will be complete unless it takes into account nutritional factors
as well as internal regulatory mechanisms.

One simple example of a growth process which appears to be controlled
principally by the total supply of nutrients is the growth of the foetus. This
is accurately described by a model in which foetal growth rate is taken to be

proportional to the rate of supply of nutrients across a surface corresponding to the foetal vascular endothelium. The area of that surface will remain in constant proportion to the total surface area of the foetus and is therefore proportional to $W^{2/3}$. Thus, the growth rate, $\dfrac{\mathrm{d}w}{\mathrm{d}t}$, is proportional to $W^{2/3}$, leading to the equation:

$$W = a\,t^3.$$

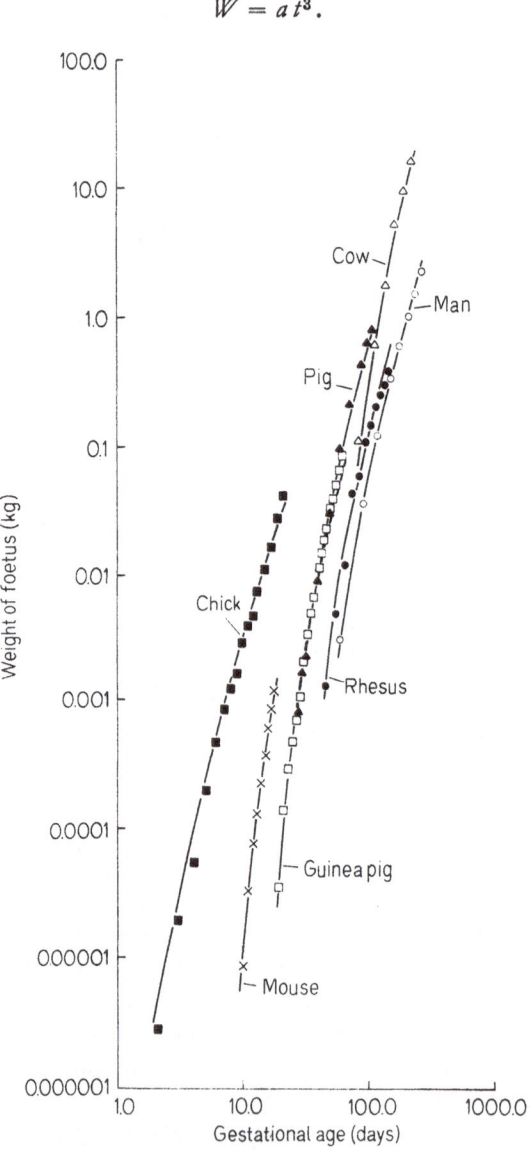

Fig. 1. Foetal weights at different gestational ages, for a number of species. The curves are calculated from the equation $W = a\,(t-t')^3$ For data see Table 1

If time is measured from the completion of a short lag phase ($t'$), associated with the process of implantation,

$$W = a(t - t')^3.$$

Figure 1 shows how well such an equation describes the growth of the foetuses of a wide variety of species.

Table 1. *Numerical values of the parameters of the equation $W = a(t—t')^3$ for foetal growth of several species.*

| SPECIES | $a \times 10^6$ | $t'$ |
|---|---|---|
| Chick | 4 | 1·1 |
| Mouse | 1 | 8·0 |
| Guinea Pig | 0.6 | 15 |
| Pig | 1 | 28 |
| Cow | 3.2 | 56 |
| Rhesus | 0.3 | 29 |
| Man | 0.2 | 36 |

In order to develop a mathematical description of growth in later life, we must start from what is known of the efficiency of food utilisation over short periods of time. MILLER and PAYNE [4] have described two equations which relate instantaneous growth rates to the balance between gain due to the anabolism of available nutrients in the diet, and losses due to the catabolism of body constituents. Thus, when the growth limiting factor in the diet is protein,

$$\frac{dw}{dt} = a(P\Theta C - 250\,W^{0.73}) \tag{1}$$

where $C$ = intake of metabolisable energy per day (kcals)

$P$ = protein calories expressed as a percentage of total calories in the diet

$\Theta$ = efficiency of protein utilisation (NPU), and,

$a$ is a constant relating body weight gain to body nitrogen gain.

When the only dietary factor limiting growth rate is the consumption of total metabolisable energy ($C$), we have:

$$\frac{dw}{dt} = a(6.8\,C - 725\,W^{0.73}) \tag{2}$$

These two equations give useful predictions of weight gain over short periods of time for animals, when $C$, $P$ and $\Theta$ are known. For long term predictions we need additional information about the relation of caloric intake to body weight and stage of growth. We have proposed a simple model which describes the relation between caloric intake $C$ and body weight $W$ as

$$C = k_1 W^{0.73} - k_2 W .$$

Here $k_1$ and $k_2$ are constants, characteristic of the particular animal concerned and represent the genetically determined aspects of growth.

When $W$ is small, $C \simeq k_1 W^{0.73}$, which agrees with the observation that in young animals the maximum observed calorie consumption is a constant multiple of the basal metabolic rate, which is in turn proportional to $W^{0.73}$. Kleiber [3] has shown that for a number of species the value of $k_1$ falls between 300 and 450 kcal/day/kg$^{0.73}$, and we have taken 400 as a mean value for all species. The term $-k_2 W$ expresses the fact that, as growth proceeds, calorie intakes per unit of body weight progressively decline towards a limiting value corresponding to adult maintenance levels. The minimum calorie intake necessary for weight maintenance is close to 1.5 x the basal metabolic rate, or $\simeq$ 107 kcal/day/kg$^{0.73}$ [4], hence for any maximum body weight ($W_{max}$), the value of $k_2 = \dfrac{400-107}{W_{max}^{0.27}} = 293 \, W_{max}^{-0.27}$

The resulting expression for caloric intake

$$C = 400 \, W^{0.73} - 293 \, W_{max}^{-0.27} W \tag{3}$$

is in very good agreement with observed food intakes at different weights of the rat, the dog and man.

Equation (3) in conjunction with (1) and (2), leads to two differential growth equations which have exact solutions and are very similar to those described by Von Bertalanffy [1].

Where protein content is the limiting dietary factor,

$$W = \left( \frac{400 \, P\Theta - 624}{k_2 P\Theta} \right)^{3.7} \left( 1 - e^{-3.99 k_2 P\Theta t \times 10^{-6}} \right)^{3.7} \tag{4}$$

When caloric intake is limiting,

$$W = \left( \frac{293}{k_2} \right)^{3.7} \left( 1 - e^{-6.8 k_2 t \times 10^{-6}} \right)^{3.7} \tag{5}$$

Figure 2 shows the general form of these equations using a value of $k_2$ appropriate to the rat. The energy limited curve, derived from (5), forms an enveloping maximum growth curve and would only be obtained if the diet contained ideal proportions of all essential nutrients. The lower curves, derived from equation (4) predict the effects of varying proportions of utilisable protein ($P\Theta$) in the diet on growth velocity and maximum body size. These curves are in good agreement with the observed growth performance of rats and dogs fed different diets; some predicted and observed values for dogs are shown in Figure 3.

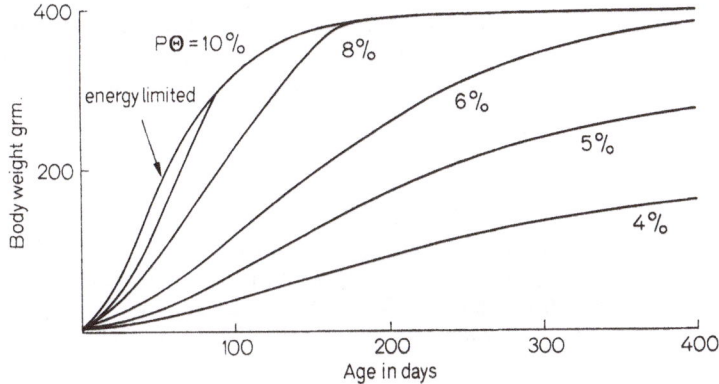

Fig. 2. Predicted growth curves of rats fed diets containing different percentages of utilisable protein

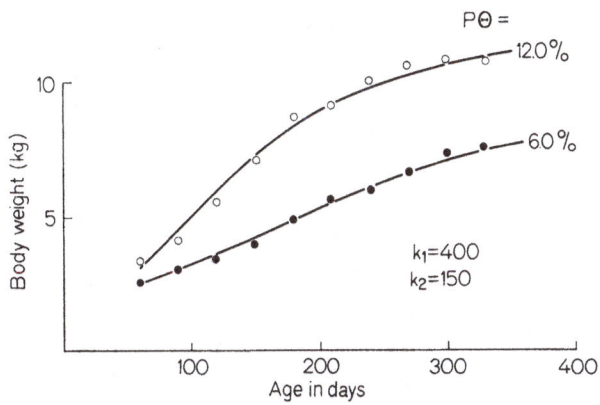

Fig. 3. Predicted curves, and measured body weights, of litter mate Beagle dogs fed diets containing 12.0 or 6.0 per cent of utilisable protein

These equations exhibit many of the features required for a successful model of the growth of animals.

1. The age-weight curves in Figure 2 illustrate an early self-accelerating phase and a self-inhibiting approach to an asymptotic adult value.

2. The point of inflection, representing the transition between these phases, occurs at between $1/3$ and $1/4$ adult size.

3. The model also shows the effect of a limiting amount of dietary protein on both growth velocity and maximum body size.

4. The equations predict a constant relationship between the maximum body weights of different species and the times required to reach a given

fraction of adult weight. The weight curves of different species could therefore be combined into a common "normalised" curve showing, as in Brody [2], the weight-growth equivalence of animals.

5. Finally, the presence of the term (P $\Theta$) representing the quality of the diet, in the time exponent of equation (4), expresses the fact that a protein deficient diet has the effect of expanding the physiological time scale. In animals fed such diets, many aspects of maturation are delayed, and the life-span is extended.

## References

1. Bertalanffy, L. von: Quart. Rev. Biol. **32**, 217 (1957).
2. Brody, S.: Bioenergetics and growth. New York: Reinhold 1945.
3. Kleiber, M.: The fire of life. London: J. Wiley and Sons 1961.
4. Miller, D. S., and P. R. Payne: J. Theoret. Biol. **5**, 1398 (1963).

### *Discussion*

Krüger:

For the description of growth of fish the formula of Ford-Walford ($l_{n+1} = a + b\, l_n$) is well suited. From this empirically found formulation one may derive the formula: $l_\tau = L_{max} + (L_{max} - L_0)b_\tau$, which agrees with the Bertalanffy-function if we put $e^{-K} = -\ln b$. Evaluations of the growth of fish done by me have shown, that the value $b$ of the Ford-Walford-formula is a better measure for the real velocity of growth than is the Bertalanffy-parameter $K$.

Payne:

In my opinion length is a most unsatisfactory parameter to use as a measure of growth, and in particular the quantity derived from the Bertalanffy-equation as $\sqrt[3]{W}$ is only loosely connected with the idea of length!

Krüger:

In man the relation $l = \sqrt[3]{W}$ is not fulfilled, but in coldblooded animals this relation covers rather exactly the observed data.

Meyer-Döring:

A formula, which describes the growth of animals, has also to take into consideration the different manner of nutrition as being related to the morphology of the animals if, for instance, ruminants are compared with other mammals.

Payne:

The relationship described by Brody, between adult weights of animals and the length of time needed to reach adult weight, is not statistically very good since, as Brody himself has pointed out, the data were confounded by variations of diets fed and by differing physiological states of the animals.

# Biomathematical Interpretation of Organismic Growth

F. Collot

*Abstract*

An example is given for the derivation of a general equation of growth assuming proportionality of growth to an energetic flow through the organism. It is also shown how, on the basis of this assumption, malignant growth can be understood.

Concepts like shape, function, cellular differentiation, rate of complexity etc. are usually thought of as scientific concepts; this, however, is incorrect since only measurable concepts deserve the term "scientific". For this reason it is necessary to give precise meaning to any of the terms used and to find a mathematical or physical equivalence for them. The example to be outlined will help us to understand the intellectual approach we recommend.

We start with the evident fact that growth of an organism needs a certain amount of energy and that the intensity of growth is proportional to the quantity of energy available at any given moment. Let us call "incident energy" (IE) the energy absorbed by an organism, such as that provided by food, oxygen or carbon dioxide for plants, after having subtracted the energy stored in the form of glycogen, starch, lipids etc. Hence, let $dq_i$ be the quantity of incident energy during the time $dt$ and $\dfrac{dq_i}{dt}$ an energetic flow that we call incident flow $\varphi_i$. Let us call "reflected energy" (RE) the energy which returns to the external world as work performed by muscle activity, as heat produced, as chemical or electrical energy, etc., and let $\varphi_r$ be the reflected flow. The energetic flow available for growth, called $\varphi_a$, is:

$$\varphi_a = \varphi_i - \varphi_r \tag{1}$$

This flow $\varphi_a$ is proportional to the intensity of the growth $\dfrac{dv}{dt}$, $v$ being the volume of the organism at time $t$ or the number of cells.

$$\frac{dv}{dt} = \varphi_i - \varphi_r \tag{2}$$

To integrate this differential equation it is necessary to write $\varphi_i$ and $\varphi^r$ in terms of $v$. We can admit reasonably that the incident flow $\varphi_i$ is proportional to $v$:

$$\varphi_i = a\,v + c \tag{3}$$

But if we want to write the reflected flow $\varphi_r$ in terms of the number of cells there arises a difficulty. Each cell, each tissue, each organ reflects a quantity of energy which depends neither on its volume nor its weight but on its degree of differentiation. Moreover, in a growing organism, this rate of differentiation varies considerably from one tissue to another and perhaps from one cell to another. Therefore, it is impossible to speak of average differentiation at time $t$; in order to make the solution homogeneous, while still assuring permanent growth, the notion of the cell has to be replaced by that of the "bioton". We shall call "bioton" some volume of living matter of arbitrary size not necessarily equal to each other but receiving a flow of IE which is equal to each other and remains constant. Our approach is similar to that of the physicist, who in the statistical mechanics renounces to enter into more details and to calculate the trajectory and kinetic energy of each individual molecule. Let us call $w$ the number of "biotons". Thus, equation (3) becomes:

$$\varphi_i = a\,w + c \tag{3'}$$

We now postulate: To any increase of biotonic volume of an organism corresponds a proportional increase of the rate of average specialization or of the rate of average differentiation of its constituent parts. This postulate can be understood intuitively by means of a sociological example. Let us imagine a primitive community in a remote island composed of ten individuals who live very distant from each other. Each of them will have to satisfy his proper needs and he will be therefore a fisherman, farmer, mason, miller, baker and so on. Here the *rate* of specialization or differentiation (we ascribe the same meaning to these words) is practically negligible, i.e. a specialization once attained will remain unchanged. Suppose now that the number of invididuals increase to one thousand. Consequently, various trades are going to appear, and the rate of specialization will increase in direct proportion to the number of individuals. If we apply this postulate here under the conditions mentioned we have:

$$\frac{\varphi_r}{\varphi_i} = \frac{1}{g - b\,w} \quad \text{or} \quad \varphi_r = \frac{\varphi_i}{g - b\,w} \tag{4}$$

By applying equations (2) (3') and (4) we get:

$$\frac{d\,w}{d\,t} = (a\,w + c) - \frac{a\,w + c}{g - b\,w} \tag{5}$$

Upon integration it results

$$t = \frac{A}{a} \log(a\,w + c) - \frac{B}{b} \log(g - 1 - b\,w) + K \tag{6}$$

The examination of this equation reveals that the representative curve is a sigmoid one with two asymptotes and a point of inflection. The position of

the inflection varies according to different and possible values of the parameters, thus giving rise to three types of curves.

It is easy to understand how, starting from this equation, we could achieve
a mathematical interpretation of malignant growth. Suppose that the ratio
$\dfrac{\varphi_r}{\varphi_i}$ expressing the rate of differentiation remains constant instead of
increasing with the volume of the organism; or even suppose that in the
adult organism it drops below its usual value. We have

$$\frac{\varphi_r}{\varphi_i} = \alpha \quad \text{or} \quad \varphi_r = \alpha\,\varphi_i \tag{7}$$

Then equation (2) becomes:

$$\frac{\mathrm{d}w}{\mathrm{d}t} = \varphi_i - \alpha\,\varphi_i = \varphi_i\,(1 - \alpha) \tag{8}$$

which yields upon integration an equation of the form

$$w = e^{\beta t} \tag{9}$$

This means that for the group of cells under consideration growth pro-
motes exponentially and one might, therefore, assert: Any (external or
internal) physical cause capable of bringing about within the living organism a constant cellular de-differentiation automatically results in cancer.

The example introduced should show how valuable an attempt at
direct mathematical interpretation of biological facts could be. Essential
for the approach here is not to enlarge a purely statistical calculation or to
apply those mathematics which are being offered by physics or chemistry.
The aim is directly to build up appropriate concepts by means of highly
abstract mathematical theories in order to reach a sensible goal.

### *Discussion*

GRIFFITH:

> It seems to me undesirable to use simple arguments, which ignore the detailed
> structure of the biological system, and then apply them just to a single example
> —the form of a growth curve. Even is this case, as Dr. PAYNE has shown in
> his discussion of the effects of varying diets, the experimental curve actually
> depends crucially on factors other than mere weight or volume.

COLLOT:

> In a preliminary approach to a theoretical interpretation of growth it becomes
> necessary to ignore secondary factors like temperature or relative proportion
> of essential nutrients in the diet, which are acting only as disturbing factors.

LOCKER:

> We are by no means justified in interpreting an exponential growth curve,
> simply derived as a consequence of a parameter variation, as an indication for
> an "automatic" occurrence of a cancer development. It is well known that,
> within shorter time intervals, also in normal embryonic development certain
> tissues undergo an exponential growth.

# Relationships Between Respiration During Imbibition and Subsequent Growth Rates in Germinating Seeds

L. W. Woodstock

With 3 Figures

*Abstract*

Differences in the respiration of germinating seeds precede differences in seedling growth. Seeds which germinate rapidly and produce vigorously growing seedlings have higher respiratory rates and lower respiratory quotients than non-vigorous seeds. Vigorous and non-vigorous seeds can sometimes be distinguished within 1 hr after the start of imbibition by differences in respiration. The following germination-inhibiting treatments affect respiration during the first 6 hrs of germination: Unfavorable long-term storage conditions in corn and barley; freezing injury and heat damage in corn; chilling injury in lima beans; desiccation treatments in sorghum; and gamma-irradiation in corn, sorghum, wheat and radish. Measurements of respiration and growth of individual undamaged and highly vigorous corn seeds revealed significant positive correlations between respiratory rates and seedling growth. The results suggest that seedling growth during the first few days of germination depends more on the overall level of metabolic activity than on the activity of particular enzymes. Several possible interpretations of the data are discussed.

## Introduction

The question may be asked whether growth is usually controlled by a single limiting reaction or whether it depends on the general level of metabolic activity. Even so "simple" an organism as *E. coli* is estimated to contain between 2000—3000 different kinds of enzymes [14]. Because of the large number of feedback control mechanisms and the speed of enzyme synthesis, as soon as any particular enzymatic reaction becomes limiting, processes may come into play which speed up that reaction. When this happens, some other enzyme or compound becomes limiting and growth may represent a kaleidoscopic series of "little crises" wherein one enzyme or compound after another becomes limiting in rapid succession. With sufficient metabolic adaptiveness to repair deficiencies as they occur, limitations in particular enzymes will be too transient to measure, and the overall level of metabolic activity will be seen as limiting for growth. Under such circumstances, a positive correlation would be expected between some measure of this activity and the growth rates.

In other cases, as in irrepairable injury and the hormonal or genetic suppression of certain reactions, the system cannot increase the limiting factor and a given enzyme or compound may be limiting for a measureable period of time. Here, growth may be correlated with the activities of particular enzymes rather than with some index of overall activity.

In an earlier study [19], we observed a direct proportionality between rates of root elongation and ribosenucleic acid (RNA) contents in the roots of *Zea mays*. In the six inbred lines studied, a striking parallel was found between the RNA content in the apical 3-mm portion and the growth rate of the entire root. The relationships, specific for RNA, were not found for deoxyribosenucleic acid (DNA) and were independent of the size of the root. Similar correlations were later reported [8] for seedling roots of sorghum cultured at various temperatures, where growth and RNA contents varied as a function of temperature and sampling period. CHANG and THOMPSON [3] recently observed a direct relationship between the growth rates of the seedling root of corn and the RNA contents of the 3-mm root tips where differences in growth and RNA were due to treatments with sodium fluoride and to time of measurement. These quantitative relationships between RNA contents and growth are similar, in many respects, to those reported for the bacteria, *Aerobacter aerogenes* [7]. Because of the known involvement of RNA metabolism in enzyme synthesis [2], we suspected that, in germinating seeds, overall levels of metabolic activity might be more important in determining seedling growth rates than the activity of any single enzyme.

Energy is an essential prerequisite for growth and metabolic energy is derived from respiration. Thus, when overall metabolic activity is related to growth, a correlation between respiration and growth is to be expected. Many such observations have been reported. Usually, however, respiratory rates have been measured on actively growing tissues and do not indicate whether differences in respiration are a cause or a result of differences in growth. Our first concern was to determine whether differences in the respiratory rates of germinating seeds could be shown to clearly precede differences in growth. We found that differences in the respiration of germinating seeds do, in fact, precede differences in seedling growth. Seeds which germinate rapidly and produce vigorously growing seedlings have higher respiratory rates and lower respiratory quotients than non-vigorous seeds. These differences may appear as early as 1 hr after the start of imbibition. This is many hours before rupture of the seed coat by the elongating radicle or other morphological evidence of growth. In the remainder of this talk a number of experiments indicating relationships between respiration and seedling growth in different kinds of seeds will be described. From these experiments, we will attempt to deduce something of the nature of the relationship between respiration and growth.

*Effect of storage conditions on respiration and growth in corn:* In this investigation, we [16] measured the respiration of samples of corn seeds having undergone different amounts of deterioration due to various conditions of temperature and humidity during long-term storage. Respiration was measured between 2 and 30 hrs after the start of imbibition as $O_2$ uptake and $CO_2$ evolution and was compared with several indices of seedling growth (Fig. 1a). There was a striking positive correlation between rates of $O_2$ uptake only 3 hrs after the start of imbibition and the shoot lengths of 5-day-old seedlings. An equally striking negative correlation between the respiratory quotients and seedling growth rates was observed (Fig. 1b). Both correlations were significant at the 1 % level.

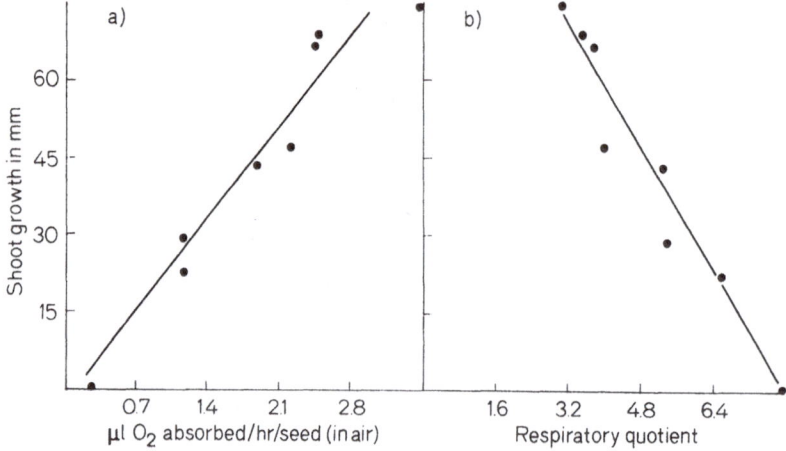

Fig. 1a. Relationship between respiratory rates of corn seeds, measured between 2 and 6 hrs after the start of imbibition, and shoot lenght of 5-day-old seedlings, where differences between seed lots were due to different storage conditions.

Fig. 1b. Relationship between respiratory quotients of corn seeds and average shoot length of 5-day-old seedlings. (From: WOODSTOCK and GRABE, Plant Physiology **42**, 1071 (1967)).

Because radicle emergence was not observed until 20 hrs after the start of imbibition, differences in respiration were assumed to precede differences in growth rates. To test this assumption, rates of water uptake in vigorous seeds and in seeds weakened by heat treatments were compared with respiration at 2 and 4 hrs after the start of imbibition [16]. The heat treatments markedly inhibited seedling growth. They also inhibited respiration and increased the respiratory quotients, but did not inhibit water uptake during the initial 4 hrs. Had there been differences between heated and control seeds in imbibition or rates of embryo elongation during this period, differences in water uptake would have been observed. Thus, differences in respiration cannot be due either to differences in imbibitional water uptake or to differences in growth rates. These results also show that differences in respiration occur within only 2 hrs after wetting, and that injury due to short-term heat treatments affects respiration in a manner similar to that due to gradual deterioration during long-term storage.

*The respiration and growth of individual corn seeds:* In the above tests, different populations of seeds were compared. The respiration and growth of individual seeds can also be compared. The respiration of an individual seed can be measured during imbibition and compared with the growth of that particular seed at a later stage of development. In these experiments, the respiratory rates of 16 single individual corn seeds were measured for 6

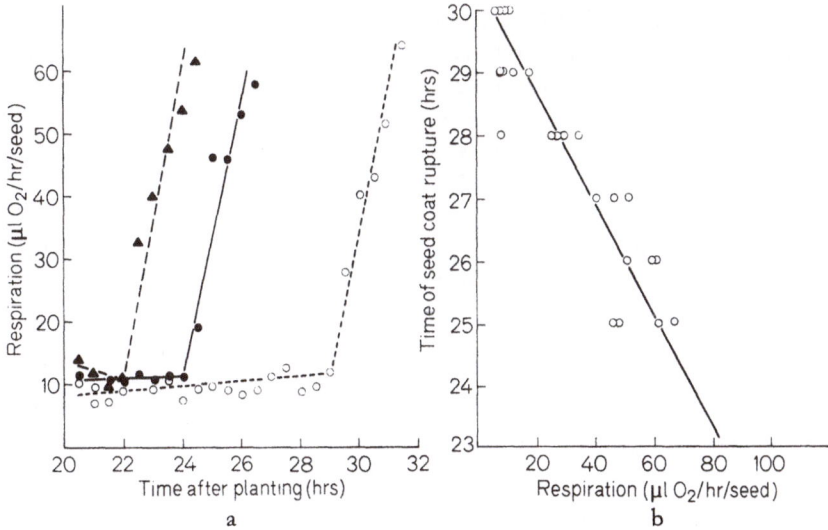

Fig. 2a. Respiratory rates of three individual randomly selected corn seeds as a function of time after planting

Fig. 2b. Respiratory rates of individual corn seeds measured between 27 and 29 hrs after planting vs. time of seed coat rupture (hrs after planting) (From: WOODSTOCK. (1965) Bioscience, 15: 783—784.)

hrs after wetting and the seeds were then placed in moist paper towels and germinated in darkness [15]. Respiratory rates of individual seeds between 2—3 hrs after the start of imbibition and seedling axis lengths 3 days later were correlated at the 1 % level of significance. Again, a striking negative correlation between RQ and seedling growth was observed. Since no known injury was involved and the seeds were of high quality, the results indicate a general relationship between early respiratory rates and later seedling growth in corn.

The time of seed coat rupture could be determined by visual observation and was accompanied by a rapid rise in the respiratory rate (Fig. 2a). This suggested that the seed coat might have restricted oxygen diffusion to the embryo. The time of seed coat rupture was negatively correlated with respiratory rates 27 to 29 hrs after planting, i. e., the sooner the seed coat was ruptured by the elongating radicle, the higher the respiratory rates (Fig. 2b). The following relationships are indicated: Initial respiratory rates are related to seedling growth rates during the first 24 hours of germination, which in turn, are related to the time of seed coat rupture (between 22 and 32 hrs after the start of imbibition). In other experiments, seedling growth during the first 24 hrs was highly correlated with seedling growth during the subsequent 2—3 days.

   *The effect of gamma-irradiation on the respiration and growth of corn, wheat, sorghum and radish seeds*: In assessing the significance of changes in respiration relative to changes in seedling growth, we wished to determine (a) whether the above relationships occurred in species other than corn, and (b) whether other types of injury would cause similar differences in respiration. We, therefore, determined the effects of gamma-irradiation on the respiration and seedling growth of corn, wheat, sorghum, and radish (Table 1). Treatments of 80 krads markedly inhibited seedling growth in all four species, but reduced seed viability only slightly [17].

   The growth-inhibiting 80-krad treatment always either inhibited oxygen uptake, increased the RQ, or did both when respiration was measured in 100% oxygen [17]. This suggests either that the mechanism

Table 1. *Effect of $\gamma$-irradiation on respiration and growth of seeds.*

| Kind of Seeds | Dose (krad) | $QO_2$ ($\mu l\ O_2\ hr^{-1}\ seed^{-1}$) | Resp. Quot. | Stem Axis (mm) | Time Meas. (Days) |
|---|---|---|---|---|---|
| Corn | 0 | 2.6 | 2.4 | 30 ± 4.6 | 4 |
| Corn | 5 | 3.9 | 1.6 | 37 ± 4.3 | 4 |
| Corn | 20 | 3.3 | 1.8 | 31 ± 5.3 | 4 |
| Corn | 80 | 1.7 | 3.3 | 9 ± 1.7 | 4 |
| Wheat | 0 | 3.1 | 0.77 | 29 ± 2.5 | 6 |
| Wheat | 5 | 3.4 | 0.88 | 38 ± 4.5 | 6 |
| Wheat | 20 | 3.0 | 0.87 | 27 ± 1.9 | 6 |
| Wheat | 80 | 2.8 | 0.96 | 13 ± 0.5 | 6 |
| Sorghum | 0 | 1.2 | 0.67 | 16 ± 2.1 | 4 |
| Sorghum | 5 | 1.0 | 0.80 | 22 ± 1.1 | 4 |
| Sorghum | 20 | 1.0 | 0.90 | 15 ± 0.8 | 4 |
| Sorghum | 80 | 0.7 | 1.43 | 7 ± 0.5 | 4 |
| Radish | 0 | 2.7 | 0.74 | 82 ± 4.0 | 4 |
| Radish | 5 | 2.6 | 0.77 | 90 ± 5.4 | 4 |
| Radish | 20 | 2.3 | 0.78 | 78 ± 6.5 | 4 |
| Radish | 80 | 2.0 | 0.85 | 52 ± 2.6 | 4 |

from: L. W. WOODSTOCK and O. L. JUSTICE, Radiation Botany 7, 129 (1967) of irradiation damage by high dosages is via the respiratory system, or that respiration reflects metabolic injury at a deeper level. A large body of data indicates that the chromosome may be the primary site. Nevertheless, damage at the chromosomal level must express itself at the metabolic level before it can affect morphology or growth. The respiratory system may be a key mechanism in this process. SKOOG [13], and GORDON and WEBER [4], and others have shown that ionizing radiation affects auxin metabolism. Auxin concentration, in turn, exerts a marked influence on the respiratory rate. Our data do not indicate whether the effects of gamma-irradiation on respiration reflect possible auxin destruction during or after irradiation or are due to some more direct effect of irradiation on respiratory metabolism. The fact of greatest interest here is that gamma-irradiation injury affects respiration during imbibition in the same way as heat treatments and adverse storage conditions: It lowers respiratory rates and increases respiratory quotients.

The stimulation of plant growth by exposing seeds to low doses of ionizing radiation has been reported by many investigators and was reviewed recently [12]. Although in most cases the reported stimulations were small and not always reproducible, SAX [12] concludes that the documentation is sufficient to demonstrate growth stimulation by small doses of irradiation. An interesting aspect of the present studies is the apparent stimulation in seedling growth produced by the 5-krad treatments. Although the stimulations were small and not always statistically significant, they were observed in all four species tested, and were apparent regardless of whether seedling growth was measured as stem axis length or as seedling fresh weight.

In corn, we found that stimulation of seedling growth was preceded by a stimulation in respiration (Table 1). In corn and sorghum, the 20-krad irradiated seeds grew at about the same rate as the non-irradiated controls. The respiration of these seeds, however, was similar to that of seeds given the growth-promoting 5-krad treatments. Conversely, in radish, the growth-promoting 5-krad treatment had no apparent effect on respiration. Since the effect of the irradiation treatments on respiration can be separated from that on seedling growth, the growth- and respiration-stimulating mechanisms are not identical.

*Desiccation-induced dormancy of sorghum:* Seeds of sorghum (*Sorghum vulgare* Pers.) dried in a forced-air dryer from an initial moisture content of 12% to either 10 or 7% exhibited physiological dormancy [9]. Dormancy was more marked in seeds dried to 7% than to 10% moisture, and was more pronounced in germination at 15° or 20° than at 25 °C. That drying to a moisture content of 7% did not cause permanent injury is evidenced by normal germination at high temperatures and by reversal of dormancy. Cutting through the integumentary membrane and slightly into the endosperm with a dissecting needle was highly effective in promoting the germination of dormant seeds at 15° and 20 °C. The site of the block in the germinative process thus appears to be the integumentary membrane. Changes in the physical nature of the membrane caused by desiccation which result in restriction of gas exchange may account for the dormancy.

Respiration during imbibition of seeds dried to 10 or 7% moisture content was inhibited [9]. The inhibition was especially evident when respiration was measured in an atmosphere of 100% oxygen. Under such conditions, differences between 7%-moisture and control seeds were detected within 2 hrs after the start of imbibition. The dormant seeds also exhibited higher RQ values than non-dormant controls.

In seeds from a given treatment, respiration increased linearly with fresh weight during the first 6 hrs after planting [9]. This suggests that in both 7- and 12%-seed lots moisture was limiting for respiration during this period. The two curves show, however, that at any one moisture content, the respiration of the previously desiccated seeds is higher than that of the controls. The initially lower moisture content of the dried seeds cannot by itself, therefore, account for inhibited respiration during imbibition. More probably, the drying treatments caused physical changes in the seeds which induced dormancy. The dormancy, so induced, is reflected by changes in metabolism (including respiration) upon reactivation of the enzymes during the early stages of imbibition.

*The effect of chilling treatments on the respiration and growth of lima bean seeds:* In lima beans, imbibition at 5—15 °C causes an immediate and pronounced inhibition in respiration and a subsequent inhibition in seedling growth [18]. Exposure to cold water after imbibition does not cause injury. Chilling injury was roughly proportional to the inhibition of respiration and when respiratory rates at 4 hrs after the beginning of imbibition were compared to seedling axis lengths after 5 days' germination, positive correlations were found. Bleached seeds were more susceptible to chilling injury than normal green seeds. Respiration of bleached was lower than that of green seeds only during the first 2 hrs of germination, i.e., during the temperature-sensitive period. These results, and observations by Pollock [11] that anaerobic conditions augment the injury, suggest that respiratory rates during imbibition may be a factor in determining susceptibility to chilling injury. In lima beans, the surface area of the seed increases about 4-fold during 2 hrs of imbibition. Subsequent rates of increase in surface area are lower and are limited to the elongating regions of the embryonic axis. Marked leaching of organic materials from the seed accompanies chilling injury [11]. The data indicate the following explanation for chilling injury in lima beans: During low-temperature imbibition, the respiratory rate is too low to supply the energy needed for orderly physiological stretching of the membranes. Under these conditions, membranes are physically torn as cells expand under the force of imbibitional water uptake. Extensive membrane damage in critical sites, such as the embryonic axis, might lead to subsequent inhibition of seedling growth, or even to death of the seedling.

Surprisingly, inhibition of respiration by the 5 °C treatment was more pronounced at 2- than at 6-hrs. This suggested that some kind of repair mechanism might be present. Upon measuring the respiration of cotyledons and embryonic axes separately, we found that respiration of the chilled cotyledons approached control levels after 4 hrs. That of the embryonic axes remained inhibited even after 6 hrs. This was interpreted as evidence indicating that the embryonic axis was the principal site of chilling injury. The data also show that correlations between respiration and growth do not necessarily improve as the time between the two kinds of measurements is

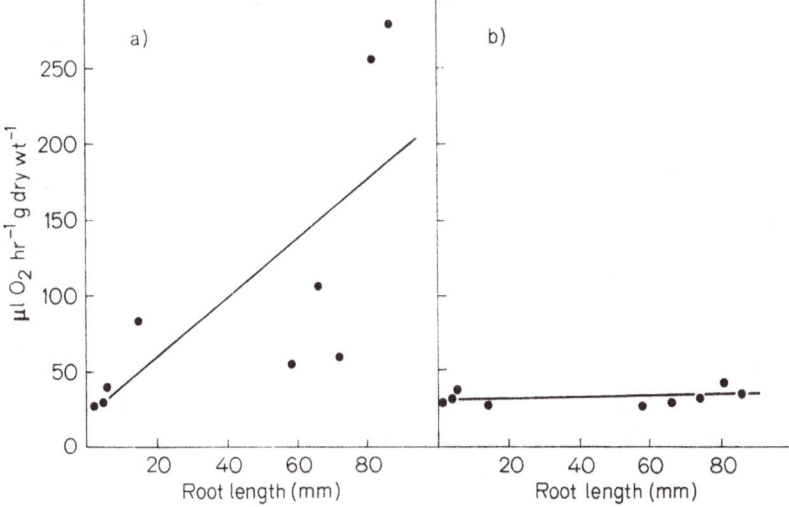

Fig. 3a. Respiratory rates in water of ground meal from different lots of barley vs. growth potential, measured as root growth of 3-day-old seedlings, of the lots

Fig. 3b. Respiratory rates in $5 \times 10^{-4}$ M 2,4-DNP of ground barley meal vs. growth potential of the lots. (Unpublished data)

diminished. In fact, in corn, respiratory rates at 3 hrs were more highly correlated with seedling growth than were respiratory rates at 12 or 18 hrs.

*Respiration and seedling growth in barley:* Relationships between respiration and seedling growth in barley are even more striking than those in corn. In barley, respiration only 1 hr after the start of imbibition was significantly correlated with seedling growth measured 3—4 days later. As with other kinds of seeds, a highly significant negative correlation was observed between the respiratory quotients and seedling growth. With barley, the seeds were graded for size, imbibition was rapid, and vigorous respiration occurred immediately after planting.

The respiratory uncoupler, dinitrophenol, failed to reduce the degree of correlation between respiratory rates and seedling growth when applied to intact seeds, even at concentrations which were growth inhibiting. When barley was ground to a dry powder, respiration per gram dry weight was initially higher than for intact seeds, indicating that possible injury to mitochondria by grinding was compensated by better aeration. In contrast to intact seeds, however, respiration of meal from low-vigor seeds declined with time. Respiratory rates of ground barley were significantly correlated with the growth potential of the seeds (Fig. 3a). Dinitrophenol (DNP) markedly reduced the degree of correlation between respiration and growth when applied to ground barley (Fig. 3b), suggesting that its failure to do so in intact seeds was due to its inability to reach active sites.

*Significance of relationships between respiration during imbibition and subsequent seedling growth:* Metabolic processes during imbibition are evidently related to the capacity of the seed for growth. The above data show that differences in seedling growth are preceded by differences in respiration during the initial 6 hrs of germination; i.e., either by changes in $QO_{2\,(seed)}$, changes in RQ, or both. What is the nature of this relationship? Four alternative interpretations of the data may be suggested: (a) Respiratory energy is limiting for the synthetic processes required for growth, as suggested by Pfeffer [10], (b) the rates of the synthetic processes set the pace, i.e., the more rapidly ATP is converted to ADP by the biosynthetic reactions, the more rapid will be the respiratory rates, (c) the respiratory and synthetic processes are mutually independent and are controlled by some other mechanism, and (d) the respiratory and synthetic processes are mutually independent, but are equally susceptible to injurious treatments and tend to "degenerate" during storage at approximately the same rates.

The fact that there was, in some cases, poor correspondence between respiration and growth (e.g., certain corn lots had similar growth rates, but differed in respiratory rates) might be taken as evidence in favor of alternative (d) above. Lack of correspondence then would be ascribed to differences in rates of degradation of the respiratory and growth mechanisms in certain lots. If alternative (d) is indeed correct, then there is no direct cause-and-effect relationship between respiration and growth and the correlations might be regarded as "fortuitous". There are, however, several difficulties with this alternative. Most important is the widespread occurrence of the correlations between respiration and seedling growth. It is not likely that similar correlations in such varied experimental systems are due to chance. Furthermore, alternative (d) offers no explanation for the negative correlation between RQ and growth. Perhaps the lack of correspondence in some of the corn lots might be attributed to differences in waxy coating or endosperm composition which would influence both imbibition and rates of oxygen diffusion. Such differences might develop during storage and would obscure relationships between tissue respiratory capacity and growth.

The arguments against alternative (d) also apply to alternative (c). Alternative (c), likewise, does not explain the inverse relationship so frequently found between the respiratory quotients and seedling growth. The wide occurrence of the correlations again would seem to argue against this alternative. One would expect that under at least some circumstances either respiration or growth would become "unstuck" from the control mechanism which this alternative assumes. Possibly the most serious objection to (c) is the effect of DNP on correlations between respiration and growth in ground barley meal. The fact that a compound which uncouples energy from respiration should reduce the degree of correlation between respiration and growth might imply that energy plays some important role in the correlation.

The present data do not permit a definitive choice between alternatives (a) and (b). Nevertheless, the negative correlations between RQ values and seedling growth may be significant. The tricarboxylic acid cycle is probably more easily injured than the glycolytic pathway or the hexose-monophosphate shunt, perhaps because it depends on the integrity of the mitochondrial membranes [1]. If the growth effects of injurious treatments were due to impaired mitochondrial activity, then accumulation of pyruvate could result in enhanced decarboxylation reactions and higher RQ values. Since mitochondria appear to be self-replicating organelles [5], injury to the mitochondria or their progenitors at a given stage of development would tend to be perpetuated. Such injury would show up in decreased respiratory rates, higher RQ values, a lowered production of ATP, and decreased rates of seedling growth. This would mean that seedling growth depends on the general level of biosynthetic activity which, in turn, depends on ATP generated by respiration, i. e., alternative (a). If alternative (a) is true, then mitochondrial activity should be closely related to seedling growth rates. Recent evidence for this comes from McDANIEL and SARKISSIAN [6], who showed that heterosis in seedling vigor, measured as seedling growth, in corn corresponds to a biochemical heterosis in mitochondria, measured as oxygen uptake.

### References

1. BONNER, W. D., JR.: In: J. BONNER and J. E. VARNER (Eds): Plant Biochemistry, p. 89. New York: Acad. Press 1965.
2. BRACHET, J.: Biochemical cytology. New York: Acad. Press 1957.
3. CHANG, C. W., and C. R. THOMPSON: Physiol. Plant 19, 911 (1966).
4. GORDON, S. A., and R. P. WEBER: Plant Physiol. 30, 200 (1955).
5. LUCK, O. J. L.: J. Cell Biol. 24, 641 (1965).
6. McDANIEL, R. Q., and I. V. SARKISSIAN: Science 152, 1640 (1966).

7.  Neidhardt, F. C., and B. Magasanik: Biochem. Biophys. Acta **42**, 99 (1960).
8.  Nilson, E. G., and A. W. Pauli: Crop. Sci. **4**, 149 (1964).
9.  Nutile, G. E., and L. W. Woodstock: Physiol. Plant., in press (1967).
10. Pfeffer, W.: The physiology of plants. London: Oxford University Press. Transl. by A. J. Ewart 1900.
11. Pollock, B. M., and V. K. Toole: Plant Physiol. **41**, 221 (1966).
12. Sax, K.: Rad. Bot. **3**, 179 (1963).
13. Skoog, F.: J. Cell. Comp. Physiol. **7**, 227 (1935).
14. Watson, J. D.: Molecular Biology of the Gene. New York: W. A. Benjamin, Inc., 1965.
15. Woodstock, L. W.: Bioscience **15**, 783 (1966).
16. —, and D. F. Grabe: Plant Physiol., in press (1967).
17. —, and O. L. Justice: Rad. Bot. **7**, 129 (1967).
18. —, and B. M. Pollock: Science **150**, 1031 (1965).
19. —, and F. Skoog: Amer. J. Bot. **47**, 713 (1960).

## *Discussion*

Klamerth:

Is there any correlation between growth rate and synthesis of rapidly labelled RNA? Could there be a possibility that preformed nucleotide triphosphates in a pool supply the whole RNA synthesis you mentioned?

Woodstock:

We have in fact found relationships between RNA contents and growth. Yes, it may be that there exists such a pool.

Payne:

Can you distinguish between your first hypothesis (a) that growth is limited by total respiratory energy, and the possibility that the rate of transport of nutrient substrate is itself the factor limiting overall metabolism?

Woodstock:

If nutrients were limiting for growth then we have to assume that their sum is also limiting for the uptake of oxygen. We cannot distinguish between the two hypotheses.

Kacser:

The growth rate was not constant; it would not be the same at 4 days or at 3 days after planting. However, when growth at one period is compared to growth at some later period, the two are positively correlated. "Growth" measured as length  increase in etiolated seedlings may not be the same as growth measured by dry weight, because under these conditions dry weight decreases. But, under normal conditions growth measured as elongation parallels protein synthesis, cell formation and the criteria of growth.

Woodstock:

We measure the growth of a seedling in the dark. If we measure the rate of growth in terms of dry weight, we measure growth rate of decrease in dry weight.

# Quantitative Parameters of Rat Liver Mitochondria in Relation to Age

D. Adamiker

*Abstract*

From 22 5-month-old and 20 24-month-old Sprague-Dawley rats in a total of 21 361 hepatic mitochondria the following parameters were determined by means of electron microscopy: (1) number of mitochondria per unit of sectioned cell area, (2) area of the individual mitochondrial section and (3) total area of mitochondrial sections expressed as percentage of the sectioned cell area. In old animals (1) was significantly higher and (2) significantly lower than in young ones, whereas the amount of (3) did not differ between young and old animals.

Mitochondria react quickly and sensitively to disturbances from internal or external causes. In electron microscopy the aspect of mitochondria is a known criterion for the degree of fixation of a tissue [5]. Also in vivo mitochondria react very sensitively to disturbing influences exerted by the organism. Very important among them are the metabolic changes occurring with increasing age [2, 3, 4]. Ageing appears to be the result of evolutionary and involutionary processes among which involution eventually prevails. Morphologically, ageing is characterized by three distinct processes: (a) atrophy of the specific organ tissues, (b) hypertrophy of the connective tissue and (c) deposition of metabolic end-products.

The important role mitochondria play in the cell and body metabolism and the decrease of vital functions with age were the reasons for the present investigation. It was decided to determine electron microscopically in young and old rats whether there are any age differences in hepatic mitochondria with respect to number, section area and total area.

11 male and 11 female 5-month-old Sprague-Dawley rats (group Y) as well as 5 male and 15 female 2-year-old rats (group O) were used for these investigations. Electron microscopical sections of the livers of these animals were prepared according to the usual method (i.e. fixation in $OsO_4$-solution, embedding in a polyester resin), sectioned and photographed. The area of certain sectors as well as that of the mitochondria contained in them were determined planimetrically and the results analyzed biostatistically:

Table 1

A) Number of Mitochondria per 56.25 $\mu^2$ of sectioned cell area

| Statistical parameter | group Y | group O |
|---|---|---|
| n | 220 | 200 |
| $\bar{x}$ | 43.5 | 59.0 |
| $s_{\bar{x}}$ | 0.81 | 1.36 |
| $P_Y$ | | ≪ 0.001 |

B) Mitochondrial section area in $\mu^2$

| | | |
|---|---|---|
| $n_m$ | 9566 | 11 795 |
| $\bar{x}$ | 0.331 | 0.253 |
| $s_{\bar{x}}$ | 0.0023 | 0.0017 |
| $P_Y$ | | ≪ 0.001 |

C) Total mitochondrial section area in $\mu^2$ per 56,25 $\mu^2$ sectioned cell area

| | | |
|---|---|---|
| n | 220 | 200 |
| $\bar{x}$ | 14.386 | 14.947 |
| $s_{\bar{x}}$ | 0.2293 | 0.3997 |
| $P_Y$ | | < 0.4 |

n   number of sectioned cell areas investigated
$n_m$ number of mitochondrial section areas investigated
$\ddot{x}$   mean
$s_{\bar{x}}$  standard error of the mean
P   probability

Analysis of variance of the results presented in table 1 (A, B, C) shows that the variance between animals was significantly higher as compared with that within animals. This reveals also significant individual differences. With respect to the size of the individual mitochondrial section area, it was found in both groups that the essential source of variance between plates was significantly higher than within plates. In the analysis of variance F-values were below 0.1 % probability in all cases (table 2, A—F).

The literature contains only a few reports on age changes of mitochondria. Samson et al. [6] also investigated rat mitochondria in connection with age. They used 1 to 50-day-old rats, however, and, moreover, did not study liver mitochondria but brain mitochondria. They were able to find an increase in the number of mitochondria from 286 to 1328 mitochondria per cell within the period of time investigated. This finding coincides in principle with our results. In hepatic rat mitochondria an increase of the number of mitochondria with increasing age has been observed. The above authors also investigated, aside from the number of mitochondria per cell, the weight of the brain mitochondria of the rats and found that the mean mitochondrial weight of 0.387 pg in the younger animals decreased to 0.288 pg in 50-day-old-animals. This supports our results in which the section area of the hepatic mitochondria of young rats was significantly larger than that of the older animals.

Table 2

A: Number of mitochondria per unit of sectioned cell area in group Y

| Source of variance | d.f. | M.S. | S.S. | F | P |
|---|---|---|---|---|---|
| between animals | 21 | 15 293 | 728.24 | 8.99 | <0.001 |
| within animals | 198 | 16 046 | 81.04 | | |
| within group | 219 | 31 339 | | | |

B: Number of mitochondria per unit of sectioned cell area in group O

| | | | | | |
|---|---|---|---|---|---|
| between animals | 19 | 37 509.2 | 1 974.17 | 9.83 | <0.001 |
| within animals | 180 | 36 138.7 | 200.77 | | |
| within group | 199 | 73 647.9 | | | |

C: Mitochondrial section areas in group Y

| | | | | | |
|---|---|---|---|---|---|
| between animals | 21 | 1 656 077.95 | 78 860.85 | 23.59 | <0.001 |
| within animals | 9 544 | 31 897 619.70 | 3 104.28 | | |
| within group | 9 565 | 33 553 697.65 | | | |
| between plates | 19 | 2 170 589.30 | 9 911.36 | 3,10 | <0.001 |
| within plates | 9 325 | 29 727 030.40 | 3 187.88 | | |
| within animals | 9 544 | 31 897 619.70 | | | |

D: Mitochondrial section areas in group O

| | | | | | |
|---|---|---|---|---|---|
| between animals | 19 | 1 678 842.34 | 88 360.12 | 39.54 | <0.001 |
| within animals | 11 775 | 26 313 179.15 | 2 234.66 | | |
| within group | 11 794 | 27 992 021.49 | | | |
| between plates | 199 | 708 644.75 | 3 561.02 | 1.60 | <0.001 |
| within plates | 11 576 | 25 604 534.40 | 2 211.86 | | |
| within animals | 11 775 | 26 313 179.15 | | | |

E: Total mitochondrial section areas per unit of sectioned cell
   *area in group* Y

| | | | | | |
|---|---|---|---|---|---|
| between animals | 21 | 66 711 907.30 | 3 176 757.49 | 5.33 | <0.001 |
| within animals | 198 | 117 992 764.90 | 595 923.05 | | |
| within group | 219 | 184 704 672.25 | | | |

F: Total mitochondrial section areas per unit of sectioned cell
   *area in group* O

| | | | | | |
|---|---|---|---|---|---|
| between animals | 19 | 298 616 660.9 | 15 716 666.36 | 17.16 | <0.001 |
| within animals | 180 | 164 867 629.5 | 915 931.27 | | |
| within group | 199 | 463 484 290.4 | | | |

A comparison of the total mitochondrial section area per unit of sectioned cell area results in no significant difference between old and young animals because of the compensation by means of different number and different sizes of sectioned areas of the individual mitochondrion. A certain compensation of the decrease in mitochondrial size by means of an increase in numbers seems to be possible. However, a complete compensation of the total mitochondrial area does not necessarily mean that it is connected with a functional compensation by an increase in numbers.

# References

1. Kuriaki, K., and Ch. J. Kensler: J. Biochem. (Tokyo) **41**, 409 (1957).
2. Lazovskaya, L. N.: Biochimia (Russ.) **8**, 171 (1943).
3. Leibetseder, J.: Z. Altersforsch. **15**, 210 (1961).
4. Linder, G.: Vet. Med. Diss. Wien (1961).
5. Novikoff, A. B., In: (J. Brachet and A. E. Mirsky, Eds.) The Cell Vol. II. New York and London: Academic Press 1961.
6. Samson, E. F., M. Balfour and R. Jacobs: Amer. J. Physiol. **199**, 693 (1960).

*Discussion*

Smith:

Were control tests performed to eliminate variance due to different degrees of shrinkage in cell areas of the tissue of the young animals in comparison with that of the older group; e.g. could there have been more lipid or water in the cells of one but not the other group, which might have caused a difference in apparent cell size and/or mitochondrial size? If your data, however, show differences obviously greater than could be attributed to these effects, it would appear that your conclusions are probably quite valid. From some early work done in our laboratory some years ago (unpublished) I can state that our results agree with your present conclusions.

Horwitz:

As an addendum to Dr. Smith's comment, I would like to say that I have observed similar changes in number and size of liver mitochondria in rats acclimated to cold (unpublished data). However, changes in the content of tissue water were also observed and these could explain—to a large degree— the increases in mitochondria number per unit area. On the other hand, if the differences you observed were relatively large (i. e., of the order of 2 or more fold) it is unlikely that a decrease in water content of the tissues could account for your changes.

Adamiker:

We have done no other quantitative experiments or control tests except the measurements of mitochondria. But so far as the normal appearance of the cell could be a point, we did not see any abnormalities between the cells of younger or older rats. Both type of cells were at the same magnification of about the same size and there was no remarkable difference in the lipid or water content. The significant difference between the values of the young and olds rats was so high ($\ll 0.001$) that water loss alone could not explain the differences.

# Enzymes as Indicators of Intracellular Reaction Conditions

E. Müller

With 1 Figure

*Abstract*

In contrast to numerous data concerning enzyme activity in vitro, little is known in a quantitative sense about enzymes inside the cell. As an example of enzyme kinetics in vivo the urea-urease-system of plant cells was investigated. The enzyme located in the groundplasm shows kinetics similar to that in water solution: the Michaelis constant and the activation energy *in vivo* are of the same magnitude as *in vitro*. — Furthermore, a specific inhibition of urease by chloramphenicol takes place *in vitro*. This effect enables the free concentration of the antibiotic in the groundplasm to be roughly estimated.

Since metabolism is mainly connected with the activity of enzymes, it is important to obtain quantitative information on the behavior of enzymes inside the living cell. The magnitudes of enzyme parameters *in vivo* and their regulatory change not only provide us with information on the enzyme molecule itself, but are concerned also with reaction conditions at the site of the enzyme activity. Much is known about reactions of soluble enzymes *in vitro* under strictly defined conditions. However, even *in vitro* it is sometimes difficult to get comparable results, since many interfering factors can appear.

Serious difficulties arise if enzyme parameters are to be investigated in the living cell, in which many enzymes are bound to a structure. Their activity *in vivo* can be followed through changes in the state of the prosthetic group (cf. [1]). Most of these enzymes are not active unless in connection with a lipid structure. The consequences of the binding of an enzyme to a structure are known incompletely [12]. Even if unbound to a structure, enzymes *in vivo* do not in general yield good information on their kinetics [6, 8, 9].

Both the concentration and the state of the enzyme substrate at the site of activity are vastly unknown; many substrates are transported by active mechanisms and unevenly distributed among the cell compartments. However, in the plant tissue (*Nymphaea alba* and *Taraxacum officinale*) [17] urea, the substrate of urease, is passively equilibrated between the cell and the surrounding solution and, within certain limits, evenly distributed inside the cell [13, 14]. In order to obtain the actual substrate concentration, it must not be overlooked that the concentration of urea inside the cell is lower than in the ambient solution in which small slices of tissue are suspended. This is a consequence of urea metabolism in the cells, especially at low overall urea concentrations [14, 15].

The determination of the specific activity or of the turnover number of an enzyme requires a knowledge of the amount of enzyme per mg protein (cf. [5]), quantities which cannot be obtained *in vivo*. Therefore, it remains only to measure the enzyme activity after extraction under standardized conditions *in vitro* [9]. But how can we be sure that the extraction of the enzyme is quantitatively complete, especially in experiments with plants?

In tackling this problem we focussed our interest onto the Michaelis constant in vivo, since here the specific activity need not be known; it is sufficient to compare the reaction rate over a wide range of substrate concentrations with the reaction rate at saturation. The great advantage is that the Michaelis constant of urease is identical with the dissociation constant of the enzyme-substrate-complex and not connected with the overall reaction rate [11]; furthermore the constant is independent of the $p_H$ ([11] and [19], but cf. [3] and [17]); consequently, it is not necessary to know the $p_H$ *in vivo*.

Our experiments have shown that the Michaelis constant *in vivo* is of the same order of magnitude as *in vitro* [16] (Fig. 1). The calculated regression lines give $K_m = 3.5 \cdot 10^{-3}$ Mol/l and $2.7 \cdot 10^{-3}$ Mol/l for two experiments. Comparable results were obtained by investigating the activation energy of the urease reaction [14, 17].

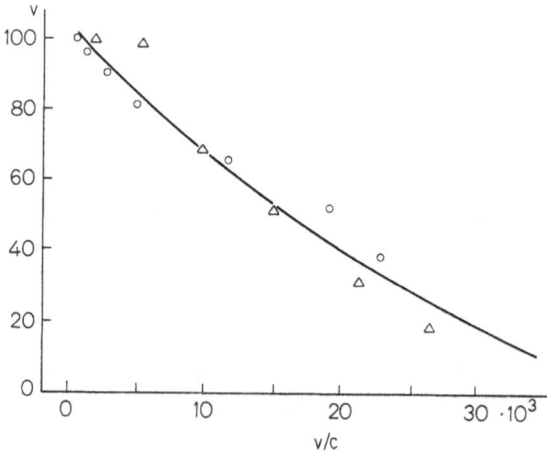

Fig. 1. Urease reaction in *Nymphaea alba*; Eadie-plot. The curve is redrawn from in vitro experiments of Kistiakowsky-Rosenberg [10]; the points from our in vivo experiments. v = reaction rate, c = concentration of urea

These results support the conclusion that the kinetics of soluble enzymes is virtually the same *in vivo* as *in vitro* [9]. Yet another conclusion can be drawn from our experiments: urease is known to be located in the groundplasm of the plant cell and not bound to a structure [17]; hence, the groundplasm must constitute a system with the properties of an aqueous solution. This conclusion is highly significant for the distribution of matter, e.g. ions, between the cells and their surroundings. Further experiments are needed; experiments with vital dyes have shown that the groundplasm may behave also as a lipid system [4, 7]. All the inferences for our present knowledge of protoplasm cannot be discussed here; but it is evident that enzyme systems are potent indicators of intracellular reaction conditions.

This fact could additionally be demonstrated with urease. We have shown that urease is specifically inhibited in vitro by the antibiotic chloramphenicol (CAP) [18]. Consequently the actual free concentration of CAP in the protoplasm can be determined by its effect on the urease reaction *in vivo*; in short-term experiments this effect is independent of the inhibition of protein synthesis by CAP. The results indicate that the effective concentration of CAP inside the cell is much lower than in the incubation medium. This obviously results from an inactivation of CAP inside the cell [2, 20], which may be one reason for the high resistance of many plant cells to this antibiotic. The effective *in vivo*-concentration of CAP must be taken into account when the sensitivity of protein synthesizing systems *in vivo* and *in vitro* are compared.

## References

1. CHANCE, B.: J. Biol. Chem. **217**, 409 (1955).
2. CZYGAN, F.-C.: Naturw. **51**, 541 (1964).
3. DIXON, M.: Biochem. J. **55**, 161 (1953).
4. DRAWERT, H.: In: Encyclopedia of Plant Physiology, Vol. **II**, p. 252, 1956.
5. GUTFREUND, H.: An Introduction to the Study of Enzymes. Oxford 1965.
6. HESS, B., and K. BRAND: Clin. Chemistry **11**, 223 (1965).
7. HÖFLER, K.: Ber. dtsch. Bot. Gesellsch. **74**, 233 (1961).
8. HOLLDORF, A., und E. FÖRSTER: In: Die Zelle, Hrsg. METZNER, p. 209. Stuttgart 1966.
9. HOLZER, H.: In: Erg. med. Grundl.forsch., Vol. **I**, S. 189, 1956.
10. KISTIAKOWSKY, G. B., and A. J. ROSENBERG: J. Amer. chem. Soc. **74**, 5020 (1952).
11. LAIDLER, K. J.: Discuss. Faraday Soc. **20**, 83.
12. McLAREN, A. D.: In: Cell Interface Reactions. Edit. BROWN, New York 1963.
13. MÜLLER, E.: Flora **148**, 529 (1960).
14. — Nova Acta Leopoldina N. F. **24**, Nr. 155 (1961).
15. — Flora **153**, 549 (1963a).
16. — Protoplasma **57**, 611 (1963b).
17. — Habil.-Schrift, Martin-Luther-Universität. Halle 1966.
18. — Naturwiss. **54**, 226 (1967).
19. MYRBÄCK, K.: Acta Chem. Scand. **1**, 142 (1947).
20. PARTHIER, B.: Naturwiss. **52**, 214 (1965).

# The Influence of Ethanol on
# Carbohydrate- and Fat-Metabolism of the Liver

H. P. T. AMMON

*Abstract*

In white mice the administration of alcohol enhances the glycogenolysis of the liver and the lipolysis of adipose tissue by means of increased liberation of catecholamines from the adrenal medulla. Liver glycolysis is accelerated via glycerol-1-P. The increased content of glycerol-1-P in the liver and of free fatty acids in the serum leads to an increased formation and deposition of triglycerides in the liver.

The liver is not only the organ frequently damaged by alcohol but also the organ where the major part of the first step of alcohol oxidation takes place. During this process considerable quantities of $NADH_2$ are formed. Furthermore alcohol leads to an increased liberation of catecholamines from the adrenal medulla [1, 9, 13, 18]. Since both the enhanced formation of $NADH_2$ in the liver and the increased liberation of catecholamines influence the carbohydrate and fat-metabolism of the liver, we investigated the influence of alcohol on the glycolytic carbohydrate breakdown and the metabolism of fat in this organ.

## Carbohydrate Metabolism

In female white mice 10 and 30 min after the i.v. injection of 4.1 or 1.5 mg/g alcohol the contents of glycogen and pyruvate in the liver decrease whereas fructose-1,6-$P_2$ (FDP), dioxyacetone-P (DAP) and glycerol-1-P increase. The lactate content is elevated only after 10 min. In the following hours the lactate content is below normal levels [2].

The diminution of the glycogen content and the increase of FDP, DAP and glycerol-1-P point to an acceleration of the glycolytic carbohydrate breakdown in the upper part of the glycolytic pathway via glycerol-1-P. The increase of glycerol-1-P in the liver is caused by the excessive formation of $NADH_2$ which carries the hydrogen from alcohol to DAP forming glycerol-1-P. The reduction of pyruvate and of lactate which is increased for not more than 10 min show that in mice the formation of the final products of the glycolytic pathway is reduced.

Since alcohol leads to an elevated liberation of catecholamines this raises the question whether the glycogenolytic effect of alcohol is caused by the catecholamines liberated by alcohol; the catecholamines activate the phosphorylase in the liver by means of increased formation of cyclic 3,5 AMP. Results from this laboratory show that glycogenolysis following administration of alcohol is due to the catecholamines because this effect could be prevented in mice pretreated with the $\beta$-sympathicolytic Kö 592 (1-(isopropylamino)-3-(m-toloxy)-3-propanol·HCl).

## Fat Metabolism

Besides glycogenolysis in the liver, lipolysis in the adipose tissue is accelerated by alcohol as is shown in a decrease of the content of fat in the epididymal adipose tissue and an increase of free fatty acids (FFA) in the serum of rats 2 hrs after the i. v. injection of 2 mg/g alcohol [5].

Nearly 30—50 % of the FFA offered to the liver is taken up and esterified by glycerol-1-P to triglycerides [15, 17]. The formation of triglycerides in the liver depends on the concentration of FFA in the blood [16]. The delivery of triglycerides from the liver seems to be limited. Therefore one finds an increase of the fat content in the liver when the uptake of FFA and the synthesis of triglycerides exceed the ability of the liver to deliver triglycerides. For this reason there is an increase of triglycerides in the liver which depends on the concentration of FFA in the serum [6, 7, 11]. The composition of the triglycerides in the liver after administration of alcohol corresponds to that found in adipose tissue [8, 14].

The increased mobilization of FFA from adipose tissue following the administration of alcohol, bringing about an elevation of FFA in the serum and of the content of fat in the liver, seems to be an effect in which the catecholamines, liberated by alcohol, play an important role. The reason for this assumption is that these effects of alcohol can be prevented by adrenalectomy [10], $\alpha$-sympathicolytics [3] and $\beta$-sympathicolytics [5].

Besides the increased mobilization of FFA, the elevated content of glycerol-1-P in the liver seems to favour the formation of triglycerides therein [12]. It is possible that the elevated offer of glycerol-1-P in the liver leads to an increased extraction of FFA from the serum. The latter opinion is supported by findings of BROWN et al. [4] who could observe an increased uptake of i.v. administered $^{14}$C-palmitate by the liver in the presence of alcohol.

## References

1. ABELIN, I., CH. HERREN and W. BERLI: Helv. med. acta 25, 591 (1958).
2. AMMON, H. P. T., C.-J. ESTLER and F. HEIM: Arch. int. Pharmacodyn. 159, 258 (1966).

3. Brodie, B. B., W. M. Butler jr., M. G. Horning, R. P. Maickel and H. M. Maling: Amer. J. Clin. Nutr. **9**, 432 (1961).
4. Brown, E. A., J. M. Hunter and H. M. Maling: Proc. Soc. exp. Biol. a. Med. **122**, 1079 (1966).
5. Estler, C.-J., and H. P. T. Ammon: Arch. int. Pharmacodyn. **166**, 333 (1967).
6. Feigelson, E. B., W. W. Pfaff, A. Karmen and D. Steinberg: J. Clin., Invest. **40**, 2171 (1961).
7. Gidez, L. I., S. Roheim and H. A. Eder: Fed. Proc. **21**, 289 (1962).
8. Horning, M. G., E. A. Williams, H. M. Maling and B. B. Brodie: Biochem. Biophys. Res. Communic. **3**, 635 (1960).
9. Klingman, G. I., and C. McGoodall: J. Pharmacol. exp. Ther. **121**, 313 (1957).
10. Mallov, S., and G. Gierke: Amer. J. Physiol. **189**, 428 (1957).
11. Nestel, P. J., and D. Steinberg: J. Lipid Res. **4**, 461 (1963).
12. Nikkilä, E. A., and K. Ojala: Proc. Soc. exp. Biol. a. Med. **113**, 814 (1963).
13. Perman, E. S.: Acta physiol. scand. **51**, 68 (1961).
14. Poggi, M., and N. R. DiLuzio: J. Lipid Res. **5**, 437 (1964).
15. Robinson, D. S.: Proc. Int. Symp. on Lipid Transport, Tennessee. Springfield Illinois: Charles S. Thomas Publisher 1963.
16. Rose, H., M. Vaughan and D. Steinberg: Amer. J. Physiol. **206**, 345 (1964).
17. Spitzer, J. J., M. Gold and B. J. Branson: Proc. Soc. exp. Biol. a. Med. **118**, 149 (1965).
18. Wartburg, I. P. v., W. Berli and H. Aebi: Helv. med. Acta **28**, 89 (1961).

# Application of Quantitative Models for the Analysis of the Mechanisms of Protection against the Damaging Effects of Ionizing Radiation

N. Arley

*Abstract*

A concrete mathematical model satisfying certain criteria for theoretical models is described for the kinetics of enzyme inactivation in solution by ionizing radiation. The model elucidates two fundamental problems in radiobiology, (a) the problem of defining a quantitative measure of radiosensitivity, (b) the problem of distinguishing between the two basic mechanisms of the effects of radio-protecting and radio-sensitizing substances, viz. their masking effects on the solute molecules, and their scavenging effects on the free radicals formed in the solvent by the irradiation.

First, I wish to deplore that the experimental part of this paper, which my experimental colleague Dr. T. Brustad from my Oslo institute was going to present, has not been completed in time for this meeting. However, for these investigations we needed about 2 mill. norw. crowns to buy the necessary experimental facility, a millionvolt accelerator in the nano-second irradiation range. As everybody knows, a large part of scientific research effort nowadays has to be spent in fighting to get money to do the research you wish to do. On this point I am happy to say that my Oslo institute has in fact now got the necessary money from the Norwegian government, the accelerator is being installed, and Dr. Brustad is at present in England learning how to use it. So we do hope that at the next symposium we shall be able to present also the experimental results of the project on which I shall report here. For I hope you will all agree that, as recently stated by the president of the University of Uppsala, "Facts without theories are blind, but theories without facts are empty".

I shall now start by repeating some general remarks I made at the Wisconsin symposium in 1963 [1] to elucidate the subject of this symposium: in what way can quantitative mathematical considerations help us to a deeper insight into Nature's structure and functioning? In our time of rapid technological progress and social change I think it is first of all well to keep constantly in mind the modest remarks about our basic knowledge of Nature which Einstein made to his friend M. v. Laue a few months before he died:

„Wenn ich in den Grübeleien eines langen Lebens etwas gelernt habe, so ist
es dies, daß wir von einer tiefen Einsicht in die elementaren Vorgänge viel weiter
entfernt sind, als die meisten Zeitgenossen glauben" (p. 397 in [10]). ("If I have
learned something from the speculations of a long life, then it is that we are a
much longer distance away from a deep insight into the elementary processes of
Nature than most of our contemporaries believe").

Actually all those basic problems in physics mentioned by EINSTEIN in
1938 as unsolved [9], are still as unsolved today as they were 25 years ago,
and one of the most fundamental problems regarding our concept of time,
discussed for more than 60 years since the publication of EINSTEIN's first
paper on his relativity theory, has only recently been cleared up [4]. It is
also a conspicuous fact that we do not yet know the basic causes of such
elementary biological phenomena as every individual Man has observed
throughout 100,000's of years and still observes today: why we are all
going to die, why the growth of every child stops at maturity, why we all
need sleep every day.

It is, therefore, no wonder that we still know so little about the basic
biological problems of both normal and pathological cells, such as why
cells divide, why cells differentiate into various tissues, how cells manage
to build themselves in their metabolism the new specific molecules they
need, and how ionizing radiation interferes with the normal processes of
the cells, the damaging effects of which may in some circumstances take
15—20 or more years to be observable (fig. XII. 1 p. 139 in [7]).

At my Oslo institute we have recently tried to give a comprehensive
review of what we know, of what we do not know, and of what experimental and theoretical methods of investigation we may today hope will be
most fruitful for extending our knowledge about some of these basic
biological problems [6]. Now the research of the last decade or so has made
it more and more probable that the reason why it has turned out to be so
difficult to fill all the gaps in our biological knowledge discussed at length
in [6] is that all the basic cell processes, both those of normal cells and those
of tumor cells, take place at a submicroscopic level, and, more specifically,
between and within single biologically important macromolecules called
DNA, RNA, and protein molecules. However, it has only recently become
possible to begin developing the experimental techniques necessary for
attacking somatic cell problems at the molecular level. Before such direct
experimental attacks become possible, the finer details can only be studied
by applying theoretical analyses.

The main problem is, consequently, to make from our present physicochemical knowledge of basic atomic and molecular processes as realistic
assumptions as possible about the primary processes. Next, to work out
from these assumptions as detailed quantitative predictions as possible
about all the results of a quantitative nature which may with present tech-

niques be obtained experimentally. It should hereby be stressed that the purpose of any theoretical studies in natural science, be it in physics, biology, cosmology or other fields, is to obtain new knowledge about Nature, in the ideal case by narrowing down the possibilities to just those which are actually realized in Nature. Consequently, only such models fulfill their purpose which allow both the deduction of quantitative predictions about all experimentally measurable quantities and the working out of explicit recipes for quantitative determination of the numerical values of the parameters of the model from actually measured quantities.

Now, in studies of biological damage due to ionizing radiations, as in most other fields of research, models may be constructed in various ways depending on which basic assumptions we start from (chap. 1 in [5]). As discussed above, it is with present experimental techniques beyond our powers to check these assumptions in direct experimentation. The best procedure is, therefore, to investigate how the phenomenon might have been formed from each set of assumptions. If one set of results looks more like the real phenomenon than another set we shall probably have discovered not only how the phenomenon is actually formed, but also which basic assumptions are correct.

At the Wisconsin symposium in 1963 I presented a review of quantitative models for the analysis of the mechanisms of carcinogenesis [1, 6] satisfying the criteria just discussed. I shall at this symposium present a short review of a quantitative model for the analysis of the mechanisms of protection against the damaging effects of ionizing radiation. For details I beg to refer you to the original paper [3] and to a short summary of it [2].

In quantitative studies of the protective, and also the sensitizing, effects of added substances on the inactivation of enzymes in dilute aqueous solution by ionizing radiation, a most fundamental problem is met with— namely, how to define a quantitative measure of radiosensitivity. If the data fit a straight line when the remaining enzyme activity, $E$, is plotted semilogarithmically versus the radiation dose, $D$, the quantity $1/D_{37}$ is usually chosen, where $D_{37}$ denotes that dose for which the enzyme activity, $E$, is reduced to $1/e = 37\%$ of the initial activity, $E_0$. However, in several experiments nonexponential dose-inactivation curves are met with [8], and the quantity $1/D_{37}$ then loses any deeper meaning as a quantitative measure of radiosensitivity. In these cases the only sensible procedure is to construct a hypothetical model of the reaction mechanisms behind the inactivation in question, from this model to deduce a mathematical formula for the dose-inactivation curves, and then to fit this formula to the experimental data in question by numerical or graphical adjustment of the parameters of the model. The radiosensitivity of the given enzyme under given experimental conditions is then expressed by these numerical values of the parameters.

Another fundamental problem in radiobiology is the problem of what are the mechanisms of the effects of radio-protecting, as well as radio-sensitizing, substances. It is well known that accumulated evidence on chemical protection does not now permit an unequivocal recognition of the underlying mechanisms (p. 51—52 in [11]). However, two mechanisms seem at the present time to be generally thought of as being the most important ones, on the one hand *the scavenging effect* of the protector molecules, $P$, on the intermediary free radicals, $X$, formed in the solvent by the irradiation, on the other hand *the masking effect* in which the protector molecules, $P$, attach themselves to the solute molecules, $E$, under investigation in such a way as to mask their sensitive sites against the inactivating attack of the free radicals, $X$, leading to the biological damage in question.

Both these problems are solved in the model and the recipe for the determination of the numerical values of the parameters of the model proposed in the paper reported here for the special case of enzyme inactivation in solution by ionizing radiation. For the details I beg to refer you to the original papers quoted. Here I shall finish by just indicating the proposed recipe for extracting the information we are looking for as discussed above from the pertinent experimental data—again stressing that these experimental data seem not yet to have been obtained, but that we hope soon to be able to obtain them in my Oslo institute.

I assume as a first approximation, which I have generalized in a second paper not yet published, that the enzyme $E$ is predominantly inactivated by one free radical $X$ formed in the water by the irradiation. Under this assumption the model contains in all 5 parameters denoting 5 kinetic rate constants of 5 chemical reactions. These 5 rate constants may be deduced from experimental data as follows. For each irradiation dose $D$ one has to calculate the time

$$t_e = \alpha\, D/I \tag{1}$$

where $t_e$ is the time of observation assumed here to coincide with the time at the end of the irradiation, started at $t = 0$; $I$ is the number of radicals formed per unit time and $\alpha$ is a proportionality constant converting dose $D$ to total number of radicals formed

$$I t_e = \alpha\, D \tag{2}$$

(The mathematical formalism of the model also covers the case where there is a time lag between the time of observation and the time at the end of the irradiation, but this case does not seem to occur in any experiments so far and it seems to be inappropriate to make the experiments in such a way.) One then has to plot

$$\ln(E/E_0) \quad \text{versus} \quad t_e^2 \tag{3}$$

where $E$ is the remaining concentration of active enzyme at time $t_e$, and $E_0$ is the initial enzyme concentration at $t = 0$, because according to the mathematics of the model $\ln(E/E_0)$ starts to vary parabolically with $t_e$, i.e. the plot (3) should for small doses asymptotically fit a straight line through the origin. If $-S_1$ denotes the slope of this line the mathematics of the model shows that

$$k_E = 2\,S_1/I \tag{4}$$

where $k_E$ is the rate constant for the reaction of the enzyme $E$ with the radical $X$ in which both enzyme and radical become inactivated. One next has to plot

$$\ln(E/E_0) \quad \text{versus} \quad t_e \tag{5}$$

because according to the mathematics of the model $\ln(E/E_0)$ versus $t_e$ should fit a straight line asymptotically for large doses $D$. Denoting this line $-S_2 t_e + R_1$ one has to calculate the quantity

$$b = [2S_1\,(1 + R_1) - S_2^2]/S_2 \tag{6}$$

and to plot

$$b \quad \text{versus} \quad E_0 \tag{7}$$

which plot should fit a straight line according to the mathematics of the model. Denoting this line $R_2 + S_3 E_0$ one has to plot

$$R_2 \quad \text{versus} \quad P_0 \tag{8}$$

where $P_0$ is the concentration of the substance whose protecting, or sensitizing, effect on the inactivation of the enzyme $E$ is studied. According to the mathematics of the model this plot should fit a straight line through the origin denoted $S_5 P_0$ with

$$k_P = S_5 \tag{9}$$

where $k_P$ is the rate constant for the reaction between the radical $X$ and the protector, or sensitizer, $P$, leading to a change of the radical.

If by varying the quality of the protecting, or sensitizing, substance $P$ a variation is found in the rate constant $k_E$ this will indicate that the protecting, or sensitizing, effect of $P$ is due to a *masking* effect of $P$ on the enzyme $E$, whereas if a variation is found in the rate constant $k_P$ this will indicate that the protective, or sensitizing, effect of $P$ is due to a *scavenging* effect of $P$ on the radical $X$.

It is therefore possible that we here have a new tool for elucidating the basic mechanisms of the effects of protecting and sensitizing substances in radiobiology. Consequently we have tentatively proposed to denote the rate constants $k_E$ and $k_P$ the *internal and external radiosensitivity coefficients*, respectively.

At this point my colleague Dr. T. Brustad from my Oslo institute should have taken over and described our experimental results, but it has taken longer time than we had thought to get the money for the necessary irradiation machine so that you will have to wait some time for the continuation of our report.

Let me conclude by repeating my concluding remarks in my paper at the Wisconsin symposium in 1963 [1]: I feel that I am no longer alone in the firm conviction that *all the basic secrets of life are hidden on the molecular level in the borderland between atomic physics and biology*. Modern quantum theory has already united the fields of physics, chemistry, and astronomy to one grand entirety of knowledge and it seems to me not unlikely that in a not too distant future the fields of biology, including both psychology and medicine, will also be embraced by this unification process thus giving us a unified picture of the universe stretching from the infinitely small world of the atoms, over the intermediary world of everyday physics, chemistry, and biology, to the infinitely large expanding world of the galaxies and their clusters.

*Acknowledgements.* The content of this paper was intended to be presented as part of lectures at the University of Copenhagen according to the royal jus docendi of the author's doctor degree and the author's royal appointment as associate professor. However, the authorities of the university of Copenhagen and the Danish minister of education have prevented by force that they were delivered in this way. — The author wishes to thank Prof. R. Schlegel, Michigan State University, and Dr. T. Brustad, Department of Biophysics, Norsk Hydro's Institute for Cancer Research, most heartily for many stimulating discussions during a number of years.

### References

1. Arley, N.: In: J. Gurland (Ed.), Stochastic Models in Medicine and Biology, p. 3, Univ. of Wisconsin. Madison: Univ. of Wisconsin Press 1964.
2. — Naturwissenschaften **53**, 276 (1966).
3. — Radiation Research **29**, 1 (1966).
4. — Naturwissenschaften **54**, 366 (1967).
5. —, and K. R. Buch: Introduction to the Theory of Probability and Statistics. Science Edition. New York: John Wiley and Sons 1966.
6. —, and R. Eker: Advances in Biological and Medical Physics **8**, 375 (1962).
7. —, u. H. Skov: Atomkraft. Eine Einführung in die Probleme des Atomzeitalters. Heidelberg: Springer-Verlag 1960.
8. Brustad, T.: Radiation Research **27**, 456 (1966).
9. Einstein, A., and L. Infeld: The Evolution of Physics. New York: Simon and Schuster 1938.
10. Seelig, C.: Albert Einstein. Leben und Werk eines Genies unserer Zeit. Zürich: Europa Verlag 1960.
11. Report of the United Nations Scientific Committee on the Effects of Atomic Radiation. New York: U.N. 1962.

*Discussion*

SMITH:

As your model is based upon single-site enzymes rather than multi-site moieties, the generality of its application is severely limited. Hopefully the theory can be developed upon a model which reflects this order of complexity. This might be facilitated by reference to the BRUNAUER, EMMET and TELLER theory.

ARLEY:

I quite agree with you that in reality the phenomenon is much more complicated than we assume in this or any other model; I also quite agree that we should try to take as many of these complications into account. However, I would like to stress that as a matter of fact the model does not assume any details about the chemical reactions entering into it.

BETZ:

In enzymology we have learned to distinguish between catalytic sites and regulatory sites, especially with allosteric enzymes. Is your model suitable to distinguish whether radiation damage will interfere with one or the other in any particular enzyme?

ARLEY:

Although it is beyond the power of any individual scholar nowadays to cover all the knowledge within even a single field such as enzymology, my personal guess is: "no", but we certainly do hope that in the experimentation of tomorrow it will become possible.

# Intracellular Recovery after Irradiation

J. Kiefer

With 4 Figures

*Abstract*

Investigations on recovery from sublethal and potentially lethal damage in cultured mammalian cells and microorganisms are reviewed and some new experimental results with yeast presented. The processes of recovery from sublethal and potentially lethal damage are not necessarily qualitatively different although they appear to be antagonistic in several repects. A model is developed, whose basic features are: after irradiation two processes enter into competition, viz. damage repair and damage fixation, both of which may be influenced differently by certain treatments. The experimental results are discussed in terms of the model suggested.

## I. Introduction

It is obvious that the manifestation of biologically significant damage may take place at different levels in the hierarchy of organismic organisation. At each of these levels reversions of the initial impact may occur. It seems important for the sake of clarity to use a clearly defined terminology, as suggested in a former paper [32] based on a scheme by Bacq and Alexander [6].

Throughout this paper recovery will mean only those processes which take place within the cell between the time of irradiation and the first subsequent cell division. A cell will be termed "lethally damaged" if it is not able to form a macrocolony. The considerations to be presented will be based mainly on work with mammalian cells in culture and microorganisms, for two reasons: 1. there are only a few data on organized systems, and 2. systemic effects may obscure the cellular phenomena or interfere with them in an unknown manner. In the classical hit-theory (cf. e.g. [54]) the course of events was considered to be determined after the absorption of the photon at the critical site. Shoulders in the survival curves were interpreted in terms of multi-hit or multi-target models. During the last decade, however, evidence has accumulated to show that the initial lesion can be modified in many respects after irradiation. One conclusion of these investigations bears out the importance of post-irradiation culture conditions. Variations in temperature, nutrient medium or the application of metabolic inhibitors may markedly affect survival level. At the present time it appears rather difficult to give a unified picture of all the effects found, nevertheless an attempt will be made to design a concept based on recent experimental work.

Recovery effects at the cellular level have been studied from two points of view, namely, recovery from so-called "sublethal" and from "lethal" damage [16]. The first phenomenon is revealed by the split-dose technique, the second leads to a modification of the survival curve. The term "recovery from lethal damage" is not completely satisfactory, because it is by definition impossible. We shall therefore prefer "recovery from potentially lethal damage". Since the techniques involved, the systems tested and the questions arising are different with the two effects, they will at first be discussed separately.

## II. Recovery from Sublethal Damage

If the survival curve of an organism shows a shoulder, this fact may be described by stating that a certain amount of damage has to accumulate before it becomes lethal, or, in other words, if the given dose does not reach this level, it is sublethal. In terms of hit-theory: one hit is sublethal if the survival curve has a two-hit shape. The question is whether a surviving cell which has suffered only sublethal damage is able to repair it. If this were the case, cells having survived a first (usually called "conditioning") dose should behave as nonirradiated cells when irradiated for a second time. If the two dose-fractions are equal, each leaving a proportion $p$ of surviving cells, the surviving fraction after two irradiations is expected to be $p^2$ if complete recovery of sublethal damage has taken place between them. To answer the question, ELKIND and SUTTON [10, 11] used the technique described with Chinese hamster cells in culture: A dose of X-rays sufficient to reach a survival level beyond the shoulder region of the survival curve was split into two fractions spaced in time by variable intervals. They found complete recovery if the time interval was sufficiently long, the cells being kept in optimal medium between the dose fractions. This result has been confirmed in the meantime for many other systems, e.g. yeast [3], plant roots [31] and the intestinal crypt cells of mice [23]. The time needed for the repair processes to be completed is of the order of the mitotic cycle time. Fig. 1 shows the kinetics of recovery for diploid *Saccharomyces cerevisiae* from early stationary phase cultures kept in glucose-yeast extract medium at 30° C between dose fractions. The peculiar shape of the curve is typical for this process and found with almost every system tested. It depends, however, on the metabolic state of the cell as discussed below.

The early experiments of ELKIND and his group stimulated further work in many laboratories, focussing interest on three main questions: 1. What is the nature of the recovery process? 2. What is the reason for the particular shape of the recovery curve? 3. How can it be modified?

We have no clear-cut answer to the first question at the present time. In summary, the investigations revealed: the first immediate rise in the survival level is not influenced by a moderate reduction of temperature [14], it is possibly delayed [8] and depressed only near 4° C [8]. It is not greatly altered by inhibition of DNA- [35] or protein synthesis [16]. Actinomycin D, which complexes to DNA, thus suppressing m-RNA synthesis [20], completely inhibits recovery in hamster cells [16], although not in L-cells [53]. All these findings refer to immediate recovery after the first dose. The

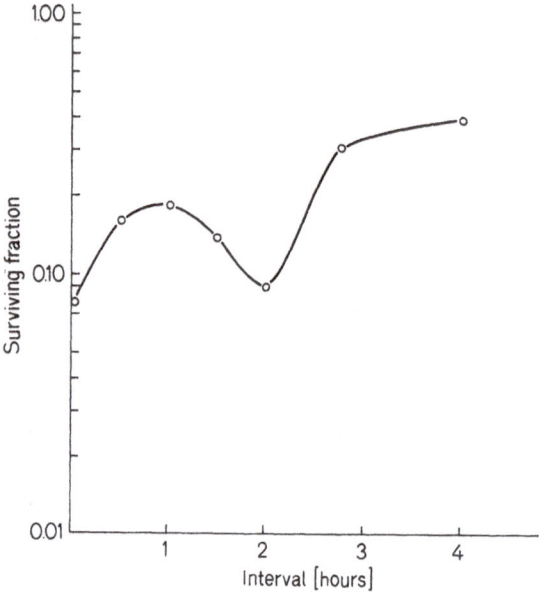

Fig. 1. Effect of split-dose exposure on diploid *Saccharomyces cerevisiae* from early stationary phase. The first dose was given at zero time, the second dose after the intervals indicated on the abscissa

matter is different with the minimum in the recovery curve: it is not found at lower temperatures [14] and is reduced after treatment with puromycin [16]. As it has been argued that the effects stated were not due to recovery but merely to selective killing of cells at a sensitive stage out of an asynchronous population [26], most of the investigations quoted have also been confirmed with synchronous cultures, as extensively reviewed recently [15, 16]. Recovery from sublethal damage has only been shown after irradiation with sparsely ionizing radiation. Attempts to demonstrate it also for UV have been unsuccessful, at least for mammalian cells [21]. In our laboratory, however, we were able to find recovery after UV for yeast although the experiments are still preliminary.

BACCHETTI et al. [3, 4, 5] studied the "Elkind-phenomenon" in diploid *Saccharomyces cerevisiae* after X-irradiation. The time-span necessary for

complete recovery is shorter because of the reduced cycletime. The shape
of the recovery curve was similiar to that found in hamster cells, if log
phase cells were used. The initial peak did not appear if stationary cells
were stored overnight at 4 °C before the experiment. Possible fluctuations
may have been obscured, however, because of the rather low dose-rate and
therefore long irradiation times used by these authors.

Recovery after UV requires a longer time than after X-rays if the same
survival levels are compared. This is in accord with our finding that division
delay is also enhanced after UV [33] and may be the reason for the lack of
success of earlier attempts although mammalian cells and yeast behave

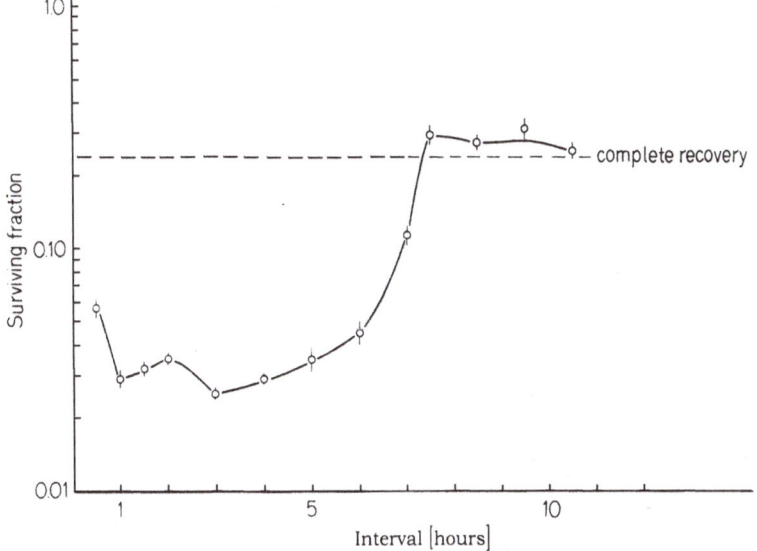

Fig. 2. Recovery from sublethal damage in starved *Saccharomyces cerevisiae*. For the
definition of "complete recovery" see text. o X-rays + X-rays

differently towards UV [9]. In a very recent paper by HARM [22] an effect of
UV-fractionation in *E. coli* has been reported which fits very well with our
results. Complete and immediate recovery from sublethal damage occurs
only if the cells are kept under optimal culture conditions between the
radiation fractions [4]; it is suppressed if only a minimal medium or saline
is present. This holds true at least for yeast.

As shown in fig. 1, the typical recovery pattern is still observable in
cells from the early stationary phase. If the same cells, however, are starved
in saline for two days at 30 °C, the shape is much altered (fig. 2). Recovery is
considerably delayed and the initial peak is missing. Our starvation procedure was different from that of BACCHETTI et al. [3]. Storing at 4°C overnight does not induce starvation, as we checked by respiration measurements, but only a resting phase.

Most of the features reviewed so far can be interpreted by ELKIND's "repair and progression" model [12]. Briefly, he postulates that the minimum is due to further progression of the irradiated cells along their normal cycle, whereas the first rise in the curve represents true recovery. The data quoted are not at variance with this model, but it offers no explanation for the nature of the process itself.

## III. Recovery from Potentially Lethal Damage

In 1953 STAPLETON and his colleagues found that the fraction of *E. coli* surviving a given dose of X-rays is a function of post-irradiation incubation temperature with each strain possessing its own optimum [49]. The authors offered an explanation which is still very attractive; it will be referred to below. Even earlier, in 1949, the phenomenon of photoreactivation was discovered by KELNER [29]: The lethal action of UV irradiation can be partly reversed if the cells are subjected additionally to visible light.

We shall restrict our considerations to these two phenomena: modification of the survival level by post-irradiation treatments after UV and X-irradiations, and photoreactivation, keeping in mind that the matter is far more complex than this, but space and time do not allow a broader discussion. In this connection we may refer to the excellent reviews given at the Cortina Congress on host cell reactivation by WINKLER [51], chromosome aberration and repair by EVANS [17], and the genetic aspects of recovery by RÖRSCH et al. [45].

Our present knowledge on photoreactviation has recently been summarized by JANE SETLOW [47]: under the action of visible light a special photoproduct of UV, thymine-dimers in the cellular DNA, may be split by an enzyme the nature of which is as unknown as the intrinsic mechanism. The protein in question has been purified and prepared in an amount which allowed recording of its absorption spectrum—no appreciable absorption could be seen at the peak of the action spectrum of photorectivation [39].

There are many postirradiation treatments which alter the survival level, the best investigated one being the so-called "liquid holding recovery" (LHR), sometimes also named "delayed plating recovery" or "dark repair", in contrast to photoreactivation, as it does not require light. The technique is basically very simple: irradiated cells are held in minimal medium (tap water in the simplest case) for 48—72 hrs before plating. By this treatment a considerable increase in survival is achieved. The resulting curves were claimed to result from "normal" ones by a simple dose modification (multiplying the dose by a constant factor) [41], but this statement could not be confirmed [4]. LHR has been studied most extensively in yeast yielding the following very significant results: The process is temperature- and pH-sensitive, showing a pronounced optimum for both. It is not altered by application of RNA-, DNA- and protein-synthesis-inhibitors, but requires oxygen and an intact respiration apparatus, as revealed by the application of KCN, azide and 2,4-dinitrophenol [42] or by the fact that "petite" mutants of yeast, being defective in cellular respiration, do not show LHR [28].

All these findings point to an enzymatic, energy-dependent process. Starved yeast (obtained by the procedure described above) also exhibits LHR, although we have indications that it is enhanced if glucose is present in the medium [34]. This agrees with the finding that exogenous ATP increases dark-recovery in unstarved yeast [42]. KOROGODIN et al. [36—38] investigated the kinetics of LHR and were able to show that the curves may be fitted by the equation

$$D(t) = D_0 (k + (1 - k) e^{-\beta t})$$

where $D(t)$ is the remaining damage at time $t$ expressed as effective dose, $D_0$ its non-reparable part (postulated by the author) and $k$ and $\beta$ are constants. $\beta$ has been found to be fairly equal for different strains of yeast tested [38]. Haploid yeast does not possess the ability to undergo LHR after X-rays where the survival curve is exponential, in contrast to its behaviour after UV where the survival curve is non-exponential [41].

The investigations on LHR must be seen in connection with the studies on post-irradiation treatments which enhance the survival level. The presence of chloramphenicol, which is supposed to inhibit protein-synthesis although its mode of action is not yet clearly understood [18], has been proved to be favourable in E. coli [18] (but to prevent repair in Rhizopus stolonifer sporangiospores [48]). Culturing at temperatures different from the optimal growth temperature also enhances the viable count [49]. On the other hand, incorporation of nucleic acid base analogues such as bromodeoxyuridine usually increases lethality (cf. e.g. [27]).

LHR has not been studied in mammalian cells because of their high sensitivity to various treatments. In 1966, however, PHILLIPS and TOLMACH [44] studied the influence of post-irradiation culture conditions after X-rays in synchronous HeLa-cells. Briefly, they found: low temperature, fluorodeoxyuridine, hydroxyurea (both supposed to be DNA-synthesis blockers) and cycloheximide (a protein-synthesis-blocker) suppressed the survival level when given within 5 hrs after irradiation. The combination of hydroxyurea and cycloheximide, however, resulted in an increased viable count. Although it appears desirable to study recovery from both sublethal and potentially lethal damage in combination, only one paper by BACCHETTI et al. [4] has yet been published on this problem. They found that LHR is no longer possible after complete recovery from sublethal damage; the lethal action of the second dose in split-dose experiments, however, could still be modified by LHR.

In our laboratory, the combination of photoreactivation or LHR and dose-fractionation has been investigated. Two doses were separated by a time interval during which either photoreactivation or LHR was permitted. A typical result is shown in fig. 3: The shoulder does not reappear if a

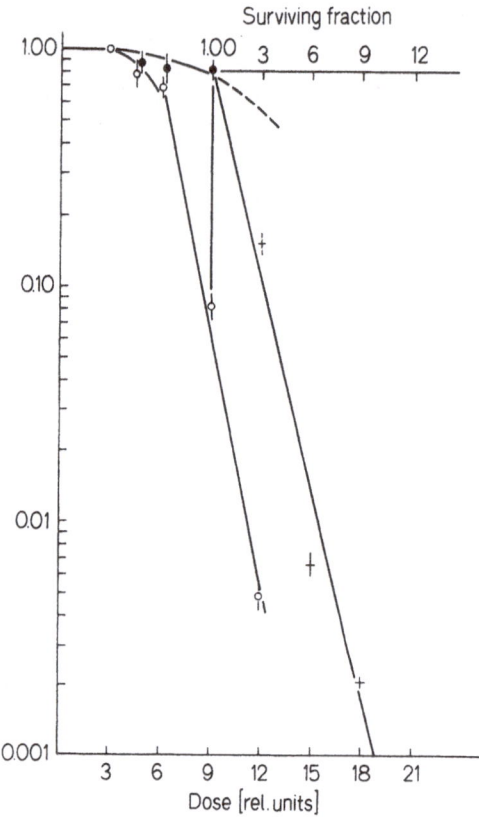

Fig. 3. The effect of photoreactivation between two doses of UV (254 nm) on the survival curve. The first dose of radiation was immediately followed by maximum photoreactivation (PR) and a second exposure to a series of doses. ○ single dose; ● single dose + PR; + single dose + PR + second dose

second dose is given after photoreactivation, simply (and, we think, too simply) speaking: recovery from sublethal damage has not occurred. Qualitatively the same holds true if LHR of starved cells is investigated instead of photoreactivation. This second result answers, by the way, a question left open in the paper by PATRICK et al. [41]—it proves that the repair process (or at least a part of it) obviously involved in LHR takes place during storage in the minimal medium and not on the nutrient agar after plating. This seems important in the light of the rescue hypothesis [19], clearly outlined by ALPER [2].

From all this it appears that recovery from sublethal and potentially lethal damage behave antagonistically in many respects. The facts are summarized in table 1.

Table 1

| Treatment | Recovery from subl. D. | Recovery from leth. D. | Object | Ref. |
|-----------|------------------------|------------------------|--------|------|
| low temperature | 0 | — | mamm. cells | 14, 44 |
| low temperature | — | | mamm. cells | 8 |
| low temperature | | + | E. coli | 49 |
| hypoxia | 0 | | mamm. cells | 14 |
| hypoxia | | — with LHR | yeast | 42 |
| inh. of protein-synthesis | 0 | — | mamm. cells | 16, 44 |
| inh. of protein-synthesis | | + | bacteria | 40 |
| inh. of protein-synthesis | | — | Rhizopus | 48 |
| inh. of m-RNA-synth. | — | — | mamm. cells | 16 |
| inh. of DNA-synth. | 0 | — | mamm. cells | 35, 44 |
| nutrient deficiency | — | + | yeast | 4, 41, 42 |
| respiration blocker | not studied | — with LHR | yeast | 42 |
| pre-irr. starvation | — | 0 | yeast | 34 |

Explanation: 0 = no effect
+ = stimulating effect
— = inhibition or delay

## IV. A Model of Recovery and Damage Fixation

The investigations reviewed so far will now be discussed in terms of a model, the basic assumptions of which are: 1. The initial irradiation damage is not harmful *per se* but may develop to a biologically significant lesion. 2. Unfixed damage may be repaired by the cell. Recovery and fixation compete under normal growth conditions. 3. There is a certain fraction of non-reparable damage, its amount depending on the dose. 4. Every fixed lesion reduces the survival probability by a certain factor, which is assumed to be independent of the number of fixed lesions per cell.

Let us first consider a cell species possessing no recovery ability, e.g. haploid yeast after X-irradiation or *E. coli* B$_s$. Let $q$ be the killing probability per lesion ("Entgleisungswahrscheinlichkeit"—HUG and KELLERER [25]) then the survival probability with $N$ lesions will be $(1-q)^N$. Assuming the initial damage to be distributed according to Poisson's law, the surviving fraction of the whole population will be

$$\text{s. f.} = \sum_{N=0}^{\infty} e^{-\nu D} (1-q)^N (\nu D)^N / N !$$
$$= e^{-q \nu D}$$

with $\nu$ as parameter.

In agreement with the experimental results an exponential survival curve is obtained.

Let us now consider another limiting case: The dose-rate of the irradiation may be low enough to allow recovery or fixation before another "hit" is received by the cell. If $\lambda$ is the fixation probability, the surviving fraction will be

$$\text{s. f.} = \sum_{N=0}^{\infty} (1-q)^N e^{-\lambda \nu D} (\lambda \nu D)^N / N !$$
$$= e^{-\lambda q \nu D}$$

Here again, an exponential survival curve is obtained but its slope is less steep. This result is also in agreement with experimental finding (see e.g. [7]). These remarks are obviously rather qualitative; the model certainly requires a more refined statistical treatment, particularly on the basis of the theory of stochastic processes, but the mathematics involved is rather complicated. Work on this problem is in progress and will be published elsewhere.

Let us now turn back to the two recovery mechanisms discussed in this paper: as the combination of LHR and split-dose experiments shows, fixation of lesions is completed if full recovery from sublethal damage has occurred [4] and, on the other hand, sublethal damage can not be repaired by the PR-procedure. It might be argued that sublethal and potentially lethal damage are qualitatively different, but this assumption is not necessary. The reasoning we should prefer is along the lines of the model suggested: under normal growth conditions recovery and fixation proceed in competition. The split-dose technique allows us to measure merely the unfixed lesions remaining at the time of the second exposure. In terms of the model, sublethal lesions are unfixed lesions. Thus, if a certain experimental treatment favours recovery as against fixation, while the rate of disappearance of unfixed damage is unchanged, the split-dose curves will remain approximately unaltered. This hypothesis may be able to explain the relative stability of the recovery pattern as studied by fractionated irradiation. Nevertheless, the experiments using metabolic inhibitors ought to be checked very carefully because of the following reasoning: If the survival level is altered by a post-irradiation treatment, the starting point for the second exposure will not be the same after different times of treatment. It appears unavoidable that this kind of investigation should have sets of differently treated controls. Fixation apparently does not occur—or at a very low rate—under LHR—or photoreactivation conditions. Thus, these techniques appear useful for further investigations of the cellular processes involved in recovery from radiation impact.

## V. Biochemical Considerations

It is not intended to treat this matter here in detail as it is beyond the scope of this lecture. Let me just set some spotlights which, I hope, are relevant to our discussion. There are many indications that DNA is the main target of radiation attack at the cellular level. It has, as everybody knows, two purposes in the cell, to serve as initial template for messenger RNA-synthesis and to replicate its own structure in order to preserve the genetic information. Nearly all post-irradiation treatments which favour recovery inhibit or delay DNA-synthesis, so that it has been proposed that damage fixation is related to DNA replication [52]. It has been found that initiation of DNA replication is connected with the breakage of hydrogen bonds in the helix [46]. This effect might also be brought about by radiation, so that the suggestion of an X-ray induced "S-phase" [30] appears plausible and might partly account for the shape of the recovery curve in split-dose experiments. DNA suffers degradation after X-irradiation (cf. e.g. [24]), but not in minimal medium [1]. This might be an indication of the validity of the explanation offered.

Recovery may take place as a normal part of RNA-synthesis. It may involve so-called "repair-replication" which is not semi-conservative, as proved by density-labelling experiments [43]. Recovery is certainly dependent on the energy metabolism of the cell. This fact fits very well into the picture of "repair-replication", as very recent biochemical investigations show that enzymes mediating the rejoining of open DNA-strands need ATP or NAD as cofactors [40, 50]. The inhibition of recovery from sublethal damage by actinomycin D [12] of of LHR by DNA-complexing dyes [42] also favours this hypothesis.

## Conclusions

Obviously we are far from a real understanding of cellular recovery processes. All the models discussed give no more than hypotheses for certain aspects. The interesting feature is the hope that we can obtain by these studies not only radiobiological knowledge but also a deeper insight into the normal functioning of the cell.

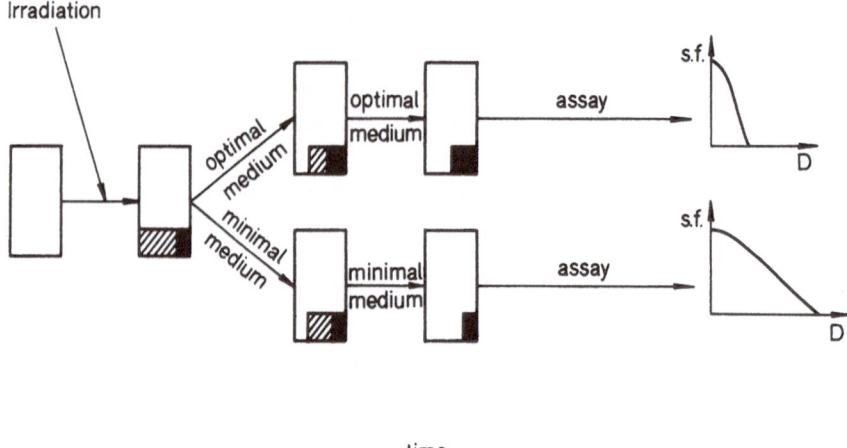

Fig. 4. Principal scheme of the model of radiation damage fixation and recovery: All damage suffered by a non-predamaged cell is unfixed just after the absorption. A certain part of it, however, is in a non-reparable state, i.e. it will be converted into fixed lesions in any case. The remaining part can, in principle, be repaired or fixed. Both processes act in competition. The probability for each is dependent on physiological parameters, e.g. medium, temperature, pH etc. Before the next cell division all remaining damage is in a fixed state, i.e. no longer reparable.

*Acknowledgement*

The author is very much indebted to Prof. Dr. A. SCHRAUB for stimulation and advice, Prof. Dr. L. RAUSCH and the colleagues at the Institute of Biophysics, Giessen, for many discussions and Mrs. Rosita VIVIANI und Miss Maria M. HLA-WICA for technical assistance during part of this work. Thanks are also due to the "Deutsche Forschungsgemeinschaft" for a grant to make these investigations possible. He would also like to thank Drs. Alma HOWARD and M. EBERT for valuable criticism and suggestions concerning this manuscript.

## References

1. ACHEY, P. M., and E. C. POLLARD: Rad. Res. **31**, 47 (1967).
2. ALPER, T.: Discussion remarks in: Repair from genetic radiation demage, (F. SOBELS, Ed.), p. 120. Oxford-London-New York-Paris 1963.
3. BACCHETTI, S.: Int. J. Radiat. Biol. **10**, 213 (1966).
4. — Rad. Res. **29**, 295 (1966).
5. —, M. CASSANDRO and F. MAURO: Exp. Cell. Res. **46**, 292 (1967).
6. BACQ, Z. M., and P. ALEXANDER: Fundamentals of Radiobiology. Oxford: Pergamon Press 1961.
7. BEDFORD, J. S., and E. J. HALL: Int. J. Radiat. Biol. **7**, 377 (1963).
8. BERRY, R. J., and R. OLIVER: Nature (London) **201**, 94 (1964).
9. CLEAVER, J. E.: Biochem. Biophys. Res. Comm. **24**, 569 (1966).
10. ELKIND, M. M., and H. SUTTON: Nature (London) **184**, 1293 (1959).
11. — — Rad. Res. **13**, 556 (1960).
12. —, G. F. WHITMORE and T. ALESCIO: Science **143**, 1454 (1964).

13. ELKIND, M. M., and T. ALESCIO, W. B. MOSES and H. SUTTON: Rad Res. **22**, 186 (1964).
14. —, H. SUTTON-GILBERT, W. B. MOSES, T. ALESCIO and R. W. SWAIN: Rad. Res. **25**, 359 (1965).
15. —, W. SINCLAIR: In: Current topics in Radiation Research (M. EBERT and A. HOWARD, Eds.), p. 165. Vol. I, Amsterdam 1965.
16. — In: Radiation Research (G. SILINI, Ed.), p. 558. Amsterdam 1967.
17. EVANS, H. J.: In: Radiation Research (G. SILINI, Ed.), p. 482. Amsterdam 1967.
18. FREEMAN, K. S., and D. HALDER: Biochem. Biophys. Res. Comm. **28**, 8 (1967).
19. GILLIES, N. E., and T. ALPER: Nature (London) **183**, 237 (1959).
20. GOLDBERG, I. H., and M. RABINOWITZ: Science **136**, 315 (1962).
21. HAN, A., N. MILETIČ and D. PETROVIC: Int. J. Radiat. Biol. **8**, 187 (1965).
22. HARM, W.: Rad. Res. **31**, 548 (1967).
23. HORNSEY, S., and S. VATISTAS: Brit. J. Radiol. **36**, 795 (1963).
24. HOWARD-FLANDERS, P., and P. T. EMMERICH: Biochem. Biophys. Res. Comm. **18**, 24 (1965).
25. HUG, O., u. A. KELLERER: Stochastik der Strahlenwirkung. Berlin-Heidelberg-New York: Springer 1966.
26. KALLMANN, R. J.: Nature (London) **197**, 557 (1963).
27. KAPLAN, H. S., K. C. SMITH and P. A. TOMLIN: Rad. Res. **16**, 98 (1962).
28. KAPULZEVICH, U. H., and V. J. KOROGODIN: Radiobiologiya (Moscow) **4**, 349 (1964).
29. KELNER, A.: Proc. Nat. Acad. Sci., U.S. **35**, 73 (1949).
30. KIEFER, J.: Naturwissenschaften **51**, 289 (1964).
31. — Int. J. Radiat. Biol. **11**, 373 (1966).
32. — Helgol. wiss. Meeresuntersuchungen **14**, 195 (1966).
33. — to be published.
34. — to be published.
35. KIM, J. H., M. L. EIDINOFF and J. S. LAUGHLIN: Nature (London) **204**, 598 (1964).
36. KOROGODIN, V. J.: Biol. Med. (Paris) **65**, 402 (1965).
37. — Problems of postirradiation recovery. Moscow 1966.
38. —, V. J. MEISSEL and T. S. REMESOVA: In: Radiation Research (G. SILINI, Ed.), p. 538. Amsterdam 1967.
39. MUHAMMED, A.: J. Biol. Chem. **241**, 516 (1966).
40. OLIVERA, B. M., and I. R. LEHMAN: Proc. Nat. Acad. Sci. U. S. **57**, 1701 (1967).
41. PATRICK, M. H., R. H. HAYNES and R. B. URETZ: Rad. Res. **21**, 144 (1964).
42. — — Rad. Res. **23**, 564 (1964).
43. PETTIJOHN, D., and P. C. HANAWALT: J. Mol. Biol. **9**, 395 (1964).
44. PHILLIPS, R. A., and L. J. TOLMACH: Rad. Res. **29**, 413 (1966).
45. RÖRSCH, R., P. VAN DE PUTTE, I. E. MATTERN, H. ZWENK and C. A. VAN SLUIS: In: Radiation Research (G. SILINI, Ed.), p. 771. Amsterdam 1967.
46. ROSENBERG, B. H., and L. F. CAVELIERI: Nature **206**, 999 (1965).
47. SETLOW, I.: in: Current topics in Radiation Research (M. EBERT and A. HOWARD, Eds.), Vol. II, p. 195. Amsterdam 1967.
48. SOMMER, N. F., J. H. GÖRTZ and E. C. MAXIE: Rad. Res **24**, 390 (1965).
49. STAPLETON, G. E., D. BILLEN and A. HOLLAENDER: J. Cell. Comp. Physiol. **41**, 345 (1953).

50. WEISS, B., T. R. LIVE and C. C. RICHARDSON: Feder. Proc. **26**, 395 (1967).
51. WINKLER, U.: In: Radiation Research, (G. SILINI, Ed.) p. 796. Amsterdam 1967.
52. WITKIN, E.: In: Repair from genetic radiation damage, (F. SOBELS, Ed.) p. 151. Oxford-London-New York-Paris 1963.
53. WHITMORE, G. F.: Paper at the 3rd. Congress of Radiation Research Cortina (Italy) 1966.
54. ZIMMER, K. G.: Quantitative Radiobiology. Edinburgh: Oliver and Boyd 1961.

*Discussion*

KLAMERTH:

(1) Is DNA synthesis essential for fixation?—(2) What happens when DNA synthesis inhibited (a) before (b) after irradiation?

KIEFER:

(1) Yes, I think it is (see ref. WITKINS).—(2) The usual technique for inhibiting DNA synthesis is to block a step in the chain of precursor syntheses. I doubt whether these techniques represent a reasonable approach to the problem.

ARLEY:

Am I right that, so far, you have not worked out the curves of your model in such detail that you are able to compare them with experimental points and if not, why not?

KIEFER:

No, I did not, but work is in progress. I think it is necessary with any model, if it is going to help one to obtain new knowledge about nature, that its predictions and conclusions can be brought to a state in which they can directly be compared with experimental data, otherwise we shall not be able to conclude backwards as to the plausibility and correctnes of the basic assumptions made in the model.

KLAMERTH:

During the symposion about repair of radiation damage in Vienna (IAEA Panel Proceedings Series, April 18—22, 1966), the question arose whether or not polymerases are involved in the repair mechanism. I was informed that inhibition of protein synthesis (by puromycin or other inhibitors) did not change the repair process.

KIEFER:

The recovery from potentially lethal damage is sometimes enhanced by treatment with protein synthesis blockers. The enzymes necessary for repair are probably present in the cell and need not be induced. As far as I know nobody has measured synthetase activities after irradiation, although this would be desirable. — *Note added at proof*: After completion of this paper, HILLOVA and DRÁŠIL (Int. J. Radiat Biol. **12**, 201, 1967) reported that recovery from sublethal damage is inhibited by iodoacetamide. In the meantime our experiments on recovery from sublethal UV damage were completed and partly published (KIEFER, J., Int. Radiat Biol., **13**, 399, 1968).

# Inhibitory Action of Radiolytic Compounds on Cell Function

O. L. KLAMERTH

With 3 Figures

*Abstract*

Glyoxal and malondialdehyde are cytotoxic for cells in culture; the primary site of attack is DNA. Its rate of synthesis is inhibited more profoundly than that of protein and RNA. m-RNA synthesis is suppressed almost entirely after a 5-hr treatment with 100 µg/ml dialdehyde. Obviously bifunctional aldehydes inhibit primarily DNA replication, whereas transcription will be stopped later.

Among the products of radiolysis of aqueous sugar solutions, glyoxal and malondialdehyde are the most important compounds because of their reactivity and high toxicity. The organo-toxic effect of glyoxal was demonstrated in rabbits [6, 8, 11] and it was found that very small amounts cause fatal hypoglycemia as a result of insulin excretion through injury and damage to the B cells of the pancreas. The toxicity of malondialdehyde on rats was reported by CRAWFORD et al. [4]. That irradiated sugar solutions in general are cytotoxic for cells in culture was first demonstrated by BERRY et al. [2], but no detailed insight was given into the mode of action.

The present study was undertaken on tissue cultures to elucidate the toxic effect of both glyoxal and malondialdehyde on the molecular level. Cells incubated with glyoxal or malondialdehyde separated from the glass wall, became transparent, and finally disintegrated. The decrease in viable cell number in relation to glyoxal or malondialdehyde concentration respectively after treatment for 5 and 24 hrs is shown in Figs. 1a and 1b.

The inhibition of cell growth and multiplication might be caused—apart from by nutritional factors—by the prevention of DNA synthesis, replication, and/or transcription due to abnormal binding and restricted strand separation. We assume that, besides the inhibitory action on protein synthesis—malondialdehyde, for example, reacts with protein, as has been demonstrated in [5]—both dialdehydes can react with DNA.

To determine the rate of synthesis of DNA, RNA and protein, pulse-label experiments were performed with the respective labelled precursors.

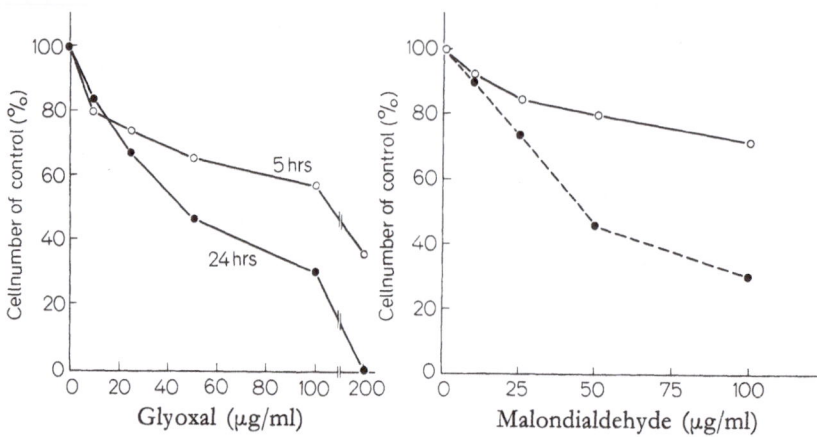

Fig. 1. The effect of various amounts of glyoxal (a) and of malondialdehyde (b) on cell number of human fibroblasts. For technical details cf. [8a]. o——o treatment 5 hrs, ●       ● 24 hrs

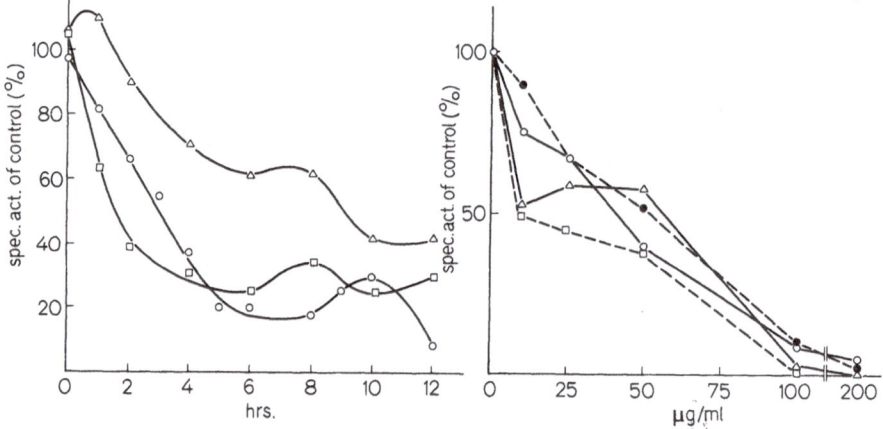

Fig. 2. Incorporation of thymidine, uridine, and amino acids in the respective cell components of human fibroblasts after treatment with glyoxal (a) and with malondialdehyde (b). For technical details cf. [3, 8a]. △——△ ³H-uridine, o——o ¹⁴C-thymidine, □——□ ¹⁴C-algaeprotein hydrolysate

The results shown in Figures 2a and 2b indicate in fact that the uptake of the label in the macromolecules is diminished with time and becomes negligible in the region of 100 μg glyoxal or malondialdehyde/ml.

The similarity of the amounts of glyoxal or malondialdehyde required to inhibit both DNA synthesis and cell division after 24 hrs by 50% permits the conclusion that DNA replication is the process most affected. This hypothesis was confirmed by the fact that the incorporation of ³H-uridine into the total RNA was diminished by only 25% after a one-hour treatment of the cell with 100 μg glyoxal/ml, an amount which reduces the synthesis of DNA by more than 50 %. The data obtained (Table 1) may indicate that the dialdehyde has only a relatively slight inhibitory effect on the synthesis of rapidly labelled RNA (for technical details cf. [8a]). It was therefore of interest to study the sedimentation profile of RNA after glyoxal treatment of the cells under different conditions. The sucrose gradient centrifugation pattern (Fig. 3a, b) shows that the major RNA components from treated and nontreated cells were practically identical in respect to size and relative position of the peaks, but the degree of label differed markedly. In the 8-min pulse

Table 1. *Effect of glyoxal pretreatment on the amount of* $^3H$-*uridine incorporated during variously timed pulses*

| Treatment with glyoxal | $\mu c/ml$ | pulse time | $^3$H-uridine incorporated | | % of control |
|---|---|---|---|---|---|
| | | | counts/min $\times 10^{-1}$ | counts/min/ $\mu g$ RNA $\times 10^{-3}$ | |
| none | 2.0 | 8 min | 2940 | 4949 | 100 |
| 100 $\mu g/ml$ | 2.0 | 8 min | 2020 | 3700 | 75 |
| none | 1.0 | 30 min | 1625 | 6271 | 100 |
| 10 $\mu g/ml$ | 1.0 | 30 min | 1490 | 6020 | 96 |
| 50 $\mu g/ml$ | 1.0 | 30 min | 1430 | 5813 | 93 |
| 100 $\mu g/ml$ | 1.0 | 30 min | 1120 | 4206 | 67 |

experiment, in which the cells were pretreated with 100 µg glyoxal/ml, the rate of incorporation hardly differed from that of the control (Fig. 3c). After a five-hour treatment, however, the incorporation of the label into the $S_{28}$ and $S_{16}$ fractions was negligible in comparison with the control, and only the $S_{4-6}$ fraction incorporated the RNA precursor (Fig. 3b). From these results it seems clear that m-RNA, regarded in the literature to be a part of the rapidly labelled RNA [7, 9], is not affected by glyoxal, at least not after short treatment.

In view of these findings, it is surprising that the synthesis of protein, as measured by the uptake of $^{14}$C-amino acids, is diminished to nearly the same degree as that of DNA (Figures 2a and 2b). A possible explanation might be a direct interaction of the bifunctional aldehydes with amino acids and proteins, as has been reported to take place *in vitro* [5]. Thymidine kinase *in vitro* however, is not affected by glyoxal in doses up to 50 µg/ml (cf. [8a]). Whether the *in vivo* change in the activity of thymidine kinase is related to the rhythmical changes observed by SACHSENMAIER and IVES [10] is not clear.

To substantiate our assumption of direct interaction of the dialdehydes with DNA in vivo, in vitro experiments were performed [3], which have shown that glyoxal and malondialdehyde react with DNA at pH's lower than 5, as demonstrated by thermal denaturation profiles, altered chromatographic behavior of the reaction product on methylated albumin, and dimin-

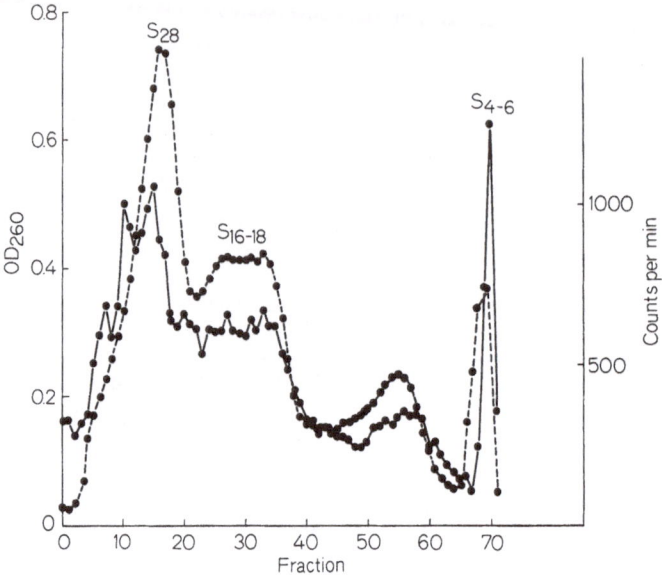

Fig. 3. a) Control sedimentation profile of uridine-labelled RNA (30 min pulse) in a 5—25% sucrose-gradient (●----● $OD_{260}$; ●——● counts per min).

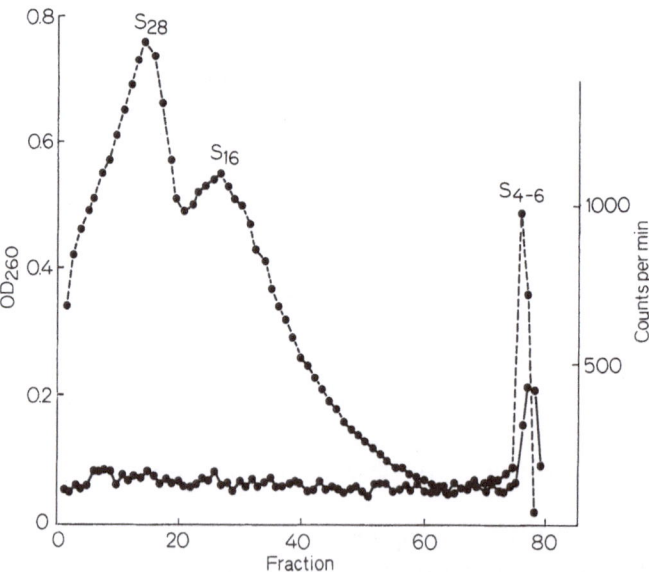

b) Sedimentation profile of uridine-labelled RNA (30 min pulse) in a 5—25% sucrose gradient 5 hrs after addition of glyoxal (100 µg/ml) (●----● $OD_{260}$; ●——● counts per min)

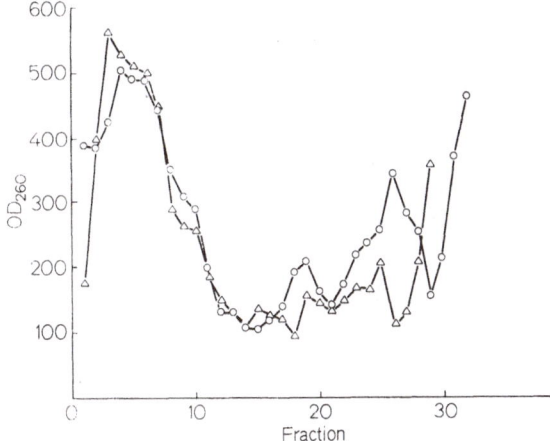

c) Comparison of the specific activity (counts/OD) of pulselabelled RNA after
1 hr treatment with 100 μg/ml glyoxal (o——o control; △——△ glyoxal)

ished susceptibility to DNAse, strongly suggesting the inducement of
cross-links by irreversible binding of the dialdehyde to DNA through
non-electrostatic bonds. Analogous observations have been made with
DNA and oxydized polyamines [1].

These observations suggested an experiment to determine the tempera-
ture-absorption pattern of DNA from cells in culture after treatment with
glyoxal and malondialdehyde and its susceptibility to DNAse. The DNA
isolated did, in fact, show an abnormal melting profile in comparison with
that of the untreated control cells and was only partially degraded by DNAse.
Since the treatment with the bifunctional aldehydes was too short (1 to
2 hrs) to interfere to any great extent with the synthesis of DNA, our results
would seem to substantiate the hypothesis of a primarily direct interaction.

## References

1. Bachrach, U., and G. Eilon: Biochem. Biophys. Acta **145**, 418 (1967).
2. Berry, R. J., P. R. Hills and W. Trillwood: Int. J. Rad. Biol. **9**, 559 (1965).
3. Brooks, B. R., and O. L. Klamerth: in press.
4. Crawford, D., R. O. Sinnhuber, F. M. Stout, J. E. Oldfield and J.
   Kaufmes: Toxicology and Appl. Pharmacol. **7**, 826 (1965).
5. Crawford, D. L., T. C. Yu and O. R. Sinnhuber: J. of Food Science **32**,
   332 (1967).
6. Doerr, W.: Verh. Deutsch. Ges. Path. **32**, 278 (1948).
7. Gros, F., W. Gilbert, H. H. Hiatt, G. Attardi, P. F. Spahr and J. D.
   Watson: Cold Spring Harbour Symp. Quant. Biol. **26**, 111 (1961).
8. Helge, G.: Verh. Deutsch. Ges. Path. **42**, 158 (1958).
8a. Klamerth, O. I..: Biochim. Biophys. Acta **155**, 271 (1968).
9. Levinthal, C., A. Keynan u. A. Higa: Proc. Natl. Acad. Sci. US **48**, 1631,
   (1962).
10. Sachsenmaier, W., and D. Ives: Bioch. Z. **343**, 399 (1965).
11. Schaack, M.: Z. ges. exptl. Med. **135**, 223 (1961).

# On the Release and Elimination
## of Liver Glutamic-Oxalacetic-Transferase

A. Fischer, and L. Takacs

With 2 Figures

*Abstract*

The activity of serum enzymes depends on their rates of entrance as well of elimination (or inactivation). Since previous observations were contradictory, we investigated G.O.T. in the blood of rats following ligature of the left liver lobe and after recirculation; an increase in S.G.O.T. occurred. Following i.v. injection of G.O.T., its excretion via the bile is considerably enhanced; however, only a small fraction of injected G.O.T. can be recovered. Addition of lung slices in vitro greatly increases G.O.T. inactivation. Some implications for the interpretation of S.G.O.T. increase are discussed.

Increased activity of serum glutamic-oxalacetic-transferase (S.G.O.T.) in liver disease is correctly ascribed to damage of liver cells; it is doubtful, however, whether cell necrosis is necessary for an increase of enzyme release. Experiments wit $CCl_4$-intoxication [5] failed to prove unequivocal relations between the toxic dose applied and the enzyme level in the blood.

In our experiments with rats the serum level of G.O.T. remains unchanged although the left liver lobe has been ligatured for 10 min., also the enzyme activity in the ligatured lobe is the same as in the non-ligatured one. Special experimental equipment allows recirculation of the ligatured lobe without additional laparatomy; then a ligature of 10 min and subsequent recirculation of 2 hrs brings about a definite increase in the serum activity and a prompt decrease of the G.O.T. activity in the previously ligatured lobe (Fig. 1). No necrosis in the liver can be detected. Thus, interruption of the liver circulation for 10 min is followed by a significant release of G.O.T. without gross morphological alteration of the liver cells.

The problem of S.G.O.T. elimination is still open [2]; some observations suggest elimination through the bile [1], but this mechanism has been questioned [3]. For the duodenal fluid a daily excretion of 20 000 units G.O.T. has been reported [4]; this value is, however, only approximate, since we observed a rapid decrease of G.O.T. activity in the bile at 37 °C, probably due to trypsin digestion. An i.v. injection of large amounts of G.O.T. (400 000—1 million units) increases the excretion through a choledochus fistula of the dog significantly, but the amount excreted during 4 hrs is only a small fraction of the total injected. The activity of S.G.O.T. diminishes by 12—41 % during the 4 hrs following i.v. enzyme injection. No significant arterio-hepatic venous difference in S.G.O.T. activity can be demonstrated in dogs prior to or following G.O.T. administration; the hepatic G.O.T. activity following G.O.T. injection never exceeds the control range. From these results it can be concluded that the liver neither eliminates nor accumulates injected G.O.T.

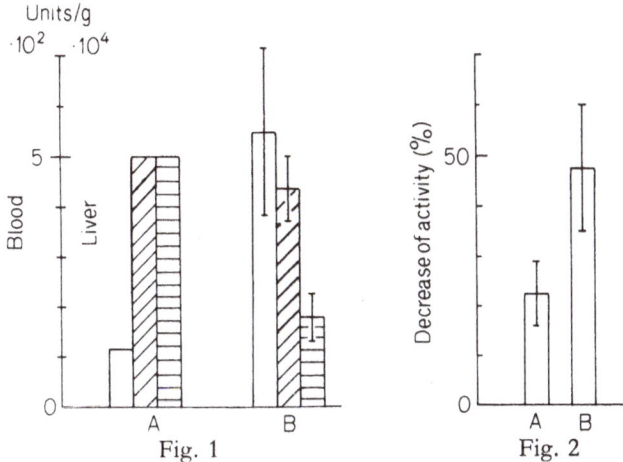

Fig. 1.
Fig. 2.

Fig. 1. G.O.T.-activity in blood and liver following 10 min ligature (A) and after recirculation of a rat liver lobe (B)

lig. liver    norm. liver    blood

Fig. 2. *In vitro* decrease of G.O.T.-activity following 4 hrs incubation without (A) and with (B) addition of lung slices

During 4 hrs *in vitro* incubation at 37 °C of G.O.T. an average inactivation of 22.4 % occurs, which is not altered by the addition of liver or lung homogenates. However, added lung slices decrease the G.O.T. activity during 4 hrs (at 37 °C) from 22.4 to 47 % (Fig. 2), a highly significant difference (t = 5.14; p < 0.001). An addition of pyridoxal-5-phosphate to the enzyme-test (following G.O.T. injection) or to the in vitro test does not exert any influence.

These results demonstrate, that a slight injury to the liver may bring about a considerable increase of G.O.T. Although the problem of G.O.T. elimination cannot yet be considered conclusively settled, it is very obvious that the presence of relatively intact tissue (e.g. lung slices) greatly enhances enzyme inactivation. This comparatively swift process is probably responsible for the rapid normalization of elevated S.G.O.T. values in many cases of myocardial infarction. Increased S.G.O.T.-activity, therefore, has to be interpreted as indicating a continuous injury to the enzyme-producing tissue, viz. myocardium or liver.

## References

1. Chinsky, M., and S. Sherry: Arch. Int. Med. **99**, 556 (1957).
2. Clermont, R. J., and T. C. Chalmers: Medicine **46**, 197 (1967).
3. Dunn, N., et al.: J. Lab. Clin. Med. **51**, 259 (1958).
4. Lorentz, K.: Acta Hepato-Splen. **9**, 403 (1962).
5. Zolton, T., et al.: Arch. Environ. Health **2**, 16 (1960).

# Metabolic Effects of Acute Hypoxia on Rat and Turtle Brain

A. T. MILLER, JR.

*Abstract*

According to one definition, hypoxia exists when, because of inadequate oxygen supply, various cellular oxidation-reduction systems must shift toward a more reduced state. This definition was tested in experiments on rat and turtle brain by comparing hypoxic changes in two oxidation-reduction systems (lactate: pyruvate and NADH-NAD+) with changes in tissue $P_{O_2}$, ATP and creatine phosphate. The results indicate that: (1) measurements of tissue $P_{O_2}$ are most useful when the electrode tip is intracellular, (2) creatine phosphate and lactate: pyruvate ratio are sensitive indicators of cellular hypoxia, (3) ATP and NADH: NAD+ ratio change significantly only in severe hypoxia, (4) tests of learning and performance show impairment at levels of hypoxia which do not significantly lower brain ATP concentration. It is concluded that intracellular hypoxia cannot be defined solely in terms of increased reduction of cellular oxidation-reduction systems.

HUCKABEE [2] has defined hypoxia as "the condition which exists when the supply of oxygen to the interior of living cells is reduced to a rate insufficient for their current metabolic needs, with the result that various cellular oxidation-reduction systems must shift toward a more reduced state". The object of this study was to test this definition by comparing changes in oxidation-reduction systems with other changes occurring in hypoxic cells. The brain was selected for study because of its vulnerability to hypoxia, and the influence of polycythemia was examined because of the conflicting opinions concerning its value to hypoxic subjects.

*Experimental:* Albino rats and water turtles (*Pseudemys scripta*) were anesthetized with urethane and pump-ventilated with gas mixtures differing in oxygen concentration. Brain oxygen tension was recorded from micro-electrodes [1] inserted into the cerebral cortex. The experiments were terminated by immersing the animal's head in liquid propane (—180 °C) and the frozen brain was removed and analyzed for ATP, creatine phosphate, lactate, pyruvate, NAD+ and NADH.

Brain oxygen tension values recorded in rats breathing air were extremely variable despite precautions to avoid inserting the electrodes near visible surface blood vessels. The electrode tip was probably extracellular in most experiments, and would therefore be influenced by its proximity to blood vessels as well as to metabolizing brain cells. The range of brain oxygen tensions was 7—40 mm Hg, with most of the values in the range 20—30 mm Hg. When rats breathed 5 % $O_2$ in $N_2$ for 3 min the final brain oxygen tension was always less than 10 mm Hg and often as low as 1—5 mm Hg. Rats made polycythemic by transfusion, cobalt administration or altitude exposure had no higher brain oxygen tension than did normo-cythemic controls when the inspired gas mixture was 4 % $O_2$ in $N_2$. When, however, 5 % $CO_2$ was added to the inspired gas, the polycythemic animals were able to maintain a higher brain oxygen tension than were their con-trols. These results are believed to reflect the opposing influences of in-creased oxygen capacity and increased viscosity of the blood, as well as the variability in oxygen tensions recorded from extracellular oxygen electrodes. Extracellular oxygen tension measurements are probably of little value in predicting intracellular oxygen tension changes because of the varying diffusion distances from electrode tips to cells. The use of intracellular oxygen microelectrodes should give more meaningful results.

When rats breathed 4 % $O_2$ in $N_2$ for 9 min, brain creatine phosphate concentration was reduced and lactate : pyruvate ratio was elevated, but ATP concentration and NADH : NAD$^+$ ratio remained virtually un-changed. Brain ATP was thus nearly normal even when the cortical oxygen tension was reduced to 1—5 mm Hg.

Cobalt-treated rats, with moderately elevated hematocrits, maintained slightly higher creatine phosphate concentration and lower NADH : NAD$^+$ ratio than did their controls, when 4 % $O_2$ in $N_2$ was breathed. Altitude-exposed animals, with much higher hematocrits, showed no such superiori-ty. When, however, 5 % $CO_2$ was added to the low $O_2$ mixture, both cobalt-treated and altitude-exposed animals were definitely superior to their controls with respect to creatine phosphate concentration and lactate: pyruvate ratio. It is suggested that cerebral vasodilatation induced by $CO_2$ mitigates the unfavorable influence of polycythemia on blood viscosity and flow, and reveals the beneficial effect of the increased oxygen carrying capacity of the blood.

When turtles breathed 4 % $O_2$ in $N_2$ for 9 min there was no change in any of the chemical constituents, and after 1 hr the changes were smaller than those observed after 9 min in rats. After 1 hr in nitrogen, severe brain hypoxia in the turtle was indicated by changes in creatine phosphate, lac-tate : pyruvate ratio and NADH : NAD$^+$ ratio. Nevertheless, the brain ATP concentration remained virtually unchanged. Thus hypoxia is mani-fested not so much by the end-result of reduction in high energy stores as by the evidence that the ATP concentration is maintained at the expense of anaerobic metabolism, which is ultimately a self-limited source.

One of the reasons for selecting the brain for studies on hypoxia is that changes in function can readily be detected and compared with metabolic changes. We observed that both learning and the performance of learned acts were impaired in rats breathing 5% $O_2$ in $N_2$. It will be recalled that no significant decrease in brain ATP concentration could be detected in anesthetized rats breathing even lower oxygen concentrations for periods of time approximately equal to those required for the tests of learning and performance. Rats acclimatized to simulated high altitude showed much less hypoxic impairment of learning and performance than did non-acclimatized animals, but rats made acutely polycythemic by transfusion of erythrocyte suspensions showed no such superiority. Polycythemia apparently must be accompanied by a compensatory increase in tissue vascularization if oxygen transport is to be increased.

*Discussion:* There is a continuing need for accurate means of detecting hypoxia and of measuring its severity. Ideal tests should be applicable to unanesthetized subjects and should differentiate the degree of oxygen deficiency in various organs and systems. The difficulties in achieving these goals are conceptual as well as technical. It is necessary first to determine the significance of the various intracellular responses to fluctuations in oxygen supply before they can be used as test items. When this has been accomplished, the next step should be the devising of functional tests for individual organs and systems which can be validated against the intracellular criteria of oxygen deficiency.

The studies reviewed in this paper were directed toward both of these goals as they apply to one organ, the brain. They lead to several conclusions which may point the way to a more penetrating analysis of this complex problem.

It is recognized that the most valuable single item of information in studies on cellular hypoxia would be the intracellular oxygen tension. At the present time this information is very incomplete and largely inferential. Oxygen microelectrodes inserted blindly into tissues measure oxygen tensions of unknown significance, since the spatial relations between electrode tip, blood vessels and cells are unknown. Substituting the term "oxygen availability" for "oxygen tension" accomplishes little except to emphasize the influence of blood flow on the measurements. It is now technically possible to insert oxygen microelectrodes, having tip diameters of 1 micron or less, directly into cells and to verify the intracellular position by recording the potential change on penetration of the cell membrane. In addition to measurements of intracellular oxygen tension it would be highly desirable to have measurements of oxygen tension gradients along the diffusion path from blood to cells; this will be a far more difficult problem, however, and it will probably require direct visualization techniques similar to those used in kidney micropuncture studies.

The measurements of metabolic changes in the brain during hypoxia highlight the semantic difficulties in defining intracellular hypoxia. When is a cell hypoxic? The ATP concentration may be well-maintained when the lactate: pyruvate ratio has shifted toward a more reduced state, and our studies on turtle brain indicate that ATP may be normal even when the NADH : NAD$^+$ ratio is greatly elevated. There are also considerable differences in the response of various cellular oxidation-reduction systems to hypoxia, some of them changing only when hypoxia is very severe. It would be of interest to extend these studies to other systems which characterize the oxidation-reduction state in cytoplasmic, mitochondrial matrix and mitochondrial cristae compartments, as suggested by KREBS [3].

Table 1. *Brain chemical changes in normal and polycythemic rats breathing various gas mixtures*

| Animals | Hct (%) | Gas breathed-9' | Cr P* | ATP* | L/P | NADH/ NAD+ |
|---------|---------|-----------------|-------|------|-----|------------|
| Transfused | 63 | air | 3.09 | 2.88 | 14 | .12 |
| Controls | 49 | air | 3.09 | 2.61 | 16 | .14 |
| Cobalt | 66 | air | 3.58 | 2.80 | 15 | .08 |
| Controls | 55 | air | 3.49 | 2.87 | 16 | .09 |
| Altitude | 76 | air | 3.64 | 2.43 | 17 | .12 |
| Controls | 54 | air | 3.66 | 2.78 | 17 | .12 |
| Transfused | 62 | 4% O$_2$ | 1.43 | 2.35 | 75 | .16 |
| Controls | 48 | 4% O$_2$ | 1.29 | 2.42 | 72 | .14 |
| Cobalt | 67 | 4% O$_2$ | 2.26 | 2.39 | 60 | .11 |
| Controls | 52 | 4% O$_2$ | 1.54 | 2.21 | 66 | .14 |
| Altitude | 81 | 4% O$_2$ | 1.99 | 2.92 | 46 | .15 |
| Controls | 55 | 4% O$_2$ | 1.72 | 2.99 | 51 | .17 |
| Cobalt | 71 | 4% O$_2$—5% CO$_2$ | 3.22 | 2.59 | 22 | .11 |
| Controls | 51 | 4% O$_2$—5% CO$_2$ | 2.29 | 2.17 | 29 | .15 |
| Altitude | 77 | 4% O$_2$—5% CO$_2$ | 3.05 | 2.59 | 14 | .14 |
| Controls | 54 | 4% O$_2$—5% CO$_2$ | 1.95 | 2.47 | 31 | .19 |

* $\mu$M/gram wet tissue

The most satisfactory indication that a cell is hypoxic might be the demonstration that it is forced to utilize anaerobic metabolism to a greater than normal degree, even though the maintenance of normal levels of high energy compounds is thereby accomplished. This would simplify the problem from the experimental standpoint provided a satisfactory solution of the interpretation of "lactate increase", "lactate : pyruvate ratio" and "excess lactate" can be achieved.

In our experiments, tests of brain function showed impairment at levels of hypoxia which did not reduce the concentration of ATP in the brain. Much needs to be learned about the chemical basis of brain function depression in hypoxia before existing information about intracellular changes in hypoxia can be used to explain changes in behavior and other higher neural activities.

Table 2. *Brain chemical changes in turtles breathing various gas mixtures*

| Gas mixture | Cr P* | ATP* | L/P | NADH/NAD+ |
|---|---|---|---|---|
| Air | 3.6 | 1.4 | 19 | .11 |
| 4% O$_2$-9 min | 4.2 | 1.4 | 15 | — |
| 4% O$_2$-60 min | 2.5 | 1.2 | 23 | .12 |
| N$_2$-60 min | 1.5 | 1.1 | 94 | .21 |

* $\mu$M/gram wet tissue

## References

1. CATER, D. B., and I. A. SILVER: In: Reference Electrodes, p. 503. Ed. by D. J. G. IVES and G. J. JANZ. New York: Acad. Pr. 1961.
2. HUCKABEE, W. E.: J. Clin. Invest. **37**, 255 (1958).
3. KREBS, H. A.: In: Advances in Enzyme Regulation, Ed. by G. WEBER, New York: Pergamon Pr. 1967.

*Acknowledgments*

Supported by grants from U.S. Army Medical Research and Development Command, the U.S. Air Force Aerospace Medical Division and the National Science Foundation.

*Discussion*

Voss:

(1) Did you carry out an investigation on the concentration of CP during brain ischemia? (2) Do you find a correlation between the decrease of CP concentration and the function of the brain?

MILLER:

(1) In the experiments reported today, no studies were made on the effect of ischemia. (2) In work which we reported at this symposium two years ago we observed that changes in the ATP concentration in the kidney were very slow. No measurements on creatine phosphate concentration were made in the ischemia experiments.

# Changes in Metabolite Concentrations in Ischemic Kidneys of Rabbits and in Transplanted Kidneys of Dogs

H. R. Schoen, and R. E. Voss

With 3 Figures

*Abstract*

The behavior of high-energy phosphates and of lactate and pyruvate was investigated in 212 kidneys of rabbits at 24° following a period of ischemia ranging from 15 sec to 10 hrs. Within the first 5 min the ATP content decreased by 70% and the AMP increased by 130%; lactate increased more slowly. The subsequent breakdown of AMP was considerably protracted and persisted for 10 hrs. This behaviour was the same in 12 transplanted canine kidneys. After only 15 min of revascularization the resynthesis of ATP occurred in the transplanted organ. The survival time and the functional duration of the transplants were determined in 22 dogs.

In order to perform renal transplantation, the renal blood flow has to be interrupted totally. During this period of ischemia, the energy for the structure-maintaining metabolism is obtained by anaerobic glycolysis only. Therefore, the amounts of reserve of glucose, glycogen and high-energy phosphates are decisive for the duration of the revival time, defined as the maximal duration of ischemia which still permits the restoration of a verifiable function. Under normothermic conditions this amounts for the kidney to 150 min. In contrast to the myocardium, the reserves of glycogen and phosphocreatin in the kidney are small. Freely available energy for the endergonic reactions is made available by the breakdown of ATP mainly.

## 1. Investigations with Rabbit Kidneys

*Method:* After extirpation the kidneys were stored at 24° in humid containers. The metabolite contents were determined at fixed intervals by enzyme-chemical and -optical methods. Control values were gained after immediate *in-situ* freezing [5] using frozen crushing forceps.

*Results:* Within 5 min. of ischemia a rapid decrease of ATP, a less pronounced decrease of ADP and a considerable accumulation of low-energy AMP occurred. In contrast to myocardium and brain, glycolysis starts slowly, as can be concluded from the gradual increase of lactate and the decrease of pyruvate. The violent alterations in the content of high-energy phosphates—within the initial 5 min—are followed by considerably slower reactions lasting 10 hrs. The constancy of the AMP content over a period of 6 hrs together with negligible loss of the total adenine nucleotides is obvious. This extremely delayed breakdown of AMP makes possible the resynthesis of ATP which rapidly takes place after termination of the ischemia. Thus the AMP content seems to be the limiting factor for the

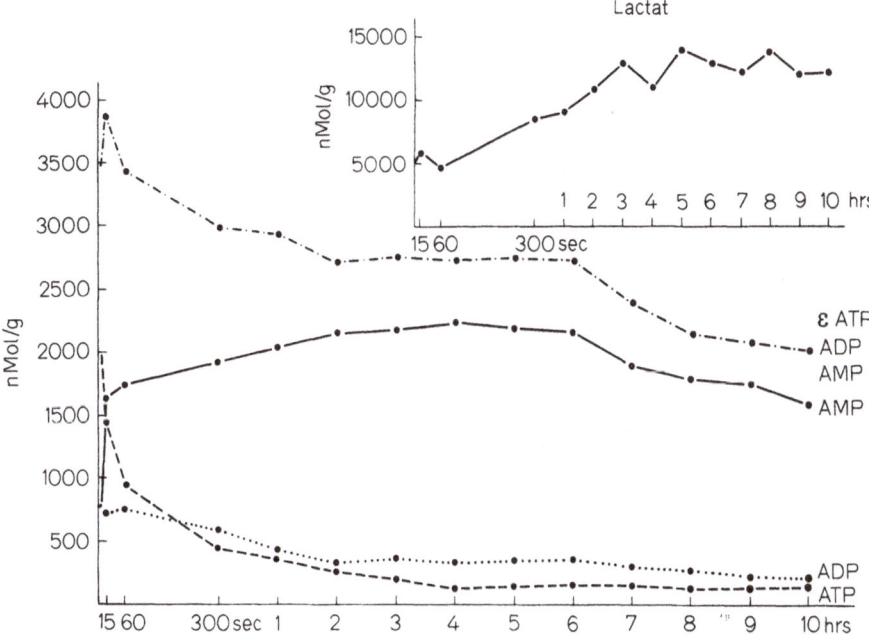

Fig. 1. Behaviour of adenine nucleotides and of lactate (insert) during renal ischemia (rabbit) from 15 sec to 10 hrs. Note the quick breakdown of ATP within the initial 5 min and the constancy of the AMP content

revival time of the kidney. Since the alterations in the energy carriers occur already during the first 5 min. it seems to be of minor importance whether an extirpation (in the clinic) is carried out 5 or 15 min after the interruption of blood supply. This fact is significant for transplantation in man, since in surgery kidneys from dead persons are being used with increasing frequency [2].

## 2. Investigations with Canine Kidneys

*Method:* The metabolites mentioned above were determined prior to, during and after transplantation in 6 dogs following pretreatment with

70 mg/kg body-weight inorganic iodine (ENDOJODIN[1]) and 3 mg/kg body-weight dipyridamol (PERSANTIN[2]), applied 12 hrs and 1 hr before the start of the investigation; 6 dogs were used as controls. The amounts of substrate were measured under three different conditions: (1) during normally functioning circulation; (2) after 40 min of ischemia and (3) after 100 min of ischemia followed by a 15 min period of recirculation in the transplant. The kidneys to be transplanted were perfused with 500 ml of Calorose-solution containing 2% glucose, 2% mannitol, 0.4% procaine and 800 i. u. of heparin. (The homologous and heterotopic transplants were located in the fossa iliaca.)

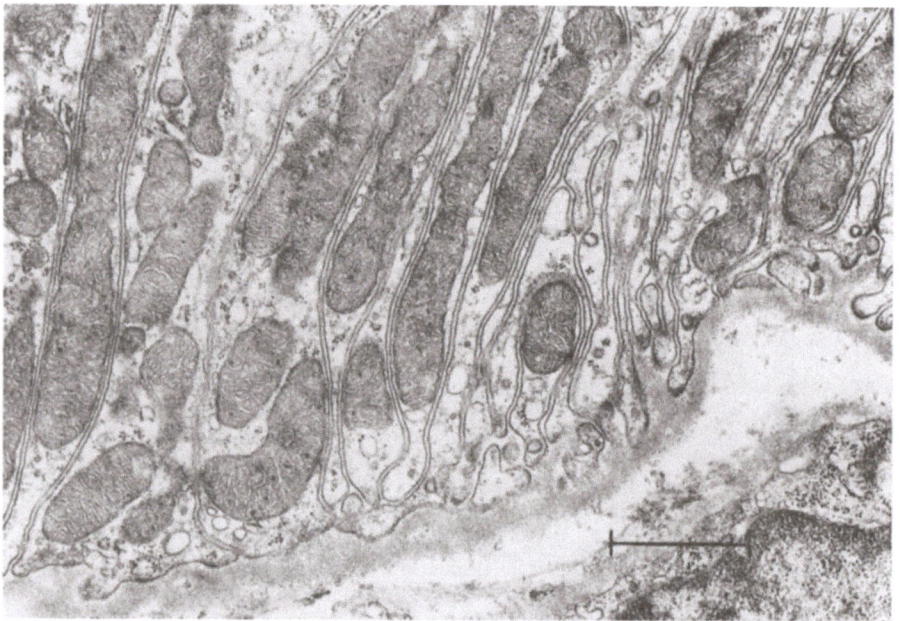

Fig. 2. Electronmicroscopical specimen of normal renal (rabbit) mitochondria of a cell of the proximal tubulus (1 : 30000)

*Results:* The dephosphorylation processes after the interruption of blood supply were similar to those in rabbit kidneys. The decrease of ATP occurred quickly; simultaneously a steep increase in AMP was recorded. Glycolytic activity developed only slowly, the inorganic phosphate increased. Following 100 min of ischemia and a subsequent 15 min recirculation period in the transplant, resynthesis of high-energy phosphates

---

[1] manufacturer: Farbenfabriken Bayer, Leverkusen/Rh.
[2] manufacturer: Dr. Karl Thomae GmbH, Biberach/Riss.

recommenced. Average ATP content at this time was already higher than that found after 40 min. of ischemia, while lactate and inorganic phosphate decreased measurably. The lactate/pyruvate ratio decreased significantly. The metabolite changes indicate the relatively quick recovery of the transplants, even after a fairly long period of ischemia. Our biochemical findings could be confirmed morphologically by electronmicroscopical investigations (together with W. VOGELL) of the transplanted kidneys. The re-establishment of function of the transplant depends upon the structural lesions caused by ischemia; these lesions proved to be extremly slight even after 2 hrs of interrupted blood flow, as may be seen in fig. 3.

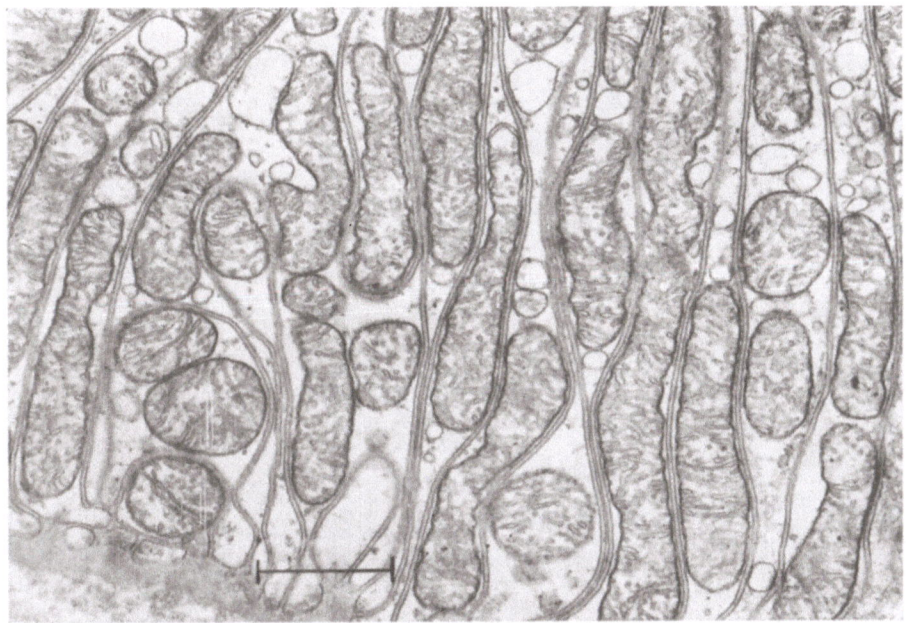

Fig. 3. Electronmicroscopical specimen. Renal tissue (proximal tubulus) after 2 hrs of ischemia. No morphological lesions. (1 : 30000)

Following pharmacological pretreatment, the ATP content in the test animals was significantly higher than in animals of the control group. After transplantation, the animals' own kidneys were extirpated. The function of the transplant was checked by determination of the serum-urea and serum-creatinine. There was no correlation between duration of ischemia and the postoperative functional duration of the transplant and the survival time of the animals. 6 dogs survived between 14 and 26 days with one homologous kidney [2, 3, 4]. The literature gives only a few examples of such a long survival time without immuno-suppressive treatment [1, 4]. Therefore, it may be concluded on the basis of our results that an improved availability of energy in the transplant possibly exerts an influence on the onset and extent of immuno-reactions after homologous transplantation.

## References

1. Largiadèr, F.: Organtransplantationen. Stuttgart 1966.
2. Schoen, H. R., R. E. Voss, K. Ruile u. J. E. Wildberger: Zbl. Chir. 92, 769 (1967).
3. Voss, R. E., H. R. Schoen und K. Ruile: Z. ges. exp. Med. 143, 255 (1967).
4. — — — u. J. E. Wildberger: Langenbecks Arch. klin. Chir. 317, 266 (1967).
5. Wollenberger, A., O. Ristau u. G. Schoffa: Pflügers Arch. 270, 399 (1960).

# The Mutual Co-Action of two Regulatory Mechanisms Realizing Optimal Autoregulation of Metabolism

G. Detchev, and A. Moskona †

With 1 Figure

*Abstract*

Autoregulation of metabolism is considered by applying certain elements of an extended thermodynamics of regulated irreversible processes. Since the application of the existing thermodynamics of irreversible processes is invalid for several reasons — chemical reactions are running far from their equilibrium; Onsager's relations do not hold true because of the existence of feedbacks; minimum entropy production does not occur — a system of equations has been derived based upon Bellman's dynamic programming and Pontrijagin's maximum principle, which overcomes these difficulties.

The existing experimental data concerning control of cell metabolism unambiguously suggest the existence of two regulatory mechanisms—the one acting swiftly and based on an allosteric modification of the enzyme pool activity, and the other going into action more slowly and based on control of the synthesis of the respective enzymes.

The investigation of the first regulatory mechanism dates back to the discovery by Novick and Szilard [5], that tryptophan inhibits the synthesis of its own precursor. It is of special interest to us that usually the activity of the first enzyme in a given chain is inhibited in this way, either by the end-product [6] or by some more or less distant precursors or even intermediates; inhibition is exerted also by end-products of simultaneously proceeding side pathways, the so called cross inhibition [4].

The second regulatory mechanism is most thoroughly studied in the β-galactosidase system of *E. coli*. By means of tracer methods it is now proved that the increase in enzyme activity is actually due to a de-novo synthesis of the enzyme. The question is raised about how the two regulatory mechanisms are correlated. The experiments of Gorini [3], who had found the effect of arginine on *E. coli*, show that the allosteric inhibition is much more sensitive to changes in arginine concentration than is repression. For this reason Davis [1] terms the first regulation "fine" and the second one "rough" adjustment (or tuning in).

We look upon the two regulatory mechanisms as co-operating to attain their ends—an optimal selfregulation of cell metabolism in such a way as to work with minimum dissipation of free energy. This common aim was formulated by us at the last symposium on quantitative biology of metabolism on Helgoland [2].

To elaborate the problem we search for the most natural method of describing the synthetic evolution of interrelations and the common functional organization of metabolism by means of appropriate mathematics. Since sufficient kinetic data about metabolic processes are still lacking, we select a thermodynamic method which adequately treats the liberation and utilization of energy in the living cell, irrespective of the underlying mechanisms of molecular interactions. The additional application of formal kinetics in considering the problem offers us, on the other hand, the possibility of a constant source of new biochemical conceptions with reference to the functional cell organization.

When making use of the thermodynamics of irreversible processes we encounter difficulties of a fundamental nature:

(1) Almost all chemical reactions which take place in biological systems run far from their equilibrium.

(2) The linear phenomenologic correlations between the corresponding rates of the chemical reactions and their affinities, as well as ONSAGER'S relations of mutuality are invalid because of the existence of feedbacks. The latter, however, are those which regulate the relations between the reaction rates.

(3) The theorem of minimum entropy production does not hold for biological systems.

By making use of the theory for optimal regulation and of PONTRIJAGIN'S maximum principle and by considering cell metabolism as a self-regulated system, we succeeded in overcoming these difficulties.

For this purpose we take into account that the rates of enzymatic reactions are not constant but changing, thus transforming the metabolic network into a subject accessible to regulation. The concentrations of metabolites turn into coordinates of the $n$-phase space of metabolism and the rate constants to regulatory parameters. The sum of the latter in the $n$-phase space of regulation constitutes the system's control.

However, the control parameters are not independent of one another; among them certain relations are to be assumed that describe the existing feedbacks and consequently the functional organization of cell metabolism. In our previous paper [2] we called these relations $\psi$-functions. Now we shall term the relations which reflect regulation by modulating enzyme activity $\psi$-functions, and the others, referring to regulation by means of synthesis of enzymes, $\Phi$-functions.

The two types of regulation are illustrated in fig. 1. Taking into consideration that the metabolic reactions are grouped into two different kinds, viz. destructive and conjugated reactions, BELLMAN's dynamic programming equation can be derived.

$$\frac{\delta G}{\delta t} = - \min \sum_{j=1}^{n} \frac{\delta G}{\delta x_j} \frac{dx}{dt} j \tag{1}$$

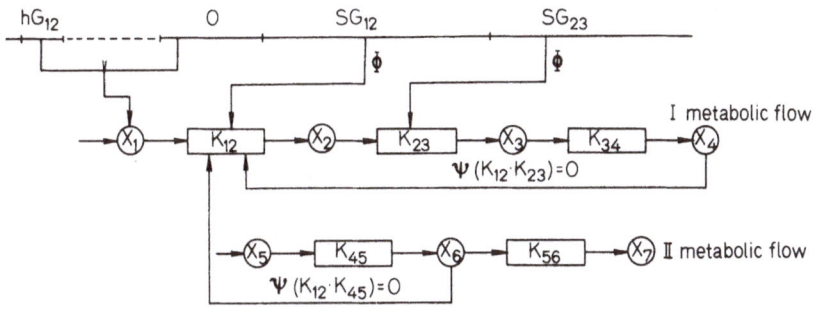

Fig. 1. Diagram illustrating both types of regulatory mechanisms: by a change in enzyme activity ($\psi_1$ and $\psi_2$-functions); and by a change in enzyme synthesis ($\Phi_1$ and $\Phi_2$-functions). $X_1 \ldots X_9$ = metabolite concentrations; $K_{11} \ldots K_{32}$ = enzyme reactions rates; O = operon; SC = structural gene

With the aid of the latter and taking into account some peculiarities of the biological objects, we found the normal functional $F_{norm}(x)$, according to which the cell mechanism shows that an optimal aim really exists; it proves to be the minimum difference between the thermodynamic potentials of destroying and synthetizing, respectively, substances in the cell, or, in other words, the minimum dissipation of free energy. It can be shown that the minimization of the entropy production in biological objects is basically different from the principle of minimum entropy production characterizing thermodynamics of irreversible processes. Whereas the latter results form purely thermodynamic reasons, the minimization of entropy production in biological objects appears to be brought about by processes leading to the evolution of cell organization together with a saturation by feedbacks.

The normal functional, as well as the relations $\varphi$ and $\Phi$ between the controlling parameters allow us to derive, with the aid of PONTRIJAGIN's maximum principle, a system of equations that makes it possible to find the optimal regulation and evolution of the system in time.

$$\text{(a)} \quad \frac{dx}{dt} = Ax + \varphi(t) \qquad x^0 = x(t^0)$$

$$\text{(b)} \quad \frac{dp}{dt} = -\tilde{A}p \qquad p^0 = p(t^0) \tag{2}$$

(c) Equations minimizing the Hamilton function

$$H\left[x\left(t\right), p\left(t\right), k\left(t\right), t\right]$$

(d)   $H\left[x\left(T\right), p\left(T\right), k\left(T\right), T\right] = 0$

Here the first system of $n$-differential equations reflects the kinetics of the biochemical reactions. With the second one are introduced $n$-auxiliary functions, the so-called impulses, corresponding in our case to the chemical potentials of the metabolites; equations (c) minimize the auxiliary functions, called Hamiltonian by analogy with statistical mechanics and corresponding to the momentary dissipation of free energy; condition (d), valid for open thermodynamic systems, makes it possible to find the time $T$, which the system needs to reach the new steady state.

Having found the optimal regulation $k\left(t\right)$ of metabolism by means of this system of equations, we are able to determine the distribution of energy flows in the cell, as well as the so called synthesis functions

$$k_{ij} = k_{ij}\left(x_1, x_2, \ldots, x_n\right) \qquad \begin{matrix} (i = 1, 2, \ldots, n) \\ (j = 1, 2, \ldots, n) \end{matrix} \qquad (3)$$

An illustration of the synthesis functions would be the following: let us assume that the concentration of a metabolite is changed. By altering the enzyme activity the regulatory mechanism enters immediately into action and changes the rates of the respective reactions. If the change in the metabolite concentration exceeds the regulatory capability of the first mechanism, and consequently a certain concentration of the metabolite penetrates into the cell nucleus, the regulation by enzyme synthesis is resorted to. This mechanism operates at the level of crude modelling of enzyme functions, whereby an approximate adjustment of the synthesis is effected, after which it switches off. The subsequent fine tuning in is effected again by the first mechanism and by a combination of the first and second regulatory mechanisms the rates of the reactions or the regulatory mechanisms are changed, due to the metabolite concentrations. In this way the synthesis functions are modulated and the optimization of cell metabolism is realized.

The synthesis functions reflect the controlling parameters as a function of the momentary state of the system and of the momentary concentrations of the metabolites. In so far as under the influence of the metabolites the goal of the control is achieved, we may call the biological entities selfregulating systems. In this manner cell metabolism is considered concomitantly as a thermodynamic and as an optimal system.

Our investigations point to some basic regularities, typical of controllable thermodynamic systems, clearly expressed by the following comparison:

·· For a thermodynamic system the $1^{st}$ principle of thermodynamics provides us only with the energy balance of the system. It can predict neither the direction of the processes nor their realizability. Thus, to one and the same increase of internal energy $\Delta U$, a strictly determined difference corresponds between the absorbed heat $dG$ and the work done $pdv$; the distribution between them could be arbitrary.

It is the second principle of thermodynamics which contains information on this distribution, thus governing the direction of the processes and their possible realizability. The second principle requires that, whenever irreversible processes occur, the every-second-production of internal entropy $T\dfrac{d_i S}{dt}$ consists of two sums; therefore, their difference should always be positive. Since this requirement could be served by various distributions among the two sums, it becomes obvious that the second principle is insufficient to determine the behaviour of the controllable thermodynamic systems.

Only equations (1) and (2) in this paper enable us to avoid this indeterminism. They place us in a position where we can find the distribution of the energy among these two sums together with the evolution of the controllable irreversible processes in biological objects.

## References

1. Davis, B.: Cold Spring Harbor Symp. $XXVI$ (1961).
2. Detchev, G., u. A. Moskona: Helgoländ. wiss. Meeresunters. **14**, 110 (1966).
3. Gorini, L.: Bull. Soc. Chem. Biol. **40**, 1939 (1958).
4. Monod, J., and F. Jacob: Cold Spring Harbor Symposia $XXVI$ (1961).
5. Novick, A., and L. Szilard: Dynamics of Growth Processes. Princeton University Press 1954.
6. Umbarger: Cellular Regulatory Mechanisms, Cold Spring Harbor Symposia $XXVI$, 1961.

*Discussion*

(This most vivid discussion of our symposium — extended until the common late dinner — could not be recovered from the tape or from discussion sheets, since most of the discussants, obviously because of being extremely excited, failed to speak correctly into the microphone, which, of course, was also not installed in the dining room. Dr. Kacser subsequently formulated some of his main arguments. The Editor).

Kacser:

The most interesting assertion of Dr. Detchev, namely, that self-regulating systems (which I take to mean those having some negative feedbacks) work with minimum dissipation of free energy, needs some elucidation.—My difficulties are two-fold: the first concerns the minimum. Is it a minimum with respect to all other possible systems? This clearly cannot be so, since I can always think of another feedback which can be added to the system. Is it, then, a minimum in the sense that when the steady state is disturbed the system will return to it? If so, then any system, if it achieves a steady state, will be at a minimum, whether it has feedbacks or not. The statement then appears tautological and is equivalent to saying that a stable steady state is achieved. — My second difficulty arises from his statement that this minimum dissipation of free energy shows "that an optimal aim really exists" and that he seems to suggest that organisms are "basically different" from physical systems. Again, if he simply means that organisms have many feedback interactions and therefore may, for example, have a smooth transient towards any steady state or that they are therefore well buffered towards many perturbations, I have no quarrel with the statement. I would not like to call such response characteristics an "optimal aim" but would consider it a somewhat dubious metaphor. Dr. DETCHEV, however, seems to imply something more, namely, that organisms have evolved in response to an *internally* given criterion, a criterion given by the thermodynamic and kinetic nature of self-regulating systems. It is my view that organisms have evolved their present complex constitution in reponse to *externally* set criteria, determined by natural selection, which has existed for the whole life span of the species. The ultimate criterion is usually described as the maximisation of fitness, which is broadly in terms of the number of viable offspring left by each generation. The elaborate feedback mechanisms, at *all* levels of organisation, are thought to "serve" this criterion insofar as organisms lacking these, or having others, have been eliminated in competition with those we find now. Thermodynamic and kinetic relations appear to place a *constraint* on evolutionary and organizational possibilities although they also provide the basis for the realization of these possibilities. — The particular "solution" which any one species has come to is a function of its particular evolutionary past and the particular ecological niche it now occupies. The "control mechanisms" exhibited are therefore specific to each of these solutions and are, for example, very different for *E. coli* and man. Dr. DETCHEV suggested that each species is at its own minimum. It is, of course, true that each species is, by definition, the "best" for its niche, an observation made many years ago by VOLTAIRE in his well known paper entitled "Candide".

DETCHEV:

The system studied by us is not a hypothetic one but rather constructed in order to reflect the processes really occurring in biological objects. It takes into consideration the existence of degrading as well as of conjugated reactions. The minimum dissipation of energy — maintained through a balance between these two kinds of reactions and thus representing the optimum functional of the system — is dependent on a set of feedbacks, exactly determined, and on the interaction between their elements. Therefore, it is not possible to arbitrarily add any further feedback to the framework of the established system. In various systems the number of minimum effective feedbacks may be different; this, however, is a question of the numerical solution on the basis of concrete data and is without significance for the general handling of the problem. — The minimization of the dissipation of free energy should not be interpreted as homeostasis. In the transition to a new steady-state — caused by any disturbance — the transition trajectory is simply determined by the optimum principle, i.e. among all the possible trajectories the one is realized which energetically is the most favorable one. — The existence of an optimum aim is not the result of a minimum dissipation, but rather conversely the latter seems to be the consequence of the former. Thus, the availability of an optimum self regulatory apparatus within living entities is the cause for the qualitative differences existing between organisms and physical systems. As against Dr. KACSER's opinion on the existence of an optimum aim as "an internally given criterion", it is especially emphasized in my paper that self-regulating and not regulating mechanisms were studied. – The regulatory loop within the system, connecting the feedback of the regulated network with the regulating enzyme system is closed; the effectiveness of this loop is a necessary condition set by the environment. Thus, in the interaction with the *external* environment during evolutionary processes, by means of which the optimal solutions achieved in the individual development have been selected, organisms succeeded in developing loops most appropriate to a well suited *internal* organization. We do not try to definitely explain evolution by thermodynamic principles; we only use a thermodynamic value such as a functional as long as intimate molecular mechanisms are vastly unknown. The maximum principle, to which this thermodynamic value is subordinated as a functional expressing the aim of an organism in the transition from one steady-state to another, is not contradictory to evolution. The control mechanisms in different organisms are universal since the fundamental biological laws are equally valid for all living beings; yet, different ecological conditions have led to a differentiation between and within the organisms.

(Further mootpoints of the discussion were: The questionability of the proposed approach to a Hamiltonian or Lagrangian formulation in classical mechanics; the question whether the formulation proposed does help to compute anything; the problem how the choice of the functional was related to the control loops; finally, the problem to predict from the physical laws how a species will survive in the evolutionary scheme. The Editor.)

# Optimization of Structure and Metabolism in Cellular Automata

A. Cristea

*Abstract*

Within a general theory of open systems isolated cells can be considered as automata with superordinate control levels, being affected by the environment as well as by interactions of the components. The equations of motion determine the changes of position, shape, reactivity and activity. An overall criterion for optimization of structure and metabolism fixes the mean trajectories, thus inducing the logical control level to make the decisions optimal. The numerical value of the maximum can be taken as an expression for biological quality, which in the course of natural evolution increases autocatalytically.

## Structure and Function of Automata

Cellular automata are open non-homogeneous systems revealing diverse and variable structures. Interactions of the structures with signals partly admit these to flow through the system according to certain drifts, and partly allow free interactions of the components. A combined application of methods derived from statistical mechanics as well as from cybernetics seems to be most appropriate to the description of biological automata.

In the complete set of parameters, i.e. dimensions of the phase space $(S)$, the equations of motion form a dynamical system, referring (a) to geometric characteristics, (b) to reactivities, i.e. symmetrical interactions inducing shifts in the phase space, and (c) to activities, i.e. asymmetrical interactions changing the signals following a prescribed program. The structure can be represented by nodes and edges of a graph, described by concentrations and flows, respectively, of signals. In this scheme the outputs depend on the inputs. The structure is furthermore disposed of two levels, one superior, exerting processing activity on the signals, which are generated in the second, inferior one by accepting information from the environment. The amount and efficiency of information and control can be visualized by the distance existing between optimal and mean trajectory. In a stochastic evolution the open system can be considered as a linear Markovian process [1, 3]; by applying a generalized Wiener-Feynman procedure [4, 6], irreversible non-stationarities as a connection of the

Lagrangian ($L$) with the transition probabilities occur. The mean trajectory corresponds to the minimum of the functional $A = \int_{t_0}^{t} L\,dt$; the probability for small derivations from the mean trajectory has a Gaussian form. Non-linearities create a sort of potential energy with many stable varieties $\Sigma_i$ diverting the trajectory from its influence zone $\Lambda_i$. At a shift normal to $\Sigma_i$ a minimum of $L$ occurs. For closed $\Sigma_i$ the motion on it could be quasi-ergodic, for open $\Sigma_i$ it is to be separated into a longitudinal part $\bar{x}_l\,(t)$ and a transversal quasi-random one $\bar{x}_{tr}\,(t)$. It is the effective Lagrangian, precised by the anisotropic part of the transition probabilities, which determines $\bar{x}_l$; playing at the same time the role of a transversal neg-entropy.

If the deviations from $\Sigma_i$ exceed the distance between neighbouring varieties then the "kinetic" energy allows to surmount the potential barriers. Thus, the equations of motion describe the interactions between the parameters of the lower level ($\bar{x}_l$), that of the upper level which (a) creates directing fields ($\bar{x}_s'$) and (b) inversely is influenced by information devices ($\bar{x}_s''$) of/for $\bar{x}_l$.

$$\frac{d\bar{x}_s'}{dt} = \bar{F}_s'\,(\bar{x}_s'/\bar{x}_s''\,\bar{c}) \qquad \frac{d\bar{x}_s''}{dt} = \bar{F}_s''\,(\bar{x}_s''/\bar{x}_l\,\bar{c})$$

$$\frac{d\bar{x}_l}{dt} = \bar{F}_l\,(\bar{x}_l/\bar{x}_s'\,\bar{c}) \tag{1}$$

which system can be somewhat simplified under stationary conditions ($\bar{c}$).

On the stable (representative) graph the equations of motion turn into equations for $C_a^k$, i.e. the concentration of $K$ signals in the $N_\alpha$ node for the $\Phi_{a\to\beta}^K$ flow on the $\Lambda \to \beta$ edge (= pathway). If the structural and functional parameters assume optimal values, the stabilizing flow terms become maximized such that opposite flows will be reduced and the correcting feed-back loops augmented. By admitting random elements a great variety and flexibility is guaranteed and the internal pattern of the automaton can be vastly modified.

## The Possibility of Extremum Values and of Optimization

The existence of stable varieties $\Sigma_i$ allows us to introduce a functional $Q = \int_{t_0}^{t} L\,(\bar{x}_l, \bar{u}_l, t)\,dt = $ max., which for one fixed time is extremal and tends to reach another extremum at changing time ($\bar{x}_l$: parameters of lower level; $\bar{u}_l$: of upper level). Therefore, $Q$ may function as an overall criterion for optimization. By generalized BELLMAN-equations [2] one is able to show how $Q$ determines the optimal evolution of $\bar{u}_l$; by removing

$\bar{x}(t)$, $Q$ is a variational principle for $\bar{u}$, whose optimum establishment may simply be approximated by the equation of a general extremum regulator

$$\frac{d\bar{u}}{dt} = \lambda^2 \frac{\delta \bar{Q}}{\delta \bar{u}(t)} \; ; \; \frac{d\bar{x}}{dt} = \bar{f}(\bar{x}_t, \bar{u}_t, t) \tag{2}$$

The current value of $Q$, indicating "biological quality", represents a criterion in dependence on errors and dispersions around the optimum and is maximal for the extremum. $Q$ increases with the ratios

$$\frac{\partial \Phi_i^+}{\partial S_j} \Big/ \frac{\partial \Phi_i^-}{\partial S_j} \quad \text{or} \quad \frac{\partial \Phi_i^+}{\partial C_j} \Big/ \frac{\partial \Phi_i^-}{\partial C_j} \qquad \Phi^+, \Phi^-: \text{terms for stabilizing flows}$$

meaning a signal to noise ratio or the ratio of order imposing to degrading forces. The $\bar{C}$-space here reflects the stability against disturbances.

It may be assumed that an adaptive (or evolutionary) transition exists such that on the $S$-space

$$\frac{\partial N}{\partial t} = D(\bar{S}, N, t) \, V_{\bar{s}}^2 N + \vec{v} \, \vec{V_{\bar{s}}} N + KN \quad (\bar{v} \cong \frac{\partial Q}{\partial \bar{S}}) \tag{3}$$

Here the dynamics $N(\bar{S}, t)$ of a population in the domain $D$ is reflecting (a) diffusive (i.e. variant) effects (= isotropic part of the transition) (b) drift towards the extremum $v$ of the "optimum quality" $Q$ (= anisotropic part of the transition) and (c) autocatalytic rate $KN$. Self-improvement occurs by reinforcement of diffusion for small $v$ and enhancement of the transition rate toward larger $v$. The upper level uses the information from environment ($M$) and interior to select reactions of optimal decision ($R$) [3]. By anchoring this decision device into the input—output relation a certain amount of memory originates. If $v$ depends mainly on its own generation rate along with a trend towards still higher maxima an autocatalytic process representing the general process of biological evolution can be imagined.

## References

1. BARLETT, M.: An introduction to stochastic processes. Cambridge: Univ. Press 1955.
2. BELLMAN, R.: Adaptive processes. A guided tour. Princeton N. J.: Princeton Univ. Press 1961.
3. BUSH, R. R., and F. MOSTELLER: Stochastic models for learning. New York: John Wiley and Sons 1955.
4. FEYNMAN, R.: Rev. Mod. Phys. **20**, 367 (1948).
5. WALD, A.: Decision processes. Cambridge: Univ. Press 1955.
6. WIENER, N.: Cybernetics. New York: John Wiley and Sons 1948.

# Oscillatory Control of Glycolysis as a Model for Biological Timing Processes

A. BETZ

With 6 Figures

*Abstract*

Oscillatory control of glycolysis, the hitherto best known biochemical oscillator, is proposed as a model for biological timing systems. It works in yeast cells, cell-free extracts of yeast cells and beef heart muscle and in rat liver mitochondria. It affects not only glycolytic intermediates like sugar phosphates, pyridine and adenine nucleotides, but also glucose incorporation in yeast and ion transport and water binding in mitochondria. The glycolytic oscillator can be reset by adding control chemicals. Its mechanism depends essentially on the control characteristics of one or a few enzymes.

## Biological Chronometry in Principle

Life itself is the realization of chemical processes in space and time. In building up their characteristic structures, organisms undergo cycles of growth, synthetic activity, reproduction etc. Even in a completely constant environment some kind of timing is necessary to coordinate all these processes. There is a system controlling the correct sequence of reactions.

Cilia, for example, are effective in moving unicellular organisms as well as in transport only because they beat in a sequence ordered with respect to space as well as time. Beating of heart muscle is a rhythmic process, too, effective in transport. Its control depends on a timing system which works on a cellular level, for even isolated cells persist in beating. Peristaltic motion, another transport system, depends equally on well-timed muscle contraction.

Chronometry is even more important for organisms living under regularly changing conditions of light intensity, temperature, humidity, gravity etc. This is the case for most organisms because our earth is neither air-conditioned nor a chemostat. Adaptation to changing conditions becomes more efficient when some regulatory processes can be initiated in advance. It is advantageous for green algae to become mobile at sunrise and to be ready for photosynthesis just as the sun is rising. Because of this advantage, it is not surprising that nearly all groups of animals and plants, except bacteria, fungi and some blue-green algae, have some endogenous chronometric systems. Circadian rythms are as effective in *Euglena* as in man. People isolated in a cave with completely constant conditions show rhythmically changing phases of activity and of quiescence. The period length of these phases differs slightly from one individual to another, within the range 22—26 hrs [1]. These slight differences are indicative of the endogenous nature of those rhythms.

Anticipating control of life activities is best done by an endogenous chronometric system whose period length is similar but not identical to that of the cosmic rhythm. Some mechanisms are needed to compare the organismic timing system with the cosmic cycle and to reset the endogenous

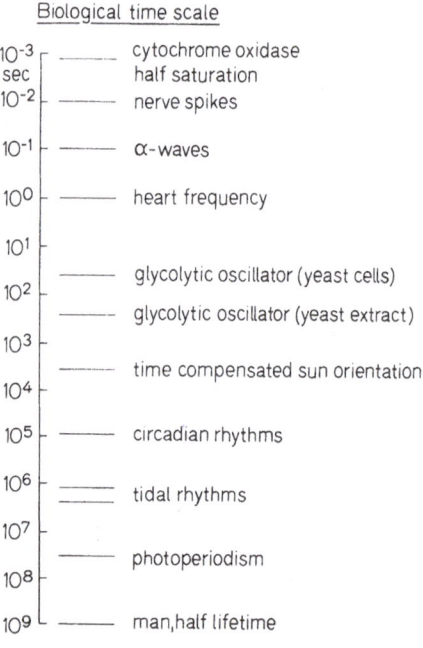

Fig. 1. Biological time scale

clock in accordance with the cosmic one. By comparing both the cosmic and the endogenous timing system, some organisms get a new type of information: retardation of sunrise in animals and an advance of sunset in the case of plants is indicative for fall time, whereas retardation of the endogenous clock with respect to those events is indicative for spring season. That is the basic mechanism of photoperiodism, and it is effective in timing blossoming of plants as well as migration of birds [2].

In marine algae and worms, tidal rhythms with a period length of 14 or 28 days are effective in controlling sexual activity, in this way improving the chance of copulation. A rather sophisticated timing system is used in the time-compensated sun orientation of birds and bees. Fig. 1 gives a rather incomplete survey on the time scale of some biological processes, most of which are oscillatory.

All timing systems which correspond to an external cosmic event are characterized by the fact that they are extremely independent of differences in temperature. Fig. 2. is a general scheme for biological chronometric systems.

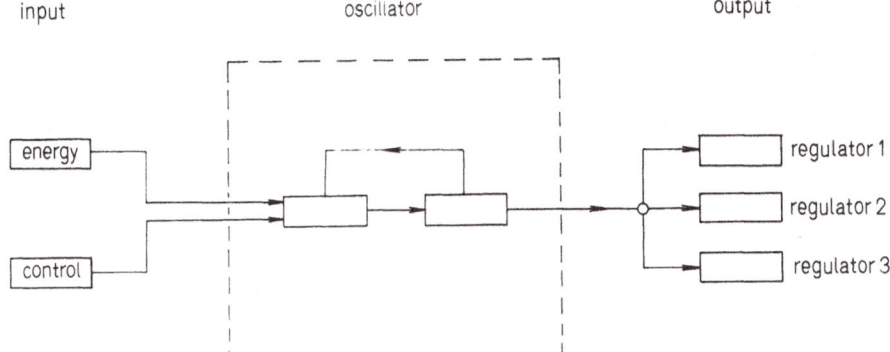

Fig. 2. Minimum requirements for chronometric systems (scheme)

The basic requirements of biological timing can be summarized as follows:
(1) constancy, by means of an endogenous self-sustaining oscillator
(2) independence of events like growth, metabolic stress, temperature
(3) ability to control biochemical processes (output from the oscillator)
(4) a mechanism for resetting and measuring the difference to cosmic events like day length, moon phase etc. (input to the oscillator).

For any absolute chronometry all these requirements are indispensable. Relative timing, like the control of beating heart muscle or the contraction in flagellae, cilia or peristaltic muscle contraction, is independent of the day/night cycle or other cosmic events. But these processes are closely connected with metabolism and therefore this kind of timing must not be independent of metabolic events. Whereas all 4 conditions have to be fulfilled for absolute timing, only the first and third are indispensable in relative timing processes.

Timing is an important feature of control. The question is whether control itself should not bear all the basic conditions for timing.

## The Embden-Meyerhof Pathway as an Oscillator

Neither statistical decay in radioactive atoms nor an hour-glass system is suitable for biological chronometry. The only realistic models are oscillators. Life is realized only in chemical processes, therefore we have to look for chemical or biochemical oscillators.

The system we are going to discuss now, the glycolytic oscillator, can be considered as a model for biological timing systems. It is surely not the basis of circadian rhythms. Whether or not it will be involved in control of the beating heart muscle or contracting cilia is still an open question, but it is one which is open to experiment, too.

Oscillatory fluctuations in glycolysis were observed independently by CHANCE [3], DUYSENS [4], MATTHAEI [5] and others later. All of them saw fluctuations in the level of reduced pyridinenucleotides during and following the transition to anaerobiosis in yeast cells. After the rediscovery of oscillating NADH concentration in yeast by GHOSH [6] in Dr. CHANCE's laboratory, the phenomenon was extensively studied there and at other places. Since we learned to prepare a soluble extract from yeast cells [7], it was obvious that control interactions between soluble enzymes acting on soluble substrates are able to form an oscillator. In the meantime FRENKEL [8], working with cell-free extracts of beef heart muscle, was able to demonstrate oscillatory control in this system too. PRESSMAN [9] and LARDY [10] observed potassium transport in rat liver mitochondria to be a pulsatory process, like glycolysis in yeast. They subsequently established a close connection between ion transport, shrinking and swelling and reduction of pyridinenucleotide in mitochondria. There are strong indications from our own work [11] as well as from HOMMES [12] and MAITRA [13] that glycolysis can continue to oscillate even under aerobic conditions. Oscillatory control in glycolysis is not merely the consequence of some disturbances in control.

The first question concerning this biochemical oscillator may be, how far will it be able to fulfill the basic requirements for a timing system mentioned above.

First, there is constancy which means a self-sustaining, virtually undamped oscillator. Certainly, glycolytic oscillations in yeast cells are far from being the undamped type. But with a cell-free extract PYE [14] was able to demonstrate more than one hundred cycles of a nearly undamped shape by adding trehalose, a substrate which is incorporated at a rather slow rate. Recently BOITEUX [15] using a constant infusion technique achieved genuinely undamped oscillations. This means that a biochemical control system can become an essentially undamped oscillator, provided that the influx of substrate is constant and of a suitable rate and further that the end-products do not accumulate.

It is not strange that a metabolic oscillator can be triggered by substrate addition. Glycolytic oscillations in cell-free extracts can be controlled not only by substrates but even by other chemicals, which are involved only as intermediates and whose concentrations are tightly balanced in glycolysis. CHANCE [16] demonstrated in yeast extract that the addition of ADP at the minimum of pyridinenucleotide reduction brought about an advance in phase, whereas pyruvate, when added at the NADH maximum brought about a retardation. FRENKEL [8], working with beef heart extract, demonstrated a complete phase shift when adding AMP at minimum NADH. The same chemical, however, caused no phase-shift when added at maximal NADH.

Chemicals like adenosine nucleotides or pyruvate are involved not only in glycolysis but in many other processes too. It is obvious that these metabolic pathways should be able to control the glycolytic oscillator. This biochemical oscillator is highly stable but there are chemical mechanisms for setting and resetting it.

As already mentioned, an oscillating control system has to be able to transfer its own rhythmicity to other metabolic pathways. In this connection PRESSMAN [9] made an instructive experiment. He was able to demonstrate that in rat liver mitochondria oscillations of NADH (under aerobic conditions) are closely connected with potassium uptake, shrinking and swelling of mitochondria and acidification of the medium. It may be that some of these events participate in the control circuit producing oscillations, but we can be sure that at least some of them are not the cause, but rather the consequence of oscillatory control.

In yeast cells the glycolytic breakdown of glucose is not only an oscillatory process, but its interconversion to UDPG, an intermediate of glycogen synthesis, is also a pulsed process. We cannot decide at present whether glycolytic oscillations are responsible for these processes or whether pulsed incorporation of glucose itself is the first cause of these rhythms.

In any case, however, there is ample evidence that a biochemical oscillator will transfer its rhythmicity to other metabolic events.

### Analysis of the Oscillator Itself

Of course, fluctuating NADH levels indicate that glycolysis as a whole will oscillate because NADH is produced as well as consumed in this metabolite sequence. The first question to be raised when NADH is oscillating is: which other metabolites will do the same and what are their phase relations? To answer this, NADH was monitored by measuring fluorescence in a batch of yeast cells, while samples were taken at different intervals as shown by the dotted lines in fig. 3. These samples had to be deproteinized and quantified for different metabolites. As the figure shows, the pool sizes of intermediates also fluctuate. Their phases can differ very greatly from the phase of NADH as demonstrated, for example, by G-6-P and F-6-P. So all intermediates have to be checked. For establishing smaller differences in phase, it is sometimes advisable to use the phase plan plot as proposed by GHOSH and CHANCE [6]. The concentration of one chemical is plotted for every sample against the concentration of another (fig. 4). NADH can be used as reference, because it can be directly measured without difficulty. In this way all metabolites can be compared among themselves [17a]. The results are the phase relations shown in fig. 5. We see there is no important control point between NADH and pyruvate, because both substances oscillate completely in phase. The 180° phase difference between F-6-P and FDP is indicative for the PFK reaction, being the most important control step responsible for glycolytic oscillations. Provided this is correct, we can

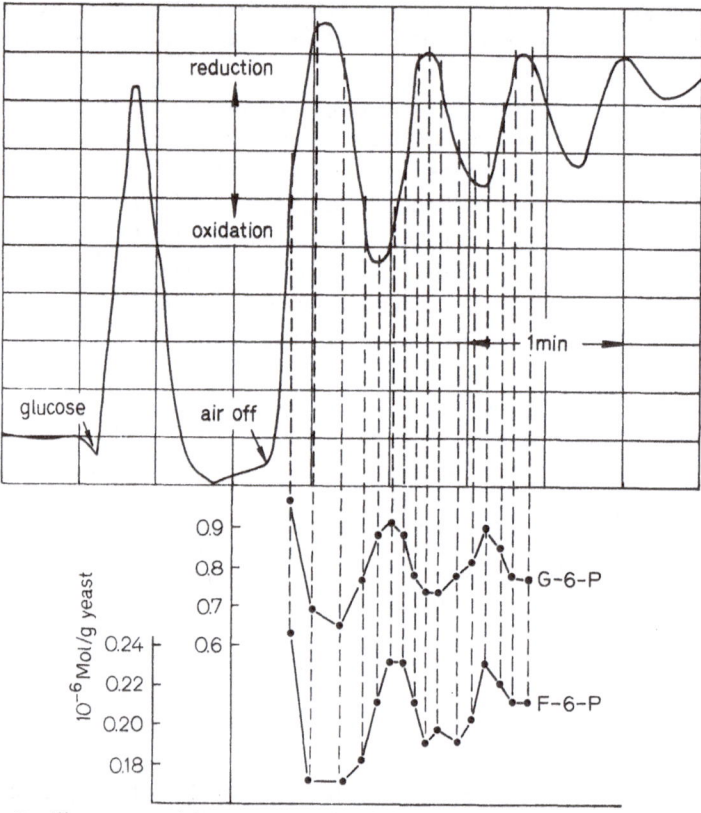

Fig. 3. Oscillatory transition to anaerobiosis in yeast cells. The reduction state of pyridine nucleotides (upper curve) was monitored directly. G-6-P and F-6-P were assayed in extracts sampled at the times indicated by the dotted lines

expect any intermediate following later in the glycolytic sequence to be in phase or retarded with respect to FDP. This is the case for DAP and GAP, which are both obviously retarded relative to FDP. But NADH and pyruvate, which follow later in the glycolytic pathway, are not retarded with respect to FDP but both evidently precede it. This means that PFK may be the control site actually producing the oscillatory metabolic flux, but it has to be controlled by some other event, obviously by some chemical which is in phase with NADH and pyruvate.

Further information about flux rates in a metabolic pathway can be derived from calculations of the apparent equilibrium constant of an enzy-

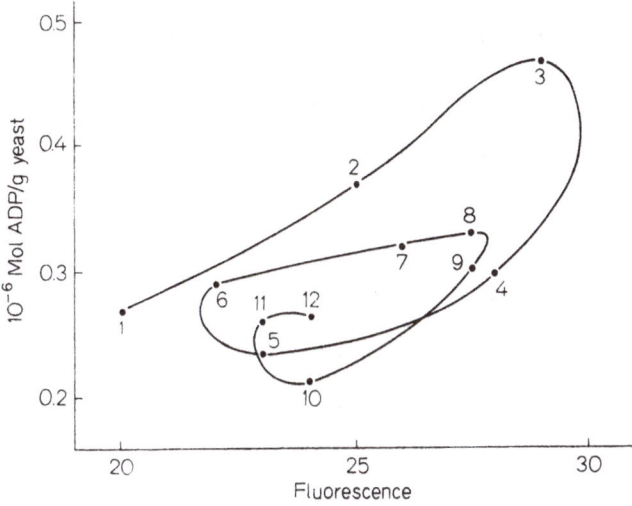

Fig. 4. Phase plan plot illustrating the phase relations between ADP and NADH (the latter monitored as fluorescence)

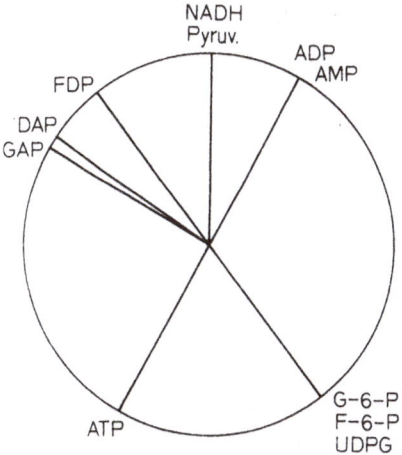

Fig. 5. Diagram of phase relations between glycolytic intermediates in yeast cells

matic step. We carried it out on the aldolase reaction [11]. The quotient $\dfrac{[FDP]}{[DAP]\,[GAP]}$ does not agree exactly with the one in the true equilibrium, for we are dealing here with a system of metabolic flux. In glycolysis, $K_{aldolase}$

14*

is not only different from its theoretical value but oscillating too. It is in phase neither with its substrate FDP nor with the products DAP and GAP. The oscillating value of $\dfrac{[FDP]}{[DAP]\,[GAP]}$ indicates that there are rhythmically changing differences between the rates of FDP synthesis and GAP consumption. Some of these metabolites oscillating with respect to their concentrations are given in fig. 6 which shows phase relations as well as amplitudes. In the lower part of this figure the fluctuations of $\dfrac{[FDP]}{[DAP]\,[GAP]}$ are demonstrated.

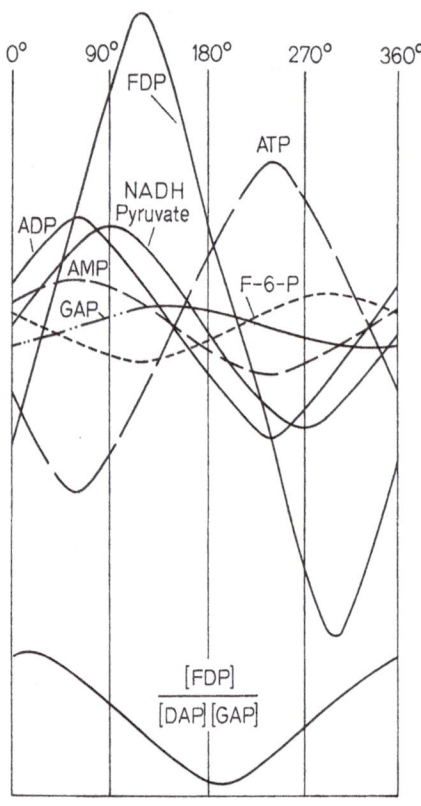

Fig. 6. Scheme on phase relations and average amplitudes of glycolytic intermediates in comparison to the quotient $\dfrac{[FDP]}{[DAP]\,[GAP]}$ as an indicator for differences in flow rates between the upper and the lower part of the glycolytic pathway

With respect to the phase relation of glycolytic intermediates and the fluctuations in the apparent equilibrium constant of aldolase, it is apparent, that there is one major control point at the PFK reaction and one more modifying control site further down which can be localized at the GAPDH reaction, because there is no further phase shift between the metabolites further down.

This indirect evidence of the alternately changing flow rates in the upper and in the lower part of the glycolytic pathway is confined by direct measurements [18]. In the glycolysing cell-free extract, inorganic phosphate is predominantly incorporated into ATP and FDP. Pulse labelling experiments show that FDP synthesis is almost completely restricted to the phase of rising NADH, whereas ATP synthesis is never completely suppressed but rather is twice as efficient during the NADH oxidation phase.

It might be of interest to mention that only 50 % of all glucose taken up by yeast cells is worked up into glycolysis. However, the uptake of all glucose is controlled in a pulsed manner and in phase with its conversion to FDP. The other 50 % of glucose which is not worked up into glycolysis is transformed into an insoluble carbohydrate, presumably glycogen. UDPG, the direct precursor of glycogen synthesis also oscillates, being in phase with G-6-P. Glucose uptake and ATP breakdown are also very closely balanced in resynthesis of ATP [19].

Oscillatory control in glycolysis was interpreted by HIGGINS [20]. He postulated that a single enzymatic step can produce oscillations when it is susceptible to substrate inhibition and product activation. PFK, the enzyme at the most obvious control site as seen by phase analysis as well as in studies of the incorporation of [32]P, actually satisfies both conditions [21, 22]. This enzyme is excessively inhibited by its substrate ATP and activated by F-6-P. It is further activated by AMP, a compound closely correlated with ADP, one of the products of the PFK reaction. Some activation is also possible by FDP, the second product. We found this compound effective only at low concentrations of F-6-P [23]; it is unlikely to be an important factor in oscillatory control. The activation by the substrate F-6-P is more important, since it does not only work by mass action but is also most efficient in neutralizing the inhibitory action of ATP. Plotting F-6-P concentration against velocity gives a sigmoid curve indicating that a cooperative process is involved which is typical of allosteric enzymes. In the Hill plot F-6-P gives a coordination number of 4, which means that 4 steps (or 4 subunits) are involved. Coordination numbers of 2 and higher are to be observed with AMP and FDP too.

Any interpretation of PFK activity is extremely difficult, because any substance acting on this enzyme has only a limited range of activating concentration, and this range depends on the actual concentration of all other substrates and effectors.

At present allosteric conditions of PFK are theoretically sufficient to explain that this enzyme makes the pathway in which it works to become an oscillator. We know that some further conditions are necessary to make this oscillator work, but we cannot decide precisely which of these accessory conditions are really indispensable*.

The glycolytic oscillator is a suitable model for the study of biochemical oscillators in general. We emphasize, however, that it is not an efficient chronometric system, since it cannot enable an organism to effect absolute timing. The frequency of glycolytic oscillations is dependent on temperature and needs a rather high energy of activation (some 14 000 cal) [17b]. This pathway producing energy by the synthesis of ATP is involved in nearly all metabolic activities of the organism. Therefore, it is not independent of other metabolic events. We should consider it as a model for biochemical oscillators and study it with respect to the principal features of similar oscillators. However, we should look out for other oscillating systems, as some of them might be the real timing oscillators of organisms.

# References

1. Aschoff, J., u. R. Wever: Naturw. **49**, 337 (1962).
2. Bünning, E.: The Physiological Clock. New York: Springer 1967.
3. Chance, B.: In: The Mechanism of Enzyme Action. (W. D. McElroy and B. Glass, Eds.) Baltimore 1954.
4. Duysens, L. N. M., and J. Amesz: BBA **24**, 19 (1957).
5. Matthaei, J. H.: (personal communication).
6. Ghosh, A., and B. Chance,: BBRC **16**, 174 (1964).
7. Chance, B., B. Hess and A. Betz: BBRC **16**, 182 (1964).
8. Frenkel, R.: ABB **115**, 112 (1966).
9. Pressman, B. C.: Federation Proceedings **24**, 425 (1965).
10. Lardy, H. A.: Federation Proceedings **24**, 424 (1965).
11. Betz, A.: Physiol. Plant. **19**, 1049 (1966).
12. Hommes, F. A.: ABB **109**, 168 (1965).
13. Maitra, P.: BBRC **25**, 462 (1966).
14. Pye, K. E.: (personal communication).
15. Boiteux, A., and B. Hess,: 4. Meeting of the Fed. of Europ. Biochem. Soc. (FEBS) Oslo 1967.
16. Chance, B., B. Schoener and S. Elsaesser: Proc. Ntl. Acad. Sci. **52**, 337 (1964).
17a. Betz, A., and B. Chance: ABB **109**, 585 (1965).
17b. — — ABB **109**, 579 (1965).
18. — Europ. J. Biochem. **4**, 354 (1968).
19. — — and R. Hinrichs: Europ. J. Biochem. **5**, 154 (1968).

*Addition at proof*: In the meantime theoretical considerations of Sel'kov (Europ. J. Biochem. **4**, 79, 1968) suggested to recheck the kinetics of yeast PFK under conditions simulating those of the cellfree yeast extract. In that case only F-6-P and most drastically AMP are efficient control chemicals (Betz, A.: Abstracts, 5th Febs meeting, 62, Prague 1968). Oscillating glycolysis in yeast extracts fits well with Sel'kov's model (Sel'kov, E. E., Abstracts, 5th Febs meeting, 240, Prague 1968; Sel'kov, E. E. and Betz, A., Symposium on short term oscillations, Prague 1969, in press).

20. Higgins, J.: Proc. Natl. Acad. Sci. **51**, 989 (1964).
21. Atkinson, D. E.: Regulation of Enzyme Activity. In: Annual Review of Biochemistry **35**, 611 (1966).
22. Moore, C. L., A. Betz and B. Chance: In: Control of Energy Metabolism (B. Chance, R. W. Estabrook and J. R. Williamson, eds.), p. 97. New York 1965.
23. Betz, A., and C. L. Moore: ABB **120**, 268 (1967).

## *Discussion*

Simon:

What about a system of differential equations which describe the process and behave oscillatory (undamped), as e.g. Goodwin's theoretical system?

Betz:

This was done, as far as I know, by Dr. Higgins in Dr. Chance's laboratory in Philadelphia as well as by Dr. Sel'kov in Moskow.

Smith:

To what extent does the uncontrolled step from DAP to α-glycerophosphate affect the oscillatory behavior of DPNH?

Betz:

The shunt from DAP to glycerol is actually participating, the flux being somewhat like 10% of ethanol production. But we could not decide whether it is oscillatory, too.

Walter:

Do whole yeast cells that are not synchronized exhibit oscillations?

Betz:

The yeast cells are actually synchronized by our treatment, by feeding or by switching over to anaerobic conditions. But from an experiment with a single cell, Dr. Chance recently told me, it is evident that damping is merely a population effect and is less efficient in single cells.

Kacser:

I would like to ask you about something I have not quite understood: the rationale of the phase shift pattern and the assignment of control point effects.

Betz:

A phase shift of 180° between F-6-P and FDP would be a perfect proof that PFK is the main regulatory point, if we could assume the rate of glucose incorporation to be constant. Since we know now that phosphorylation of glucose is a pulsatory process, too, we have to consider that PFK may well be controlled by F-6-P-concentration rather than by product activation.

Arley:

As a physicist I am most interested in your experiments because, as you know, the concepts of time and ageing are fundamental and for more than 40 years it has been thought that relativity theory could elucidate the biological aspects of these problems, predicting that space travellers at relativistic velocities should get younger as compared with their twin brothers resting on the earth during the time of the space flight, a prediction that has only recently been shown to be incorrect (N. ARLEY, Naturwissenschaften **54**, 366 (1967)). Is it not actually the first time that anybody has been able to give a concrete chemical explanation of organic time oscillations?

BETZ:

There are many observations on inorganic oscillatory systems. As I mentioned in my lecture, fluctuations in NADH level have been observed independently by CHANCE, DUYSENS, MATTHAEI and rediscovered by GHOSH in Dr. CHANCE's laboratory. It is in this laboratory that the work I have reported was actually begun.

# Evaluation of the Coordinated Control of Activated Glycolysis and Respiration by Orthophosphate and ADP in Ascites Tumor Cells

E. L. Coe, Mary H. Coe, and In-Young Lee

With 2 Figures

*Abstract*

During the first minute after addition of glucose to respiring ascites tumor cells, there are two internally regulated glycolytic units: hexokinase through phosphofructokinase; and triose-P dehydrogenase through lactate dehydrogenase. The velocity of the former, $V_{PFK}$, and of the latter $V_{LDH}$, as well as the velocities of respiration and ATP hydrolysis are linear functions of both ADP and intracellular $P_i$. Expressions for $d/dt(ADP)$ and $d/dt(P_i)$ are solved for ADP, assuming various combinations of ADP and $P_i$ dependence for the velocities. Comparisons of theoretical and experimental ADP curves indicate that neither ADP nor $P_i$ control both $V_{PFK}$ and $V_{LDH}$, but that $P_i$ probably controls $V_{PFK}$ while ADP controls $V_{LDH}$.

Velocities of glycolytic enzymes and of respiration in Ehrlich ascites carcinoma cells have been correlated previously with intracellular ADP and $P_i$ concentrations [3]. During the first minute after glucose addition, glycolysis divides into two internally regulated segments: the "head end" including hexokinase through phosphofructokinase; and the "tail end", from triose-P dehydrogenase through lactate dehydrogenase [1]. The intermediate segment, aldolase and triose-P isomerase, appeared to equilibrate fructose diphosphate and the triose phosphates without much effect on either the head or tail. The velocity of the phosphofructokinase step, $V_{PFK}$, was chosen to represent the head, and that of lactate dehydrogenase, $V_{LDH}$, the tail, since all enzymatic rates within a segment are nearly equivalent. The velocities of respiration, $V_R$, and of overall ATP hydrolysis, $V_S$, as well as $V_{PFK}$ and $V_{LDH}$ are shown as functions of ADP and esterified phosphate ($P_{est}$) in Fig. 1, for a representative experiment (Exp. A., ref. [3]). Assuming a slow rate of extracellular $P_i$ entry [4] intracellular $P_i = (C - P_{est})$, where C is the initial intracellular concentration [3]. $V_s$ is calculated from the combined theoretical ATP yield of glycolysis and respiration [1, 2]. All the velocities are approximately linear functions of both ADP and, by deduction, intracellular $P_i$ from 10—60 sec. Hence, further considerations are necessary to separate ADP from $P_i$ dependency.

Assume energy metabolism may be divided into 4 internally regulated blocks:

$$G \xrightarrow{\ V_{PFK}\ } F \xrightarrow{\ V_{LDH}\ } L$$

$$2RH_2 + O_2 \xrightarrow{\ V_R\ } 2H_2O + 2R$$

$$ATP + X \xrightarrow{\ V_S\ } ADP + P_i + Y$$

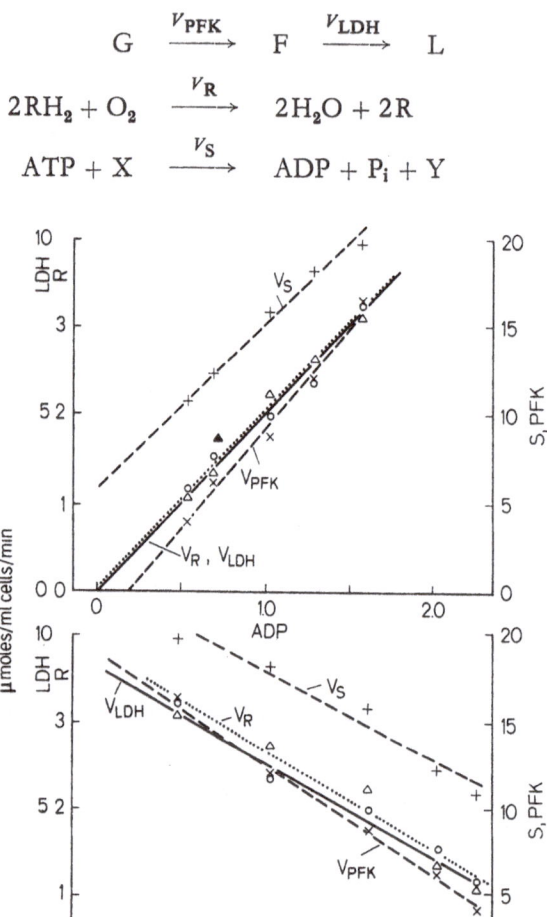

Fig. 1. Velocities of respiration ($V_R$), the head end of glycolysis ($V_{PFK}$), the tail end of glycolysis ($V_{LDH}$) and ATP hydrolysis ($V_S$) as functions of ADP and esterified phosphate ($P_{est}$).

Units for $V_R$, in μmoles $O_2$ and $V_{LDH}$, in glucose equivalents, are given at left; units for $V_{PFK}$, in glucose equivalents, and $V_S$, in μmoles ATP, are given at right. The points cover the first 60 sec after glucose addition, the highest point in each series being at 10 sec. Conditions: 37 °C, 54 mM phosphate buffer, 2.9% $V/V$ cells, 0.77 mM glucose (cf. ref. [3] for further details). Symbols: +——, $V_S$; △····, $V_R$; ○——, $V_{LDH}$; ×——, $V_{PFK}$; ▲, $V_R$ prior to glucose addition. $P_{est}$ estimated by summation of changes in fructose diphosphate, triose phosphate, and other phosphate esters where F includes fructose diphosphate and triose phosphates and $V_S$ includes AMP formation and conversion to ADP. Assuming complete coupling of both oxidative and glycolytic phosphorylation:

$$dD/dt = 2V_{PFK} - 4V_{LDH} - 6V_R + V_S \tag{1}$$

$$dP/dt = -2V_{LDH} - 6V_R + V_S \tag{2}$$

where $D$ = ADP and $P$ = intracellular $P_i$ concentration, and $V_{LDH}$ is in glucose equivalents ($\frac{1}{2}dL/dt$). Since there are 2 possible variables for each of the 4 velocities, there are 16 possible combinations and solutions. To limit the choices, it will be assumed that $V_R = K_{RD}(D)$ and $V_S = K_{SD}(D) + S_0$ in all cases, as suggested by independent evidence (for $V_R$ cf. [5]). $V_S$ contains an additive term, $S_0$, since it extrapolates to a substantial rate at zero ADP; the relationship between $V_S$ and ATP is more consistent, but ATP may be considered a linear function of ADP (1). The remaining combinations are then: *case* (*a*), $V_{PFK} = K_{FD}(D)$, $V_{LDH} = K_{LD}(D)$; *case* (*b*), $V_{PFK} = K_{FD}(D)$, $V_{LDH} = K_{LP}(P)$; *case* (*c*), $V_{PFK} = K_{FP}(P)$, $V_{LDH} = K_{LD}(D)$; and *case* (*d*), $V_{PFK} = K_{FP}(P)$, $V_{LDH} = K_{LP}(P)$. If either ADP or $P_i$ are unevenly distributed within the cell, the same equation will apply so long as the concentration in a given compartment is proportional to the total. These can be substituted into eq. 1 and 2 and the resultant solved for ADP.

Case (a) requires only eq. 1:

$$D = 1/r\left\{(rD_0 + S_0)\,e^{rt} - S_0\right\} \tag{3}$$

where $r = (2K_{FD} + K_{SD} - 4K_{LD} - 6K_{RD})$, and $D_0$ = ADP at $t = 0$. Cases (b)–(d) require both eq. 1 and 2 and reduce to the same general form:

$$\frac{d^2D}{dt^2} + A\frac{dD}{dt} + BD + C = 0 \tag{4}$$

$$(A^2 > 4B)\qquad D = S_1\,e^{-\alpha t} + S_2\,e^{-\beta t} - C/B \tag{5}$$

where $\quad \alpha = \frac{1}{2}\left(A + \sqrt{A^2 - 4B}\right);\qquad \beta = \frac{1}{2}\left(A - \sqrt{A^2 - 4B}\right)$

$$(A^2 < 4B)\qquad D = S_3\,e^{-\frac{1}{2}At}\sin\left(\frac{1}{2}\sqrt{4B - A^2}\right)t - C/B \tag{6}$$

$S_1$–$S_3$ are constants of integrations.

Note that $(A^2 < 4B)$ produces an oscillating system. In case (b), $A = (6K_{RD} + 2K_{LP} - 2K_{FD} - K_{SD})$; $B = 2K_{LP}(K_{SD} - 2K_{FD} - 6K_{RD})$; $C = 2K_{LP}S_0$. In (c), $A = (6K_{RD} + 4K_{LD} - K_{SD})$; $B = 2K_{FP}(2K_{LD} - 6K_{RD} - K_{SD})$; $C = -2K_{FP}S_0$. In (d), $A = (6K_{RD} + 2K_{LP} - K_{SD})$; $B = 2(K_{FP} - K_{LP})(6K_{RD} - K_{SD})$; $C = 2(K_{LP} - K_{FP})S_0$. From Fig. 1, in $(min)^{-1}$: $K_{RD}$, 2.10; $K_{SD}$, 9.1; $K_{LD}$, 5.0; $K_{FD}$, 11.1; $K_{LP}$, 0.67; $K_{FP}$, 1.57. $S_0 = 5.9$ $\mu moles/ml$ cells/min, and $D_0 = 0.72$ $\mu moles/ml$ cells. Solutions are as follows:

case (a)  $D = 2.56 - 1.84\ e^{-2.30t}$

(b)  $D = 0.36\,e^{-1.80t} + 0.37\ (D_0 = 0.72)$

(b')  $D = 1.4\,e^{-1.80t} + 0.37\ (D_0 = 1.77)$

(c)  $D = -1.22e^{-21.5t} + 1.50e^{-1.98t} + 0.44$

(d)  $D = 10\ e^{-2.42t}\sin(0.665)t + 1.69$

where $t$ is in min, $D$ is $\mu$moles/ml cells and $S_1 - S_3$ are chosen empirically. Cases (a) and (d) deviate greatly from the experimental points and are therefore improbable (Fig. 2). Case (c) gives the closest approximation although case (b') could apply, provided that a separate mechanism is postulated for the initial elevation of ADP.

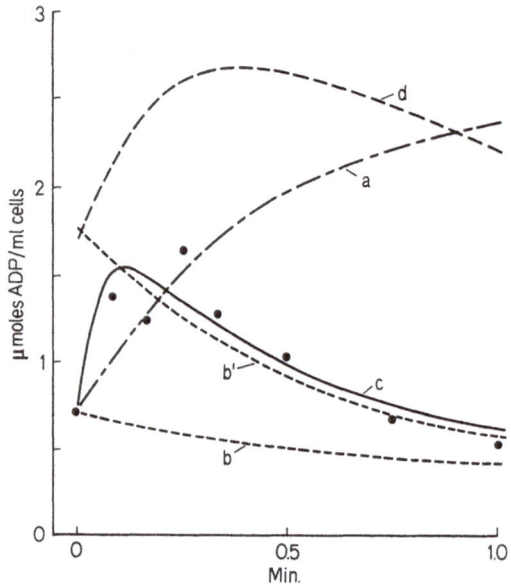

Fig. 2. Experimental and predicted changes in ADP level after glucose addition. Symbols: •, experimental ADP points; ———, curve based on assumption a. ------, curves based on b; ———, curve based on c; and — —, curve based on d; Curves b and b' assume different initial ADP values (see text)

## References

1. Coe, E. L.: Biochem. Biophys. Acta **118**, 495 (1966).
2. Lee, I.-Y., and E. L. Coe: Biochim. Biophys. Acta **131**, 441 (1967).
3. —, R. C. Strunk and E. L. Coe: J. Biol. Chem. **242**, 2021 (1967).
4. Levinson, C.: Biochim. Biophys. Acta **120**, 292 (1966).
5. Maitra, P. K., and B. Chance: In: B. Chance, R. W. Estabrook and J. R. Williamson (Eds.) Control of Energy Metabolism, p. 157. New York: Academic Press 1965.

This work was supported by U.S. Public Health Research Grant No. CA-10723.

# Profiles of Substrate Levels of the Central Metabolic Pathways in Frog Heart

P. Arese, A. Bosia, and L. Rossini

With 3 Figures

*Abstract*

Stepwise discriminant analysis (according to Dixon's BMDX 67/BMD O7M program) allows highly significant discrimination between groups of mutually correlated little-differing metabolic parameters. The analysis was applied to groups of 18 and 35 parameters (18 glycolytic-, citric-acid-cycle-, high-energy-compounds plus 17 derived values, such as *in vivo* equilibrium constants, redox ratios, etc.) determined in frog heart. The following conditions were chosen: wash-out effects, and short-term effects of ionically modified Ringer-solutions and of ouabain.

The first purpose of this communication is to investigate how functionally comparable dynamic situations correlate with the behavior of the glycolytic-, citric-acid-cycle- and energy-rich-substrates. The second aim is to describe further patterns of metabolite movements in complex cellular systems changing from one steady-state to another, in order to make the formulation of computer models [7, 8] more general and realistic. Frog myocardium was chosen because of its homogeneity—owing to its lack of connective tissue and coronary circulation—and because of its great ionic mobility [3]. Injection with calcium-enriched or ouabain-containing Ringer solutions had no effect on heart dynamics after 30 sec, caused a moderate positive inotropism after 60 sec, and a sustained contracture after 120 sec. At these times, the hearts were quick-frozen and 18 substrates were measured. Owing to the great number of parameters, computer-assisted stepwise discriminant analysis was necessary. The analysis was performed separately on the 18 substrates, 17 derived values or on the resulting sum of parameters. The discriminating differences between normal and treated hearts were significant at the 0.01 F level. The statistical procedure we adopted identified little-differing but functionally relevant parameters, and changes of the whole pattern.

## Materials and Methods

Frogs of both sexes, weighing 15 to 30 g were used. Control animals were injected in the vena cava inferior with Ringer-frog solution. Treated frogs were injected with ionically modified ($4 \times Ca^{++}$, minus $K^+$) or ouabain-containing (0.1 %)

isotonic Ringer solution. Injection rates of 0.5 ml/30 sec, 1 ml/60 sec or 1 ml/120 sec were chosen. After 30, 60 or 120 sec the ventricles were cut, quick-frozen, and stored in liquid nitrogen. The following substrates were determined in the neutralized perchloric acid extracts, according to Bergmeyer [2]: glucose (GLU), glucose-1-P (G1P), glucose-6-P (G6P), fructose-6-P (F6P), fructose diphosphate (FDP), dihydroxyacetone-P (DAP), glyceraldehyde-3-P (GAP), α-glycero-P (α-GP), pyruvate (P), lactate (L), oxaloacetate (OAA), malate (M), α-ketoglutarate (α-KG), P-creatine (PC), creatine (Cr), ATP, ADP, and AMP. Beside the 18 substrates, the following 17 derived parameters were also taken into account: mass-action ratios of the phosphoglucomutase (G1P/G6P), phosphohexose isomerase (F6P/G6P), triosephosphate isomerase (DAP/GAP), phosphofructokinase (1. $FDP \times ADP/F6P \times ATP$; 2. $FDP \times Cr/F6P \times PC$), aldolase (1. $DAP^2/FDP$; 2. $GAP \times DAP/FDP$), adenylate kinase ($AMP \times ATP/ADP^2$), creatine kinase ($Cr \times ATP/PC \times ADP$); redox ratios L/P, α-GP/DAP, M/OAA; ratios PC/Cr, ATP/ADP; the sums PC + Cr, ATP + ADP + AMP, and L + P. The 18 substrates or the 35 parameters were subjected to stepwise discriminant analysis [5], carried out on an IBM 7094 computer. Three main variance factors were considered: different times of injection, varying amounts and composition of injected fluids. Every group ($4 \times Ca^{++}$, 0.1% ouabain and related controls at 30, 60 and 120 sec) included 125 hearts. As 20—25 frozen hearts were pooled for each extract, every group was the result of 5 to 6 distinct experiments.

## Results

*Control conditions* (Fig. 1) Comparison of control steady-state levels showed that most glycolytic and citric-acid-cycle intermediates, as well as AMP, are 3 to 6 times higher in the frog than in the rat [12, 13]. Close correspondence was found only for α-GP, DAP and ATP. The redox ratios L/P, α-GP/DAP and M/OAA are apparently not kept at equilibrium through a common NAD-pool. Whereas the phosphoglucomutase reaction was near to thermodynamic equilibrium (mass-action ratio *in vivo* 0.09, $K_{eq}$: 0.05) clear shifts were observed for the two isomerase-catalyzed steps. Owing to the compartmentation of ATP in muscle tissues [9, 10], the phosphofructokinase equilibrium constant was also computed using the PC/Cr-ratio [4]. Both results showed a clear shift from thermodynamic equilibrium. With respect to mammalian heart, ATP level in the frog was constantly higher than PC; the steady-state equilibrium constant for the creatine-kinase reaction was lower than in mammalian heart (frog: 3.06; rat: 8.1 [13]) and virtually the same as in rat skeletal muscle [9]. Multivariate analysis on hearts injected with different amounts of Ringer-frog solution for 30, 60 or 120 sec showed clear wash-out effects for glucose, pyruvate, α-ketoglutarate, α-glycero-P, and creatine.

*Effect of ionic modifications* (Fig. 2). A 30 sec infusion with calcium-enriched Ringer solution ($Ca^{++} = 7.2$ mM) caused shifts in the metabolite pattern. The following changes were relevant: (1) slowing down of glucose entry into the cell, (2) activation of phosphofructokinase, (3) increase of two NAD-linked redox ratios, and (4) negative PC and ATP balance. No activation of phosphorylase could be seen. After a 60 sec infusion, phosphofructokinase activation became more marked and was accompanied by phosphorylase activation, as shown by the increase of G1P and G6P. Glucose entry was further inhibited. The increase of the glycolytic flux at 60 and 120 sec is evident if we consider the strong increase of lactate and pyruvate.

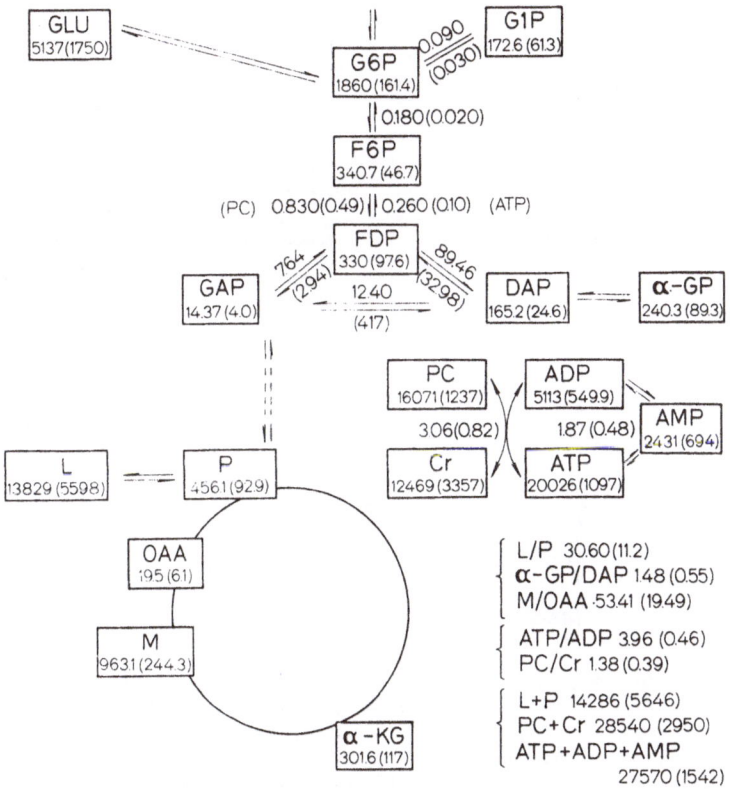

Fig. 1. Network of substrate levels and derived parameters in control frog hearts. Inside the squares: mean values, mμmoles/g dry weight. Abbreviations used: see Materials and Methods. Between the squares: mass-action ratios. For details, see Materials and Methods. Standard deviations are given in brackets

*Effect of ouabain* (Fig. 3). After 30 sec, a clearly positive balance of energy-rich phosphates was evident (increase of ATP, of ATP/ADP- and PC/Cr-ratios). Glycolysis was activated at the expense of glycogen: in fact, glucose, G6P, pyruvate, lactate, and the sum: lactate + pyruvate were increased. No activation of phosphofructokinase was evident. As already reported [1], after a 60 sec injection a crossover point may be observed within the glucokinase-phosphoglucomutase steps. After a 120 sec infusion, the crossover is located at the phosphofructokinase-aldolase step. The direction of the apparent control steps shift is opposite to that of the calcium-treated hearts. The parallel decrease of glucose and G6P may be explained by an activation of the G6P-shunt, in accordance with known data on dog heart [14]; whereas the phosphofructokinase (ATP) and creatine kinase *in vivo* constants increase with time the calcium-treated hearts, after ouabain infusion they showed fluctuating profiles.

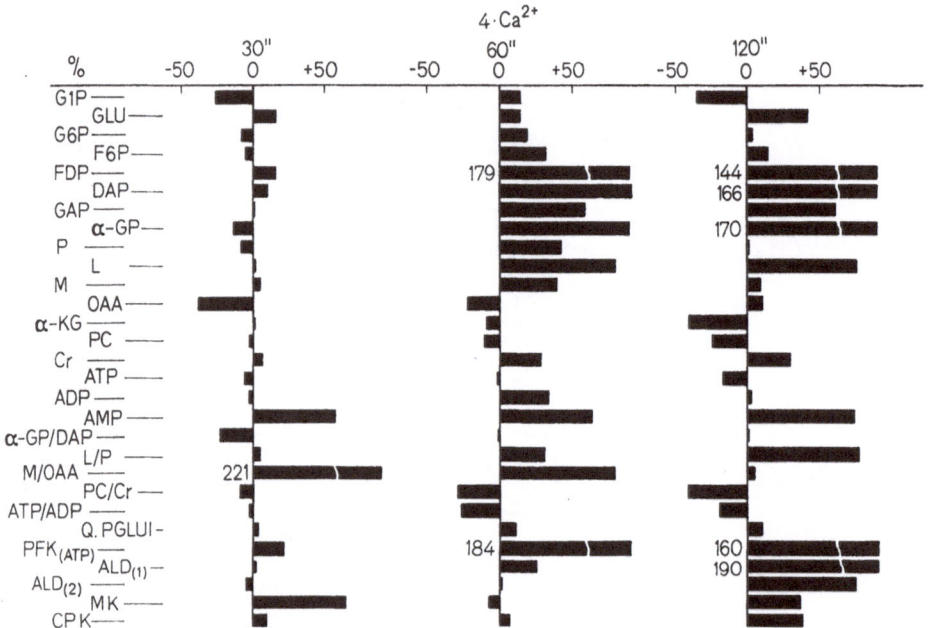

Fig. 2. Percent changes, with respect to the control groups, of the parameters which discriminate the Ca-treated ($4 \times Ca^{++}$) from the control hearts. The hearts were quick-frozen 30, 60, and 120 sec after the beginning of the infusion. Abbreviations used: substrates, see Materials and Methods; mass-action ratios (Q): PGLUI, phosphohexose isomerase; PFK(ATP), phosphofructokinase 1.; ALD(1), aldolase 1.; ALD(2), aldolase 2.; MK, adenylate kinase; CPK, creatine kinase

## Discussion

The following tentative conclusions may be drawn:

(1) Multivariate analysis is necessary to discriminate little-differing metabolic states with a great number of components.

(2) The activation of the phosphofructokinase step after calcium injection is in contrast with the $Ca^{++}$-inhibition of frog muscle vesicle-bound phosphofructokinase [11]. Our finding, which could be related to the increase of AMP and the decrease of ATP, shows that control *in vivo* of allosteric enzymes is much more complex than clear-cut results from simplified systems or purified enzymes would seem to indicate.

(3) Increased glycogen phosphorolysis, on the contrary, confirms the well-known *in vitro* effect of calcium on phosphorylase-*b* kinase [6].

(4) The same effects on heart dynamics, when attained by ouabain injection, are accompanied not only by a very different metabolite pattern, but also by different interrelationships within the treated groups; (for example, at 60 sec, phosphofructokinase activation by decreased AMP; fluctuation of many parameters after ouabain injection, undirectional behavior after calcium treatment).

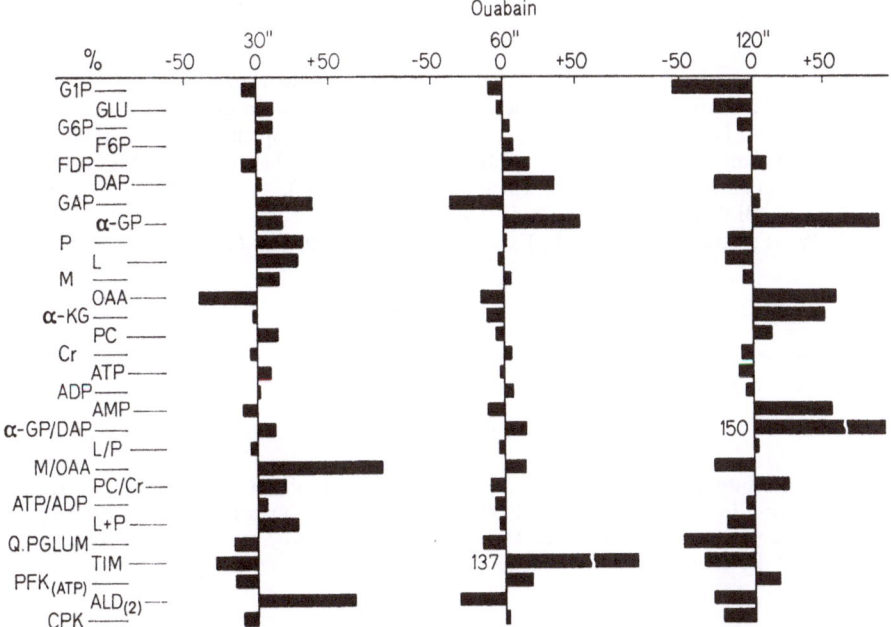

Fig. 3. Percent changes, with respect to the control groups, of the parameters which discriminate the ouabain-treated from the control hearts. The hearts were quick-frozen 30, 60, and 120 sec after the beginning of the infusion. Abbreviations used: substrates, see Materials and Methods; mass-action ratios (Q), see Fig. 2 and: PGLUM, phosphoglucomutase; TIM, triosephosphate isomerase

(5) The evalution of minimum metabolic deviations and the well differentiated kinetic response of intermediary metabolites to similar stimuli should help to improve the fitting of future computer models to the real behavior of cell systems.

## References

1. Arese, P., A. Bosia and L. Rossini: Biochem. Biophys. Res. Comm. 27, 138 (1967).
2. Bergmeyer, H.-U.: Methoden der enzymatischen Analyse. Weinheim/Bergstraße: Verlag Chemie 1962.

3. Brady, A. J.: In: Physiology of the Amphibia (A. J. Moore, Ed.). New York-London: Academic Press 1965.
4. Bücher, Th., u. W. Rüssmann: Angew. Chem. **75**, 881 (1963).
5. Dixon, W.: Biomedical Computer Programs, Dept. of Preventive Medicine and Public Health. Los Angeles: University of California Medical School 1964.
6. Drummond, G. I., and L. Duncan: Biol. Chem. **241**, 3097 (1966).
7. Garfinkel, D.: In: Instrumentation in Biochemistry, (T. W. Goodwin, Ed.) p. 81, Biochem. Soc. Symp. Nr. 26. New York-London: Academic Press 1967.
8. Higgins, J. J.: In: Computers in Biochemical Research (R. W. Stacy and B. Waxman, Eds.), Vol. 2, p. 101. New York-London: Academic Press 1965.
9. Hohorst, H. J.: Habilitationsschrift. Marburg/Lahn 1962.
10. — In: Funktionelle und morphologische Organisation der Zelle (P. Karlson, Ed.), p. 163. Berlin-Göttingen-Heidelberg: Springer-Verlag 1963.
11. Margreth, A., C. Catani and S. Schiaffino: Biochem. J. **102**, 35C (1967).
12. Williamson, J. R.: J. Biol. Chem. **240**, 2308 (1965).
13. — J. Biol. Chem., **241**, 5026 (1966).
14. Wollenberger, A.: Arch. Exp. Path. Pharm. **219**, 408 (1953).

*Aknowledgement*

This work was supported by the Italian Research Council (Consiglio Nazionale delle Ricerche), Rome.

*Discussion*

Walter:

In the interconversion of F-6-P and FDP phosphofructokinase catalyzes the reaction in one direction and FDP phosphatase catalyzes the reverse reaction. Have you any idea, perhaps from tracer experiments, whether the source of F-6-P buildup is G-6-P or FDP?

Arese:

The presence of FDP phosphatase in skeletal muscle is well established (M. Salas, E. Vinuela, J. Salas and A. Sols, Biochem. Biophys. Res. Comm. **17**, 150 (1964)). However, till now, no evidence of this enzyme in heart tissue of several species has been found (H. A. Krebs and M. Woodford, Biochem. J. **94** 436 (1965)). We think therefore that the observed effects are caused by the interaction of ions or of regulative metabolites with phosphofructokinase.

# Analysis by Computer of some Oscillatory Features of the Red Cell System

J. KIRK, and J. S. ORR

With 3 Figures

*Abstract*

The oscillatory nature of the red blood cell and bone marrow stem cell system, which is composed of two interacting feedback control loops, is studied. A mathematical description of the system is formulated as four simultaneous non-linear difference-differential equations in four variables, and the constants derived from a wide range of experimental data. The equations are evaluated numerically by daily increments using a digital computer. The behaviour of the formulation is compared with experimental data. Important features of the oscillations are described, and the entrainment of the frequencies between the loops is discussed from the point of view of the synchronisation of non-linear oscillators, in the light of an analogy with electronic oscillatory systems.

## Introduction

The oscillatory nature of the red blood cell system has been demonstrated in several experimental studies; of which those of PORTEOUS and LAJTHA [3], HULSE [1] and ORR et al. [2] are of particular interest. The system can be broken down naturally into two interacting feedback control loops, as shown in figure 1; the red cell loop in which bone marrow stem cells are differentiated under the stimulus of erythropoietin to form reticulocytes and ultimately mature red cells, and the stem cell loop in which stem cells lost to the differentiation channels are replenished by mitosis, controlled by the stem cell specific mitotic inhibitor, chalone. Erythropoietin and chalone are catabolized in proportion to their own level, and mature red cells disappear from the system after a fairly well defined lifetime. Erythropoietin production is dependent on the reciprocal of a high power of the size of the red cell population, while chalone production is proportional to the stem cell population. The reticulocyte fraction of the red cell population depends on the lifetime of a reticulocyte before becoming a mature red cell. Other differentiation channels are considered as a constant drain on the stem cell population.

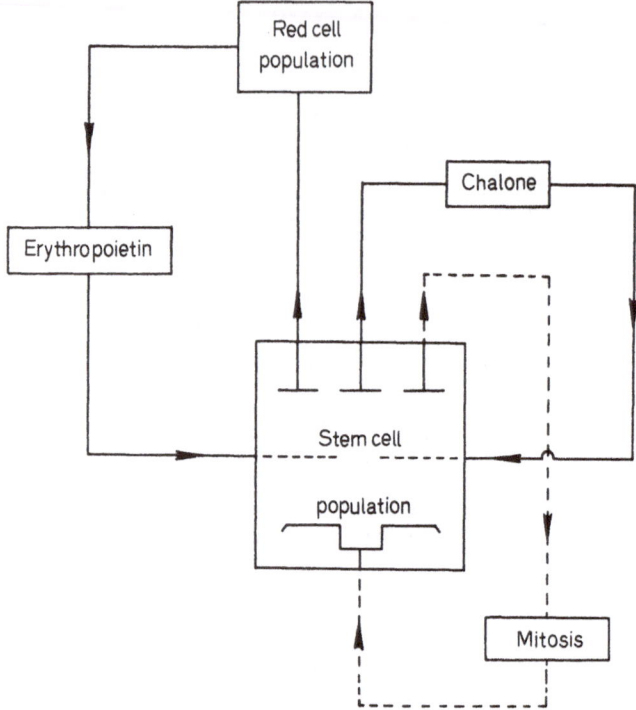

Fig. 1. Block diagram of the oscillatory system

## The Mathematical Formulation

The system is described mathematically by four simultaneous non-linear difference-differential equations in the four variables; the red cell population ($R$), the stem cell population ($S$), erythropoietin ($E$) and chalone ($C$). Since these equations cannot be solved analytically, the daily values of the above variables are calculated by numerical methods using a digital computer. The evaluation was carried out using an integration step of one day. The equations representing the addition of the daily changes of the variables to their previous values are:

$$E_r = \frac{a}{1 + R_{r-1}^z/\delta} + \beta E_{r-1}$$

$$C_r = \gamma \left[ S_r - S_{r-1} \left(1 - m C_{r-1}^n\right)\right] + \rho C_{r-1}$$

$$S_r = S_{r-1} \left\{2 - m C_{r-1}^n \left[1 + \left(\omega + p\, b\, E_{r-1}^q\right)/m C_s^n\right]\right\}$$

$$R_r = R_{r-1} + k\, p\, b\, E_{r-x}^q\, S_{r-x}\, C_{r-x}^n/C_s^n - F(R)_r$$

The constants in these equations are determined from a wide range of experimental data.

## Comparison of the Formulation with Experiment

The behaviour of the mathematical formulation, with the interacting loops and the single loops alone, has been compared with the three sets of experimental data listed above.

In the paper by PORTEOUS and LAJTHA [3], experiments are described in which mice were first hypertransfused to halt erythropoietin production, so that the red cell differentiation channel is cut off. Under these conditions, the stem cell mitotic loop is essentially operating alone, with negligible interaction from the red cell loop. With the mice in this hypertransfused condition, they were irradiated, reducing the stem cell population to 10% of the norm. The recovery of the stem cell population with time was recorded. After an initial dip of variable length, the stem cell population rises rapidly to reach the norm in 7 days and continues to reach a maximum of about 220% at about 13 days and then falls to cross its steady state value again. The behaviour of the formulation has been compared with this experimental situation, and the general features of the experimental data are found to be well reproduced.

The second set of experimental data is found in the paper by HULSE [1]. In this case, rats under normal steady state conditions were given various doses of radiation and the recovery of the reticulocytes with time was studied. A typical set of data shows a marked drop in the reticulocyte population, followed by a rapid rise to overshoot the norm and then an approach to the steady state level with a damped oscillatory motion. The effect of the radiation is a reduction of the stem cell pool, and the response of the mathematical formulation with both interacting loops to this initial perturbation is found to be in good agreement with the observed experimental data.

The third set of experimental data is furnished in the paper by ORR et al. [2], which describes the oscillations set up in the red cell system when rabbits were given a course of red cell iso-antibody. Under these conditions, the red cell lifetime is reduced due to the premature death of red cells under the action of the iso-antibody. This drop in the mean red cell level causes an increase in the production of erythropoietin, stimulating an increase in the number of stem cells differentiated per day into the red cell channel. This, in turn, gives rise to an increased production of reticulocytes and so to an increase in the mean reticulocyte level. The experimental results, two sets of which are shown in figures 2(a) and 2(b), are presented as the variation of red cells and reticulocytes with time, and the corresponding behaviour of the mathematical formulation is shown in figure 2(c). A dominant factor in satisfactorily reproducing the oscillations of the system under the above conditions is the choice of the value of the stem cell population, $S$, and this is found to be $\sim 6 \times 10^6$ stem cells per gram of body weight.

Although no experimental data are available for comparison, it is necessary to examine the behaviour of the formulation when the stem cell level is held constant. In this case, the mitotic replenishment loop is inactive, and only the red cell loop is in operation. The red cell population was given a small perturbation to set up oscillations. Under these conditions, periodic

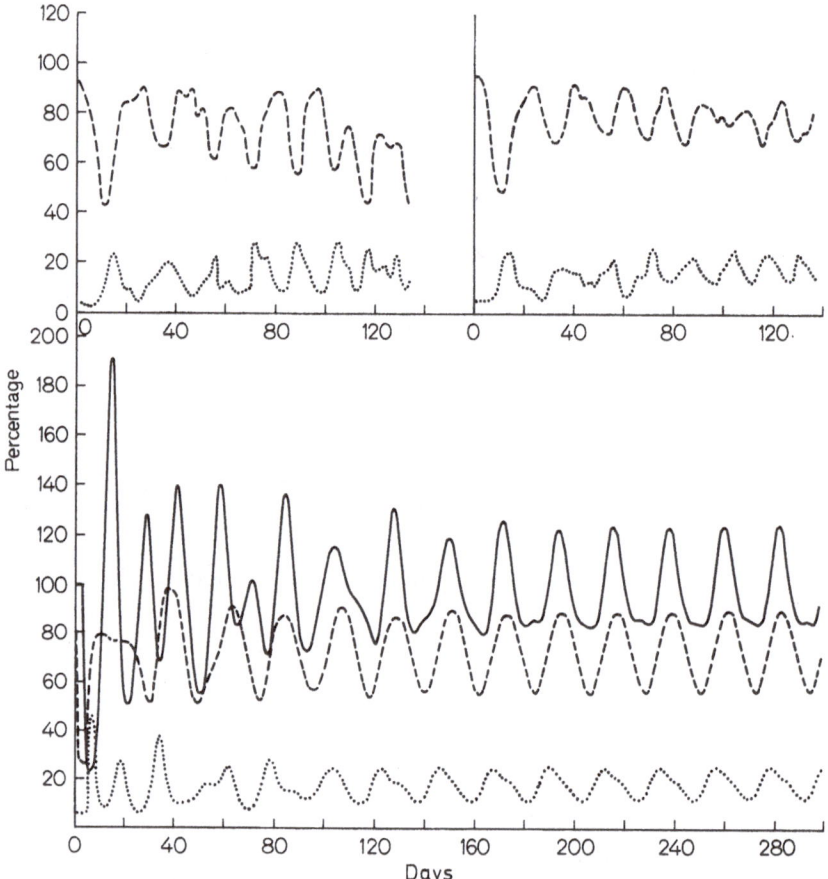

Fig. 2. Variation of cell populations with time (a, b) and the corresponding behaviour of the mathematical formulation (c).
(———— stem cell population, ------- red cell populations, ······· reticulocyte population)

variations are found which damp out in a similar way to those of the stem cell mitotic loop. It thus appears clear that neither the stem cell mitotic loop nor the red cell loop can independently maintain oscillations, and that continued oscillations can only be maintained under conditions of interaction between the loops.

## Discussion of the Mathematical Formulation

It has been shown in figure 2(c), in comparison with the experimental data of ORR et al. [2], that uniform continued oscillations are generated from the interactions between the natural frequencies of the loops. An important feature of the formulation is shown in the period of entrainment of the frequencies during the first 80 days, leading to the final uniform oscillations, in phase. Oscillations of this type only occur for small differences in frequency between the loops. The entrainment of one frequency with a natural reference frequency is well recognised in the synchronisation

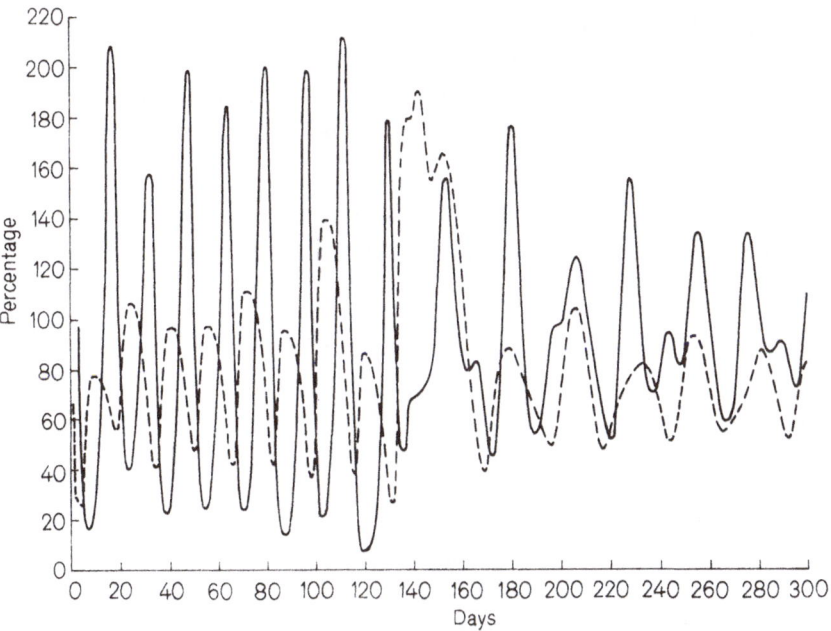

Fig. 3. Complex oscillations for some ratios of the loop frequencies.

of oscillators in non-linear systems. For the red cell system, this is perhaps an oversimplified view since a mutual entrainment occurs, so that the final frequency of the stable oscillations lies somewhere between the natural frequency of each loop.

For some ratios of the frequencies of the loops, the more complex oscillations shown in figure 3 occur. In the first 130 days, an effort at entrainment is made, leading to moderately regular but unstable oscillations, because the irregularities build up to a major disturbance at about 150 days, followed by an entirely different mode of oscillation. This mode is stable although rather complex initially, but eventually attains a greater uniformity after about 500 days, and exhibits features of beating. The disturbance at 150 days, and the neighbouring oscillations, are a manifestation of the

Ziehen, or frequency jump, effect. In a system of coupled electronic oscillators, the resonance condition for the system as a whole occurs when the
reactance is zero, and this condition is satisfied for two frequencies. The
system will be stable in one or other of the two frequencies, depending on
its energy. As the energy changes through some critical value, the system
can jump from one meta-stable state to the other. The two modes of oscillation shown in figure 3 can be compared with the two meta-stable states of
the electronic analogue, and the disturbance at 150 days represents the
frequency jump or transient involved in the transition from one mode to
the other.

The analogy between this electronic system of coupled oscillators, and
the behaviour of the mathematical formulation of the red cell system with
its two interacting control loops, appears to be very close, and an analysis
of the latter system in the light of this analogy may prove of considerable
interest and value.

## References

1. Hulse, E. V.: Brit. J. Haemat. **9**, 365 (1963).
2. Orr, J. S., J. Kirk, K. G. Gray and J. R. Anderson: Brit. J. Haemat. **15**, 23
   (1968).
3. Porteous, D. D., and L. G. Lajtha: Brit. J. Haemat. **12**, 177 (1966).

*Discussion*

Kiefer:

Is there experimental evidence for a stem-cell specific chalone?

Kirk:

Not exactly, but Bullough and Laurence (1964) described an experiment
on the inhibitory effects of chalone in mouse epidermal tissue. The form of
this inhibition effect has been carried over and applied to stem cells.

Kiefer:

How would the behaviour of your model system be changed if the other ways
of differentiation were accounted for by more feedback loops?

Kirk:

The behaviour would not be basically changed, but other differentiation channels would cause more fluctuations in the stem cell population and compete
to some extent with red cell differentiation channels.

Kiefer:

The integration step of one day is in the order of the mitotic cycle time of the
stem cells. Is it not likely to introduce artefacts by such a long integration step?

Kirk:

The one-day integration step has a great deal to recommend it owing to
diurnal effects in the system. Calculations on a small section of the system
with an integration step of $1/5$ of a day introduced a difference of less than
6%; also the integration effect is of the order of $1/20$ of the frequency of the
loops. Thus, it is very unlikely that any artefacts are introduced.

BETZ:

In every case when in your figures the frequencies of stem cells and red cells are different, there seems to be some interference of two oscillations in the case of the stem cells. Perhaps there would be no difference in frequency if we could account for this second, less pronounced mode of oscillations.

KIRK:

In figure 3 the second mode of oscillation is composed of at least two components of different frequencies and this accounts for the interference or beating effects observed.

BETZ:

In one of your figures there is a phase difference between stem cells and red cells. May this difference be due to the time difference needed for differentiation of the red cells?

KIRK:

Yes, this difference in phase represents the time for a differentiated stem cell to pass through the normoblast series and become a mature red cell.

WALTER:

What is the range of $z$ in your equation?

KIRK:

$z$ is of the order of 9.

# Memory and Cellular Control Processes

J. S. GRIFFITH

*Abstract*

The possible physical basis of memory is discussed, mainly in relation to our present knowledge of the biochemical organization of cells. Various matters are considered, which include the question of the relevance of estimates of memory capacity, the significance of attempts to transfer learning between animals, and the experimental evidence about the effect on memory of compounds which interfere with protein or RNA synthesis. — An account is given of the possible nature of the regulatory controls responsible for the formation and maintenance of differentiation in the nervous system. This is based upon the views of MONOD and JACOB, as they might apply to neurons, and it is shown that long-term memory might be dependent upon similar controls.

## 1. Introduction

I shall discuss various aspects of the problem of the physical basis of memory in man and other animals, mainly in relation to our present knowledge and hypotheses about the nature of the genetic regulatory mechanisms in cells. I believe that at the present time this is the best way to arrive at clear-cut suggestions for memory mechanisms which have, then, some hope of experimental test. Furthermore, a great deal of current experimental work is already producing results whose interpretation is best attempted in these terms. I conclude by showing that presentday ideas about differentiation [7, 42, 44] lead to a possible simple hypothesis about the nature of memory.

## 2. Survey of Various Relevant Topics

### a) Types of Memory Theory

The material basis of "long-term" memory might be anything between two rather extreme limits, one of which can be described simply at the macroscopic cellular level and the other at the molecular level. An example of the first kind is the circulating impulse theory, in which memory is based upon continuing trains of action potentials moving endlessly around closed circuits of neurons. This mechanism was first suggested by FORBES [25] as a basis for spinal reflex afterdischarge. It is not now thought likely to be a basis for long-term memory. A cogent objection to it is the small effect on well-established memories of electroconvulsive shock treatment or grand mal epileptic fits. One would expect either of these to disrupt orderly cycles of circulating action potentials and hence, if the theory were correct, to erase all memory.

Other mechanisms of a macroscopic kind involve permanent alterations of cellular membrane properties, growth of new axonal or dendritic branches and growth of new synaptic knobs or alteration in the size of old ones. However, in each of these examples, although the final result can be described only in macroscopic terms, the alteration might be triggered and possibly maintained by a simple molecular mechanism [16, 51].

At the molecular limit it is possible to distinguish between two broadly different kinds of theory. Both would suppose that memory is represented by some lasting change in the metabolic balance of the cell but one would implicate the control processes operating at the DNA-RNA-protein synthesis level [16, 30, 47, 51], and the other would not. An example of the second would occur if, after each time a nerve cell fired, the restoration of the membrane potential was coupled biochemically to the synthesis of a small quantity of a compound $X$. If $X$ were metabolically stable and only synthesized in these circumstances, then the total quantity of $X$ in a nerve cell would record the total number of times that cell had fired. The first kind of molecular theory is probably the most interesting at the present time, both because of current concern with genetic regulatory mechanisms in cells and because it gives one a framework within which one can think precisely and constructively. Accordingly we shall concentrate our attention upon it as giving potentially the most fruitful approach although, in our present state of ignorance, we should not entirely ignore the other possible types of theory.

### b) Quantity Requirements

Various authors have made attempts to give approximate estimates of the memory capacity in "bits" of the human brain. There are two fundamentally different assumptions upon which such estimates can be based. One is that the brain stores a record of the sequence of action potentials arriving from every single input cell. The other is that the storage system is arranged so that only pieces of information which would be of interest at the "psychological" or "whole organism" level are extracted from the input and stored. The difference is rather like that between the storage in picture form of a letter, such as A, on video tape and its retention in coded form as one character on a five hole paper tape record. Obviously, much more capacity is needed in a store operating in the first way than in the second.

I do not think we can be sure at the moment which of these extremes is nearer the truth for the human brain. However, the second seems more plausible on account of its greater economy and the lack of any evidence for the first. Also, experiments on rates of information acquisition and transmission in man are consistent with it. With the first assumption, the human memory capacity has been estimated as $10^{16}$—$10^{19}$ bits [28, 55] and with the second, $10^8$—$10^{11}$ bits [28, 46, 54].

These estimates are very approximate and depend upon assumptions which can reasonably be questioned. Have they any value? I believe they have, because they can be compared with theoretical calculations of capacity derived from various possible mechanisms for memory and used to give some guidance as to the plausibility of those mechanisms. For example, if memory is stored in nerve cells and their connections, then if the first assumption is correct there are $10^6$—$10^9$ bits per nerve cell. This would create insuperable difficulty for a theory which asserted that changes of threshold were responsible for memory, and considerable difficulty for one which assigned it to changes in synaptic connection. On the other hand, with the second assumption, we need only $10^{-2}$—$10$ bits per cell, which would be acceptable on either theory.

It has been suggested that we should introduce a unit of memory, to be called the mnemon, which is to be the "minimal physical change in the nervous system that encodes one memory" ([11], see also [58]). One should beware of the assumption, implicit in this definition, that there *is* a minimal unit. If memory involves the accumulation of some substance, $X$, or synaptic growth, there may be no minimum above the level of a single molecule, which might be ridiculously small in comparison with the smallest amount of storage normally achievable in practice. In this context, one should note that in spite of its "quantized" sound, a bit is not necessarily the smallest amount of information that can be stored [36].

### c) Transfer Experiments and "RNA Theories of Memory"

Many experiments have been reported which claim transfer of learning between animals (see, e.g., [33, 49, 50]) and some of these also claim support for an "RNA theory of memory". We shall not enter here into the complicated question of the extent to which these claims are justified [9, 31, 34], but shall make a few theoretical comments on the problem of their possible interpretation (see also [15, 28]).

We should replace vague talk about "RNA theories of memory" with some definite hypothesis, probably one of the following three. First, memory is stored in a coded form in the nucleotide sequence of specially-synthesized RNA, which is *not* transcribed from the inherited DNA. There is no evidence for this, and it must surely be regarded as highly improbable. Second, the physical representation of memory is in the form of the quantities of one or more types of mRNA present in each of the cells (neurons probably, glia possibly), that mRNA being partly or wholly metabolically stable and having been transcribed from inherited DNA. Third, memory resides in the quantities or conformations of some cellular constituent other than RNA, for example structural or enzymic proteins, but the process of laying it down involves the synthesis of some new, metabolically unstable, mRNA. These last two hypotheses, or combinations of them, are perfectly plausible, but there are also possible mechanisms for memory which use only previously synthesized mRNA or protein (see, e.g., [29]) or even none directly at all.

Then we have to decide how specific the transfer is [15]. For example, suppose that memory is "set" by the release of some agent $S$ whose quantity depends on the degree of reinforcement [29]. Then an extract from a brain in which much reinforcement had recently occurred might contain a large amount of $S$, which would therefore act as a "setting agent" in a brain into which it was injected and thus improve memory. Such improvement would presumably be primarily interpreted as "savings" rather than "retention", but if there were several different setting agents, $S_1, S_2, \ldots, S_n$, say, acting specifically on different cells, or synapses, then effects mimicking learning, and having some degree of specificity, could be produced.

Finally, if the essential effective material transferred is RNA, this fact by itself still does not enable one to distinguish between the three hypotheses described above. For, even in the third, the transfer agent could be RNA, namely that newly-synthesized, metabolically unstable, mRNA mentioned there.

### d) Axonal Protein Synthesis

If the laying down of long-term memory involves an action on the genetic regulatory elements of cells, this would imply an influence at some stage on the process of transcription from DNA to RNA or on protein synthesis. It is natural to suppose that this must take place in or near the nucleus of a cell, but this is not necessarily the case. It is now known that mitochondria contain DNA and, almost certainly, a complete protein-synthesizing apparatus, including ribosomes [2, 8, 45]. As each synaptic knob usually contains a mitochondrion [48], it follows that it is possible to envisage a control involving transcription or translation in the mitochondrion. The level of activity of the respiratory enzymes in the mitochondrion would be expected to affect the size and efficacy of its synaptic knob and, therefore, inducers or repressors acting on the mitochondrion would give a convenient method of control over the synapse. Whether this is so or not remains, of course, to be seen, but the possibility should be borne in mind in considering the effects on memory of puromycin and the like.

There is definite evidence for local protein syntheses in various axons [18, 37] and, at least in one case, this appears to be outside the mitochondria [38]. Although electron microscopic techniques have failed to show ribosomes in axons [17, 18], there does seem to be ribosomal RNA [43] and perhaps also DNA [19] which may be directing this local synthesis.

### e) Compounds Which Interfere with Protein or RNA Synthesis

If the laying down of new long-term memory involves, as an essential part, the synthesis of new protein or new mRNA, then it must be inhibited by any compound which prevents that synthesis. For this reason, several investigators have injected puromycin and other compounds into the brains of various animals, before and after learning, to determine the effect on memory. The results favour the view that the early phases of memory storage do not involve such new synthesis [1, 3].

However, the effects on the later phases of storage were soon found to be less simple. Puromycin and acetoxycycloheximide (AC) both inhibit protein synthesis, although by different mechanisms [20]. In the goldfish, they both appear to prevent long-term memory [1], but in the mouse only puromycin does, but not AC or cycloheximide [4, 21, 24] (actinomycin D was found to have no effect on memory in the mouse [5, 12]). Furthermore, AC injected together with puromycin can actually protect against the influence of the latter upon memory [21].

It is becoming apparent now that the situation is even more complicated than this. The effect of puromycin seems to be dependent also upon environmental factors [14] and even upon the subsequent injection of saline [22], whilst AC does seem to have a transient effect in mice after all [23]. Also it has been suggested that puromycin may work by producing an increased susceptibility of the animal to seizures [13] and that some of its effects may be caused by the brain lesion arising from the puncture by the injecting needle [6]. Evidently no clear-cut picture emerges at the present time, and it is to be hoped that workers using this sort of approach will not be discouraged from continuing what is surely, even now, a very hopeful experimental procedure.

### f) The Nerve Growth Factor

The nerve growth factor (NGF) is a protein which stimulates growth of embryonic sensory and sympathetic, and mature sympathetic, nerve cells [40]. Its properties are interesting in connection with memory because of the possibility that the latter is based upon growth stimulated by similar proteins [29]. Its effects are inhibited by actinomycin D [41], which suggests that it is operating at the level of transcription and may be acting as an inducer. In view of the discussion in section d, it would be of great interest to know its exact site of action in the cell.

## 3. Control Processes and Differentitation

MONOD and JACOB [44] have pointed out that certain interactions between genetic regulatory elements could result in a cell having two or more different stable sets of concentrations of mRNA, protein, etc. A particular example which they give can best be described as a cellular flip-flop in which there are two genes, $G_0$ and $G_1$ say, the product of either of which catalyzes the synthesis of a metabolite which represses the synthesis of the messenger from the other. Thus only one gene can be "on" at a time, the other being held "off". They suggest that differentiation arises through the operation of mechanisms of this or similar kinds, which seems highly plausible (though see [52]), and I shall suppose here that it is correct. I have discussed in a recent book [30] the relevance of this to the organization of the nervous system, and the present discussion, although self-contained, is largely complementary to my previous one and a development of it.

MONOD and JACOB's flip-flop mechanism may be symbolized by

$$
\begin{aligned}
G_0 &\dashrightarrow M_0 \longrightarrow S_0 \longrightarrow P_0', \\
G_1 &\longrightarrow M_1 \dashrightarrow S_1 \longrightarrow P_1',
\end{aligned}
\tag{1}
$$

where the $G_i$ are genes, $M_i$ messengers, $S_i$ enzymes and each of the $P_i'$ is a metabolite whose production is catalyzed by the corresponding enzyme $S_i$ ($i = 0,1$). $P_0'$ acts as a corepressor for $G_1$ and $P_1'$ as a corepressor for $G_0$. This mechanism involves the participation of enzymes and metabolites and could be switched from one state to the other by alterations in the levels of these. For example, if $G_0$ is "on" and the cell has an excessive demand for $P_0'$, the gene $G_1$ might become derepressed and set in train the sequence of reactions which lead to $P_1'$. The latter would then hold $G_0$ off, and the flip-flop would be switched. Whether this represents a serious instability in such a mechanism cannot, perhaps, be said at present. However, it is interesting to note that it can be evaded by a slight reformulation, as I shall now show.

Suppose we have the situation

$$
\begin{aligned}
G_0 &\dashrightarrow M_0 \longrightarrow S_0, \\
G_1 &\dashrightarrow M_1 \longrightarrow S_1,
\end{aligned}
\tag{2}
$$

with the same notation as before, except that the $S_i$ are no longer enzymes but merely repressors. $S_0$ is assumed to repress $G_1$, and $S_1$ to repress $G_0$. Again this is a flip-flop, but it is uninfluenced by any metabolic processes outside the machinery for synthesizing mRNA and proteins. It can, however, control the activity of the cell very well for we have merely to suppose that $S_0$ and $S_1$ also act as repressors (or inducers) for certain structural genes. Such a flip-flop is protected from being switched by changes in metabolic levels within the cell. However, it can still be controlled by further repressors, either coded for by other genes in the cell or coming in from outside the cell.

For purposes of exposition, it is convenient to consider such simple mechanisms as these flip-flops. The actual controls involved in differentiation of mammalian cells may very well be much more complicated, but certain general features should still be present. There should be a finite number of stable states for the entire set of controls within the cell, which means that there is this finite number of different kinds of differentiated cell altogether in a particular animal. We shall call each of these kinds a "category" [30], and can say that there are a finite number of categories of nerve cell which may be enumerated as $T_1, T_2 \ldots, T_n$.

At this point it may be objected that we have ignored the possibility of oscillatory solutions to the cellular control equations, and that the existence of such solutions might invalidate the statement that there are only a finite number of cell categories. This is true, but probably unimportant. It is just possible that control systems leading to oscillations may be used by nerve cells as internal clocks. For example, HORN [32] has found quite regularly firing cells in cat visual cortex and suggested that this firing may be the expression of some intrinsic biochemical process within each cell. However, one would only expect a finite number of different stable oscillations, provided they are of the limit cycle type [39]. GOODWIN has argued that oscillations can occur which have indifferent stability, which would mean the existence of an infinite number of different possible categories and hence extra complexity. We shall not consider this possibility here, except to remark that the equations on which he bases this conclusion are not entirely satisfactory because they allow the concentrations of cellular constituents, initially positive, to become sometimes negative [26].

There is a very superficial analogy between the existence of different categories of nerve cell and the existence of different eigenstates for an atom. True to this analogy, we can distinguish between spontaneous and induced transitions between categories. The induced transition of a cell between two categories, $T_i$ and $T_j$ say, could arise from the passage into the cell of a suitable repressor. Spontaneous transition could occur if the number of mRNA or repressor molecules, involved in a control situation, is small. Random variations in these numbers could occasionally lead to an "accidental" switching taking place. We would then have a set of quantities $p_{ij}$ ($i \neq j$), where $p_{ij}$ is the probability per unit time that a cell in category $T_i$ is changed to one in category $T_j$. There is no reason to assume that $p_{ij} = p_{ji}$.

Different categories would be expected to differ biochemically and also, probably, morphologically. As I have discussed elsewhere [30], part at least of the rules governing which pairs of neurons can synapse together might be expected to depend upon which categories the pair of cells are in. In this respect it is interesting to note the existence in invertebrates of cells, identifiable from one individual to another, and having fixed relations of connection between each other [35]. A fascinating possibility, which seems to me by no means impossible, is that, apart from the effect of geometrical considerations of the proximity and accessibility of pairs of cells, the extent of connection between any pair of nerve cells throughout the entire nervous system may depend upon the categories of the cells concerned and upon nothing else. Assuming that long-term memory depends upon changes of synaptic linkage [10], this would mean that both the embryonic development of the nervous system and the whole of long-term memory would depend upon the laws governing the developmental formation of categories, and the spontaneous and induced transitions between them. Furthermore, a fairly complete description of a brain at any moment, including all its long-term memory, would be given by saying which category each nerve cell is in, throughout the nervous system.

Let us conclude by considering the plausibility and the implementation of such a hypothesis. As far as quantity requirements are concerned, it would give a maximum memory capacity of $C = N \log_2 n$ bits, where $n$ is the number of categories and $N$ the number of nerve cells. This is not very sensitive to the value of $n$, which enables us to say that $C$ would be of the order of $10^{11}$—$10^{12}$ for the human brain. The possibility of glial cells being also involved in some similar way cannot be excluded. The hypothesis also allows for forgetting, through the existence of spontaneous transitions between categories, although forgetting could also occur in other ways (ref. [30], page 37). Clearly the rate of spontaneous forgetting could vary with the material learnt (because different categories would be involved) and with the animal (cf. the finding of almost "perfect" retention in certain experiments with goldfish [1]).

The control over synaptic connection might be effected by a differential synthesis of various growth substances, $Y$, which pass across the synaptic clefts in suitable circumstances, as discussed previously [29]. Induced transitions would be expected to be a result of the passage of materials, $Z$, across synaptic clefts from one cell to another. These materials, $Z$, would not necessarily be the same as the growth substances, $Y$. Because of the action of the NGF at the transcription level, either or both of $Y$ or $Z$ might be of a similar nature. I have argued that $Y$ is likely to pass from the soma or dendrites of one cell to "feed" the axonal tips which are synapsed to them [29, 30]. $Z$ might pass in either direction, but the fact that the axoplasm of neurons flows continuously outward from the soma and down the axons might suggest that it is impossible for any material to pass upstream to the nucleus from the axonal tips. There may, however, be specialized passages in a nerve cell for transport, apart from this general flow [27, 57], and, if this is so, there is no reason why some of it should not be in the opposite direction. This possibility seems worth investigating experimentally. In this context it is worth noting that it seems likely that muscle cells which are artificially caused to be innervated by inappropriate motoneurons are capable of transforming the character of the latter into that of the appropriate ones (Weiss [56], though see ref. [53]). This phenomenon, which is called "specific modulation" or the "myotypic response", could have a very similar mechanism to the one I have hypothesized above.

In relating memory at the cellular level to the overall nervous integration it is easiest, though not necessarily correct, to suppose that it involves a change in the strength of connections which already exist, rather than the formation of completely new ones [10, 30]. This would be possible according to the present hypothesis. One need only suppose that the switching involved in memory is, usually at least, between categories which produce identically the same set of growth substances, but in different amounts. This could occur if the structural genes were repressed to different extents, or if several structural genes coded for the same mRNA and different numbers were switched on.

Finally, note that although it is appealing to hope that the whole basis of long-term memory might depend upon a single mechanism which has a simple description at the molecular level, it may well be that Nature is more complicated than this. Even if the present hypothetical mechanism were correct, it might well act alongside others, such as the servomechanism I have discussed previously [29, 30].

## References

1. Agranoff, B. W., R. E. Davis and J. J. Brink: Brain Research 1, 303 (1966).
2. Barnett, W. E., D. H. Brown and J. L. Epler: Proc. Nat. Acad. Sci. 57, 1775 (1967).
3. Barondes, S. H., and H. D. Cohen: Science 151, 594 (1966).
4. — — Brain Research 4, 44 (1967).
5. —, and M. E. Jarvik: J. Neurochem. 11, 187 (1964).
6. Bohdanecka, M., Z. Bohdanecky and M. E. Jarvik: Science 157, 334 (1967).
7. Bonner, J.: The Molecular Biology of Development. Oxford: University Press 1965.
8. Borst, P., A. M. Kroon and G. J. C. M. Ruttenberg: In: Genetic Elements. Properties and Function, (Shugar, Ed.) p. 81. Academic Press 1967.
9. Byrne, W. L., and 22 others: Science 153, 658 (1966).
10. Cajal, S. R. Y.: Histologie du Système Nerveux. Paris 1910.
11. Cherkin, A.: Proc. Nat. Acad. Sci. 55, 88 (1966).
12. Cohen, H. D., and S. H. Barondes: J. Neurochem. 13, 207 (1966).
13. — — Science 157, 333 (1967).
14. Davis, R. E., and B. W. Agranoff: Proc. Nat. Acad. Sci. 55, 555 (1966).
15. Dingman, W., and M. B. Sporn: Science 144, 26 (1964).
16. Dixon, K. C.: Lancet 7 Jan., p. 27, (1967).
17. Edström, A.: J. Neurochem. 11, 309 (1964).
18. — J. Neurochem. 13, 315 (1966).
19. — J. Neurochem. 14, 239 (1967).

20. Ennis, H. L., and M. Lubin: Science **146**, 1476 (1964).
21. Flexner, L. B., and J. B. Flexner: Proc. Nat. Acad. Sci. **55**, 369 (1966).
22. − − Proc. Nat. Acad. Sci. **57**, 1651 (1967).
23. − − and R. B. Roberts: Proc. Nat. Acad. Sci. **56**, 730 (1966).
24. − − and E. Stellar: Science **141**, 57 (1963).
25. Forbes, A.: Quoted in: Burns, B. D., The Mammalian Cerebral Cortex. Edward Arnold, Ltd. 1958.
26. Goodwin, B. C.: Temporal Organization in Cells, p. 32. Academic Press 1963.
27. Grafstein, B.: Science **157**, 196 (1967).
28. Griffith, J. S.: In: Molecular Biophysics, (Pullman and Weissbluth, Eds.) p. 411. Academic Press 1965.
29. − Nature **211**, 1160 (1966).
30. − A view of the brain. Oxford: University Press 1967.
31. Hartry, A. L., P. Keith-Lee and W. D. Morton: Science **146**, 274 (1964).
32. Horn, G.: Nature **194**, 1084 (1962).
33. Jacobson, A. L., C. Fried and S. D. Horowitz: Nature **209**, 599 (1966).
34. Jensen, D. D.: Animal Behaviour, Supp. No. 1, p. 9 (1964).
35. Kennedy, D.: Scientific American **216** (5), 44 (1967).
36. Khinchin, A. I.: Mathematical Foundations of Information Theory, p. 3. Dover 1957.
37. Koenig, E.: J. Neurochem. **12**, 343 (1965).
38. − J. Neurochem. **14**, 437 (1967).
39. Leimanis, E., and N. Minorsky: Dynamics and Non-linear Mechanics, p. 122. John Wiley, Inc. 1958.
40. Levi-Montalcini, R., and S. Cohen: Ann. N. Y. Acad. Sci. **85**, 324 (1960).
41. Liuzzi, A., P. U. Angeletti and R. Levi-Montalcini: J. Neurochem. **12**, 705 (1965).
42. Maruyama, M.: Am. Scient. **51**, 164 (1962).
43. Miani, N., A. di Girolamo and M. di Girolamo: J. Neurochem. **13**, 755 (1966).
44. Monod, J., and F. Jacob: Cold Spring Harbour Symp. Quant. Biol. **26**, 389 (1961).
45. Reich, E., and D. J. L. Luck: Proc. Nat. Acad. Sci. **55**, 1600 (1966).
46. Richter, D.: In: Aspects of Learning and Memory, (Richter, Ed.) p. 91. W. Heinemann 1966.
47. Roberts, E.: Brain Research **1**, 117 (1966).
48. Robertis, E. de: Science **156**, 907 (1967).
49. Rosenblatt, F., J. T. Farrow and S. Rhine: Proc. Nat. Acad. Sci. **55**, 548, 787 (1966).
50. −, and R. G. Miller: Proc. Nat. Acad. Sci. **56**, 1423, 1683 (1967).
51. Smith, C. E.: Science **138**, 889 (1962).
52. Strehler, B. L., D. D. Hendley and G. P. Hirsch: Proc. Nat. Acad. Sci. **57**, 1751, (1967).
53. Szekely, G.: J. Embryol. Exptl. Morphol. **7**, 375 (1959).
54. Turing, A. M.: Reprinted in: Minds and Machines. Prentice Hall 1964.
55. von Neumann, J.: The Computer and the Brain. Yale: University Press 1958.
56. Weiss, P.: In: Analysis of Development, (Willier, Weiss and Hamburger, Eds.) p. 383. W. B. Saunders Co. 1955.
57. Weiss, P., and Y. Holland: Proc. Nat. Acad. Sci. **57**, 258 (1967).
58. Young, J. Z.: The Memory System of the Brain, p. 42. Oxford: University Press 1966.

*Discussion*

SIMON:

In my opinion the JACOB-MONOD flip-flop is too slow, for rapid memory, at least. To apply the flip-flop here, one has to have a sensitive decay of one repressor protein available as well as of its messenger templates. The known decay times for mRNA are min to hrs, for proteins only hrs.

GRIFFITH:

It is generally thought that memory storage may involve more than one phase, with different mechanisms. My talk has referred to the later phases and, as I discuss in reference [29], the first phase may be dependent upon residual transmitters.

# On the Steady State Nature of Evolution, Learning, Perception, Hallucination and Dreaming

R. Fischer

With 3 Figures

*Abstract*

Evolution, learning, perception as well as hallucination and dreaming will be treated as sequential steps of a single ongoing process which can be visualized as a logarithmic spiral. Each adaptational step displays an increasing time rate of change and represents a steady state with less and less space-like and more and more time-like characteristics, thus implying the involvement of less energy and more information [9].

## Organismic Steady States

The maintenance of steady state between a system and its environment is the most efficient, orderly state of an irreversible, smoothly-running, open system. The steady state can, therefore, be defined as a nonequilibrium state of an open system in which all forces acting on the system are exactly counterbalanced by opposing forces, in such a manner that all its components are stationary in concentration although matter is flowing through the system [26].

The steady state relationship offers advantages and imposes limitations. The most visible limitations on the macroscopic level are size, rate of change and duration, an interrelated triad which can be dealt with systematically especially in the case of homeothermic mammals. Within these organisms the production of heat or the $O_2$ consumption per unit surface area is approximately constant "from the shrew to the whale", since the metabolic rate relatively decreases with increasing organismic size. It is, therefore, the *metabolic rate* at which $O_2$ consumption per unit surface area proceeds which determines the chronological life-span of the organism. After Huxley [19],

$$M = k \cdot W^n \tag{1}$$

where $M$ = metabolic rate, $k$ = constant, describing metabolic activity for one weight-unit ($W = 1$) and $n$ = allometric constant as an exponent

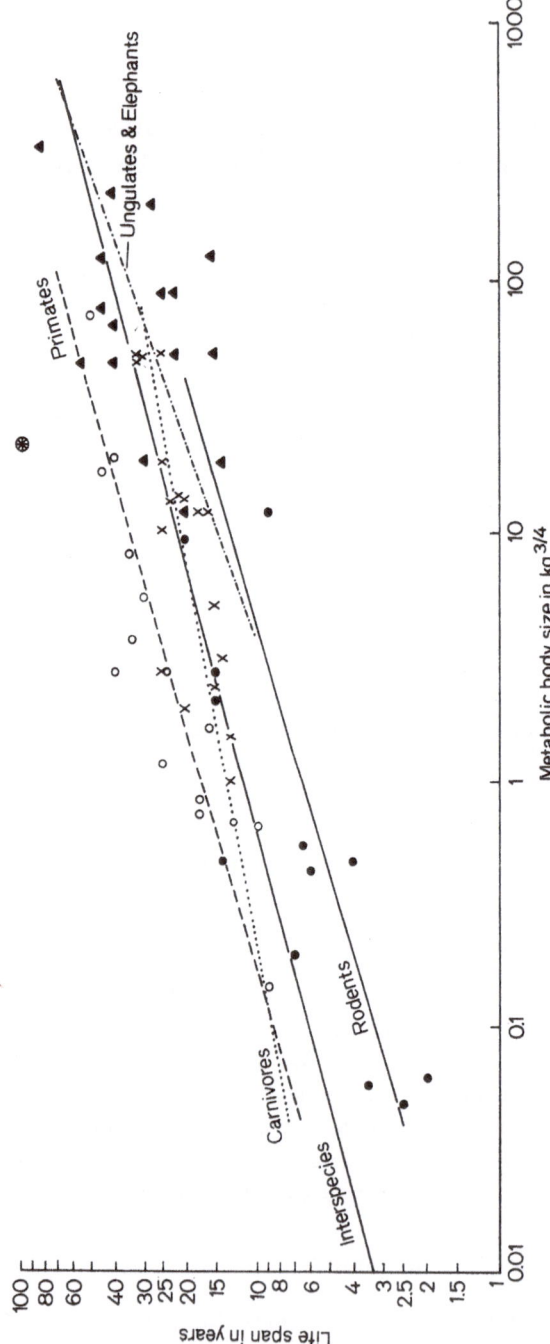

Fig. 1. The linear relationship between log lifespan and log metabolic body size in kg³/⁴ for 63 species of mammals replotted after Sacher [31]. Regression lines are for all species (interspecies) and four groups of species. Open circles — primates and lemurs; solid circles — rodents and insectivores; crosses — carnivores; solid triangles — ungulates and elephants; star in circular field—man. (For details see [13])

of the weight; specifically, $n = 1$ expresses weight proportionality; $n = {}^2/_3$, surface proportionality; and $n = {}^3/_4$, metabolic body size within homeothermic mammals[1].

Figure 1 illustrates a highly significant linear relation between logarithm of lifespan ($y$) and logarithm of metabolic rate ($x$) in kg$^{3/4}$ for 63 species of mammals (replot after SACHER, [31]). The interspecies relationship expressed by the regression line is:

$$y = .270\,(x) + 1.0649 \tag{2}$$

The standard error of estimate ($S_{x \cdot y}$) is equal to .209, which implies that the ratio between log (actual lifespan) and log (calculated lifespan) is generally less than 1.618. This means that most of the predicted lifespans are within 60% of the actual lifespans (for details see [13]).

The above relation between metabolic rate and lifespan also holds true on the single *organ* level, for instance, when plotting brain weight in kg$^{3/4}$ against lifespan in the same manner as in Figure 1 (see also [13]).

The approximately two-fold superiority of metabolic brain size over metabolic body size as a predictor of lifespan is shown by the tighter clustering to the regression lines and specifically man's closer fit within the relationship. Since man lives three times longer than expected in terms of his body size [32], the closer fit may indicate that man's brain may have something to do with the prolongation of his lifespan. That metabolic brain size can be related to performance is also indicated by VANDENBERG's [41] studies in which it was found that in identical twins, the larger the head, the higher the intelligence test score. This statistically significant relationship is obscured by other factors in the general population.

It may be of interest to note that indirect calorimetry of the whole brain of man has yielded high values of energy metabolism [36], an average of 12 mWatts per gram brain tissue [1]. There is also evidence that the significant correlation between regional cortical blood flow and EEG frequency content, shown in man and other animals, may reflect the dependency of both variables upon the oxidative metabolic activity of the nervous tissue [20].

Mitochondrial *oxidative metabolic activity* in relation to unit of homeothermic mass has been discussed elsewhere earlier [5, 11, 34]. Spectroscopic observations of intact, respiring mitochondria have shown that the electron carrier molecules of the respiratory chains participate in an exquisitely balanced dynamic steady state [26].

---

[1] For practical purposes KLEIBER [24] recommends the kg$^{3/4}$ as a useful generalization expressing the metabolic unit of body size, although VON SCHELLING [33] favors, on mathematical grounds, the 0.73 power instead of 0.75.

## The Sensory Steady States of Learning and Perception

Evolution, we have shown, is a steady state represented by a straight line on a log-log plot relating metabolic body size (space) and lifespan (time). By analogy, learning, another adaptational process, is also a steady state, using less energy, proceeding at a faster rate of change. Learning can also be characterized by higher complexity and less stability in energetic terms or, in other words, it is a less space-like and more time-like step than evolution. Notwithstanding these differences, the overriding similarity of the two should result in cumulative learning curves which also follow a straight line in log-log coordinates since they can be described by power functions of the general form:

$$P = k t^n \tag{3}$$

where $P$ is a cumulated measure of performance and $t$ is practice time or number of trials [37]. The similarity between this and equation (1), HUXLEY's [19] allometric relation pertaining to evolution, is quite apparent. From the many cumulative learning curves in log-log coordinates [37] we have selected one example as an illustration in Figure 2, labelled "learning";

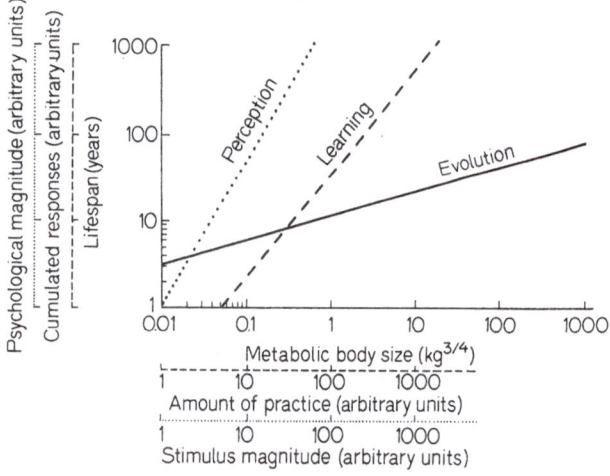

Fig. 2. Display of similarity between evolution, learning and perception as steady states. The straight line plot labelled "evolution" is the interspecies regression line from Fig. 1 and is replotted here only for comparison. The line labelled "learning" is a replot of KIENTZLE's data [23], whereas, the line labelled "perception" is replotted after [39]

it illustrates a writing-upside-down task with the abscissa containing the amount of practice from 1—70 min and the ordinate, the number of letters [23]. For comparison, Figure 2 also recapitulates the "interspecies generalization" of Figure 1 labelled "evolution".

What has been said about learning in relation to evolution, can be said in principle about perception in relation to learning. The similarity between the two is, therefore, expected to result in a straight line plot in log-log coordinates if we plot stimulus magnitude against psychological magnitude. This is indeed the case as shown in one curve of Figure 2 labelled "perception", where we have selected apparent length as an example. The relationship expressed in the straight line implies that the sensation $\Psi$ grows as a power function of the stimulus magnitude $\Phi$, or

$$\Psi = k\,\Phi^n \tag{4}$$

where the constant, $k$, depends on the unit of measurement and the value of the exponent, $n$, may vary from one sensory continuum to another [38]. The similarity between equations (1) and (3) is apparent again in equation (4). The common characteristic of *evolution*, *learning* and *perception* is evidently their steady state nature expressed in these equations and their corresponding graphic representations. It is LOCKER's [28] merit to have called attention to what he calls "inhibition reactions in higher systems" when referring to the similarity of the WEBER-FECHNER Law, the kinetics of enzyme inhibition and WILDER's law of initial values.

## Steady States of Hallucination and Dreaming — Illustrated *Geometrico Modo*

Let us begin by illustrating the interrelationship between:

(1) *invariant stimulus configuration* (i.s.c.) — commonly called the "object" in physical space — represented by the distance between two fixed points, the foci of an ellipse;

(2) the amount of *in-formation*, i.e. the reflection of the "object" in the sensory space of the system, a function of the:

(3) *matching response* (m) in cerebral space, which in turn is a function of the rate of change — i.e. state of excitation or tranquilization — of the system.

Since we assume invariant stimulus configurations to be constant, information and matching response are interdependent parts of a steady state; an increase in one must result in a decrease in the other and vice versa.

Figure 3a depicts the relationship between (1), (2) and (3) during a state of systemic equanimity while Figure 3b is an illustration of the psychoanaleptic drug-induced *excitation state*, displaying the invariance of the "object" (i.s.c.), and the experience of an increased data content[2] — or chronosystole — since

---

[2] from the observer's point of view.

more data can be experienced within a chronological time unit. Remember that during the psilocybin-LSD-mescalin produced excitatory states, sensory acuity is increased, i.e. one needs less sensory in-formation (see Table 1). These findings [10, 12, 18], are in line with the observations of an increase in neural firing in sensory pathways independent of sensed inputs during the psilocybin or LSD-produced excitation [30, 40]. The shortening

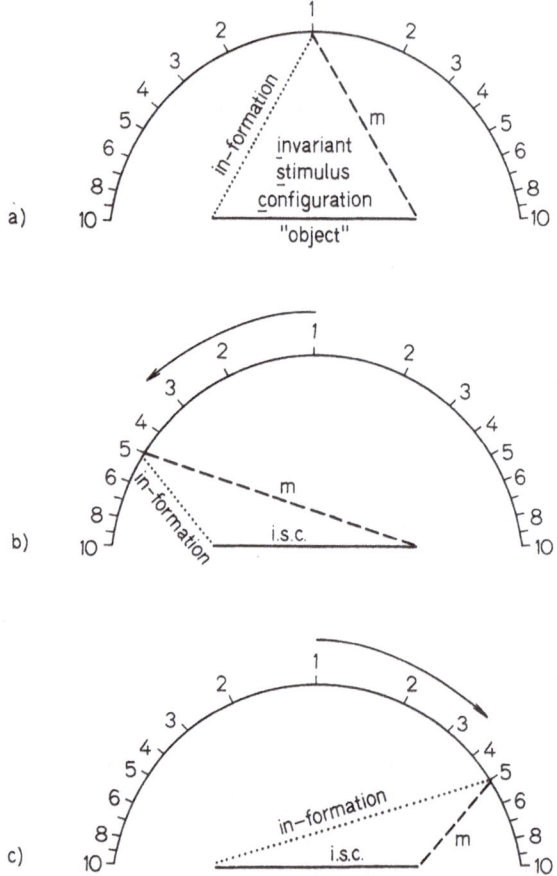

Fig. 3. Graphical representation of a steady state between invariant stimulus configuration, in-formation and matching response (meaning) during excitation and tranquilization; model of a control system with non-linear transfer functions

of the in-formation side of the triangle (in Fig. 3b) illustrates the need for less sensory in-formation. The arrow along the left logarithmic scale of the ellipse depicts states of higher and higher excitation which, again, imply the need for less and less in-formation together with an increasingly larger matching response, i.e. an amplification of sensing, knowing and attending [15] by the organism, an experience of increased meaning or in GELPKE's words "a torrential flood of inner sensation" [14][3].

Table 1. *Organismic changes as well as changes in space-time coordinates under drug-produced excitation and tranquilization. For details see* [9, 13]

|  | Psychotomimetic central *stimulants*: LSD, mescaline psilocybin, etc. | *Tranquilizers* of the phenothiazine type: Chlorpromazine, etc. |
|---|---|---|
| **Biological manifestations** | Increased rate of oxidative phosphorylation | Uncoupling of oxidative phosphorylation |
|  | Body temperature raised | Body temperature lowered |
|  | Metabolic rate increased | Metabolic rate decreased |
|  | Hyperglycemia | Hypoglycemia |
|  | Tachycardia | No Tachycardia |
|  | Piloerection | No Piloerection |
|  | Pupillary dilation | Pupillary contraction |
|  | Speed of nervous conduction* raised, 3 msec/1 °C | Speed of nervous conduction lowered, 1.84 msec/1 °C |
| **Chronological (or conventional) Time** | Overestimated | Underestimated |
| **Experimental manifestations** | Increase in freely chosen finger tapping rate | Decrease in freely chosen finger tapping rate |
|  | Increase in frequency and amplitude of saccades in involuntary micro-nystagmoid eye movements |  |
|  | Tendency to raise critical flicker fusion value | Lowering of critical flicker fusion value |
|  | Increase in handwriting rate, space and pressure | Decrease in handwriting rate, and space |
|  | Distant space diminished |  |
|  | Sensory acuity increased as measured in terms of gustatory and visual just noticeable differences | Sensory acuity decreased |
|  | Shortening of mean pause duration during reading, with unchanged total reading time |  |
|  | Increased ability to correctly read texts delected vertically up to 74% |  |

* may be—as in fever—

[3] During excitation and youth, the operations of the system are speeded up — higher rate of nervous conduction and metabolism — whereas tranquilization or old age are characterized by a lower rate of nervous conduction and metabolism [9].

Table 1 continued

| Chronological (or conventional) Time | Overestimated | Underestimated |
|---|---|---|
| Experiential manifestations | "Torrential flood of inner sensation" | |
| | "A hundred years would not suffice to describe the fullness of experience contained in a single minute" | "Time appears to pass by like magic" |
| | Hofmann's impression of "not getting ahead (on the bicycle), whereas my escort stated that we were rolling along at a good speed". | |
| | Arriving early to appointments | Arriving late to appointments |
| | Increase in data content per chronological time unit (chronosystole or time contraction) | Decrease in data content per chronological time unit |
| | Increase in meaning | Decrease in meaning |

The converse relations are valid under tranquilization (see Figure 3c as well as Table 1) [10, 12], resulting in a substantial decrease in the matching response or meaning. The well-known example of the patient who was frightened by his image of the devil illustrates this point. After tranquilization, this patient can still see the devil but he is no longer frightened [8].

The *geometrico modo* representation of the steady state between invariant stimulus configuration, in-formation and matching response during excitation and tranquilization (Figure 3b and c) implies an inverse logarithmic relation between matching response and meaning and is modelled as a control system with non-linear transfer functions.

## The Unity of Sensorimotor Performance

The fundamental interdependence of sensory and motor neural mechanisms in purposive movement was well recognized by earlier investigators, such as Munk [29], Bastian [2], and Hughlings Jackson [21], who repeatedly emphasized the sensory component of motor activity [39]. Goody [16] logically develops, from the neurologist's point of view, the postulate of the unity of sensorimotor performance. Witness the unity of a sensorimotor performance: there is no seeing without eye movements, no tasting without tongue movements, and there can be no "feeling" of touch without hand or finger movements toward the source of sensation. Goody insists that any scheme of explanation of sensorimotor performance should cover the whole range of performance, from the least to the most complex; movement is sensation [17].

We may redefine, then, the unity of sensorimotor performance within an organismic context: the subjective (sensory) and objective (peripheral motor) facets of our nature manifest themselves in our experiencing the Universe in sensations *and* in our displaying the ability to change that Universe through sensorimotor integration. It is the Yogi and Faust the man of contemplation and the man of action—who symbolize the artificial dualism. This dualism is most distinct in man, and it is based on *this* dualism that we can distinguish sensory to motor ratios within integrated sensorimotor operations.

In the light of the foregoing, matching response within our model of a non-linear transfer function (Figure 3) assumes a dual character. On the one hand it stands for the experience of meaning during states of equanimity, excitation and tranquilization, on the other hand, it also denotes a specific sensory to motor ratio. High or low sensory to motor ratios within a sensorimotor performance—behavioral state—imply, in extreme cases, either a hallucinatory state or a running race, to give examples. There is, of course, an infinite variety of behavioral states between these two extremes. Within this context and in discarding the Cartesian[4] split of universe into self awareness and an outside world, we postulate that sensations, eidetic imagery and entoptic phenomena in dreams and hallucinations are the result of inhibition of the motor component within the steady state of sensorimotor performance. Therefore, we redefine *Hallucinations as intensely active sensations with blocked gross motor manifestations*[5]; including the mainly visual alcoholic hallucinosis and the mainly auditory and tactile hallucinations of schizophrenics especially those of the excited, hallucinating catatonics.

Another excited state most closely resembling the psychoanaleptic drug-induced hallucinations is the cyclically recurring REM or D-state of dreaming during sleep [3, 4, 7]. There is no parallel increase in gross motor activity during the excitatory drug-produced sleepless dream state. Similarly, the REM dream state represents an organismic condition comparable in intensity and variability of internal events only to the most aroused wakefulness, with unit activity in the visual cortex greater than during most waking situations [6], but distinguished from waking by differences in the functional organizations of nervous activity and massive, though incom-

---

[4] For the Cartesian definition of hallucinations see DESCARTES' "Dioptrik" [27].

[5] Hallucination, as an adaptive event, is comparable to another adaptive process. KOHLER's [25] subjects when wearing strongly distorting lenses soon see "correctly" even in a new environment and thus experience what could be called a re-adaptive hallucination with no corresponding "object out there". We suggest that the difference between no object at all "out there" and no corresponding object "out there" may be indicative of the amount of gross motor performance associated with the sensation. As a matter of fact, the amount of gross motor performance may be a measure of "reality".

plete inhibition of motor expression [35][6]. We emphasize that both states are comparable through a combination of intensely active central sympathetic stimulation with a portion of it expended in blocking the peripheral-motor-manifestations. We hypothesize that the similarities between the two states may justify our extending the steady state concept, visualized in Figure 3, to the REM dream state.

*Acknowledgements*

Part of the experimental data on which the concepts are based stem from continuing studies supported by Grant 66-341 of The (Ford) Foundations' Fund for Research in Psychiatry; NIH General Research Support Grant Project No. 587 administered by the College of Medicine, The Ohio State University; and the Psychiatric Research Foundation of Columbus.

# References

1. ABRAMS, R. M., J. A. J. STOLWIJK, H. T. HAMMEL and H. GRAICHEN: Life Sci. **4**, 2399 (1965).
2. BASTIAN, H. C.: Brain **32**, 327 (1909).
3. BENOIT, O.: J. Physiologie **56**, 259 (1964).
4. BIZZI, E., O. POMPEIANO and I. SOMOGYI: Science **145**, 414 (1964).
5. DRABKIN, D. L.: J. Biol. Chem. **182**, 317 (1950).
6. EVARTS, E.: Fed. Proc. **21**, 351 (1962).
7. — J. Neurophysiol. **27**, 152 (1964).
8. FISCHER, R.: In: R. G. GRENELL and L. J. MULLINS (Eds.), Molecular Structure and Functional Activity of Nerve Cells. Washington, D. C.: Amer. Inst. Biol. Sci. 1956.
9. — In: J. T. FRASER (Ed.), Voices of Time. New York: Georg Braziller 1966.
10. —, F. GRIFFIN, R. C. ARCHER, S. C. ZINSMEISTER and P. S. JASTRAM: Nature **207**, 1049 (1965).
11. — — and L. LISS: Ann. N. Y. Acad. Sci. **96**, 44 (1962).
12. —, and R. KAELBLING: In: (J. WORTIS, Ed.) Recent Advances in Biological Psychiatry **9**. New York: Grune and Stratton 1967.
13. —, and M. A. ROCKEY: In EHRENPREIS-SOLNITZKY (Eds.), Neurosciences Research. New York: Academic Press (1968).
14. GELPKE, R.: As quoted by A. HOFMANN, Die Erforschung der Mexikanischen Zauberpilze und das Problem ihrer Wirkstoffe. Sonderabdruck aus dem Basler Stadtbuch, 1964.
15. GIBSON, J. J.: In: E. BERNARD and M. KARE (Eds.), Biological Prototypes and Synthetic Systems. New York: Plenum Press 1962.
16. GOODDY, W.: Brain **88**, 753 (1965).
17. —, and M. REINHOLD: Brain **75**, 472 (1952).
18. HILL, R. M., R. FISCHER and D. WARSHAY: Am. J. Optometry **45**, 454, (1968).
19. HUXLEY, J. S.: Problems of Relative Growth. Lincoln MacVeigh. N. Y. The Dial Press 1932.

---

[6] Strangely, but fortunately, the REM dream state depends upon the same neural structure in cat and man [22].

20. INGVAR, D. H., M. BALDY-MOULINIER, I. SULG and S. HÖRMAN: In: Regional Cerebral Blood Flow (An International Symposium), p. 197. Acta Neurol. Scand. (Suppl. 14), 1965.
21. JACKSON, J. H.: In: J. TAYLOR (Ed.), Selected Writings of John Hughlings Jackson. London: Hodder and Stoughton 1931 and 1932, 2 vols.
22. JOUVET, M., and D. JOUVET: In: R. H. PEÓN (Ed.), The Physiological Basis of Mental Activity. New York: Elsevier Publ. Co. 1963.
23. KIENTZLE, M. J.: J. Exptl. Psychol. 36, 187 (1946).
24. KLEIBER, M.: The Fire of Life: An Introduction to Animal Energetics. New York: John Wiley and Sons, Inc., 1961.
25. KOHLER, I.: The formation and transformation of the perceptual world. Psychol. Issues, III: Monograph 12, 1964.
26. LEHNINGER, A. L.: Bioenergetics: The Molecular Basis of Biological Energy Transformations. New York: W. A. Benjamin, Inc., 1965.
27. LEISEGANG, G. (transl.): Descartes Dioptrik, p. 105. Meisenheim am Glan: Westkulturverlag Anton Hain 1954.
28. LOCKER, A.: Helgol. wiss. Meeresunters. 9, 38 (1964).
29. MUNK, H.: Über die Funktionen von Hirn und Rückenmark, Gesammelte Mitteilungen. Berlin: Hirschwald 1909.
30. POLLARD, J. C., C. BAKKER, L. UHR and D. F. FEURFILE: Compr. Pyschiat. 1, 337 (1960).
31. SACHER, G. A.: In: G. E. W. WOLSTENHOLME and M. O'CONNOR (Eds.), Ciba Foundation Colloquia on Aging, Vol. V, p. 119: London: The Lifespan of Animals. J. A. Churchill, 1959.
32. — In: A. M. BRUES and G. A. SACHER (Eds.), Aging and Levels of Biological Organization, p. 276. Chicago: Univ. of Chicago Press 1965.
33. SCHELLING, H. v.: Ann. N. Y. Acad. Sci. 56, 1143 (1954).
34. SMITH, R. E.: Ann. N. Y. Acad. Sci. 62, 403 (1956).
35. SNYDER, F.: Amer. J. Psychiat. 122, 377 (1965).
36. SOKOLOFF, L.: Handbook of Physiology, Vol. III, p. 1843. Washington, D. C.: American Physiol. Soc. 1960.
37. STEVENS, J. C., and H. B. SAVIN: J. Exptl. Anal. Behav. 5, 15 (1962).
38. — Daedalus 88, 606 (1959).
39. TWITCHELL, T. E.: J. Neurophysiol. 17, 239 (1954).
40. URSIN, H.: Psychopharmacologia 3, 317 (1962).
41. VANDENBERG, S. G.: The genetics of human behavior. National Institute of Mental Health Research Project Summaries, No. 2, 31 (1965).

## Glossary

*Chronosystole* or *time contraction*. From the observer's point of view: a subject's experience of an increased data content within a chronological time unit unaccompanied by an increase in data processing. (In GELPKE's experiential interpretation: "the torrential flood of inner sensation"; or in an objective terminology: the subject overestimates chronological time).

*Cybernetic ontology*. The discipline concerned with *being* when it is viewed from without as an interaction between a self-referential system (organism) and its environment. The resulting steady state is regulated by inverse feedback which involves the connecting of output to input in order to control the output by means of comparison and correction. The interacting components of the steady-state are: (1) invariant stimulus configuration—commonly called the object—in physical space-time, (2) in-formation—the image—in sensory (visual) space-time and (3) thought—the meaning—in cerebral space-time.

*Invariant stimulus configuration.* Refers, in biology, to the invariant features of the stimulus. *Invariant* means, in physics, the unchangeability of a magnitude or formula when transposed from one into another system of coordinates; the charge of electron, e, Planck's quantum, $h$, and velocity of light, $c$, are such invariants. The invariants of perception are what the man on the street calls "real things".

*In-formation.* The irreflexive imprinting of structure on structure and the ability to discriminate a just noticeable difference (Fischer, 1967).

*"Less space-like and more time-like".* In *physical,* Euclidean space-time, we are aware of three dimensions. *Sensory,* especially visual, space-time is experienced within two-dimensional imagery. *Cerebral* space-time has no spatial characteristics; it is experienced as time.

*Psychotomimetic drugs.* Drugs which are also referred to as phantastica, psychotica, hallucinogens, psychodysleptics and psychoanaleptics, and which produce "a flight of the soul from the body" or *ekstasis* in Greek. In general, these drugs accentuate the already existing personality; visionaries become more visionary; bores become more boring.

*Weber Ratio,* specifically in gustation, is a dimensionless—or signal per noise— ratio $\Delta S/S$, where $\Delta S$ denotes the concentration difference between the lower and upper limits of a just noticeable difference [jnd] in taste and $S$ denotes the lower limit.

## Discussion

Bremermann:

The "flower children" are having difficulties in obtaining LSD. Could they use the motor immobilization that you described to have a "trip".

Fischer:

Yes, motor immobilization under condition of normal visual input results in hallucinations. Apparently the age-old Yoga practices are also based on this principle.

Bremermann:

Is the experience induced by the drug popularly called "STP" which recently appeared in the San Francisco Bay Area, comparable to the LSD experiences?

Fischer:

I think it is the same type of experience but extended for 72 hrs.

Bremermann:

In one of the papers published in "Science", it was claimed that the LSD dose that was given to rats was proportional to one average dose taken by LSD users; I do not recall whether this was a paper about chromosomal breaks or another paper about malformed and still-born young that resulted from LSD injection in early pregnancy.

Fischer:

The results of both types of papers to which you are referring could not be reproduced (see among other papers, e.g. Science **159**, 731 (1968) and **160**, 1343 (1968).

Meyer-Döring:

In the state of maximum concentration of the Yoga are there found S-waves in great number and are their spikes very high? Is the number of $\alpha$ and $\beta$ waves diminished and also the height of their amplitudes?

FISCHER:

Both Zen and Yoga practitioners shield their cerebral space against stimulation from sensory and physical space-times. Habituation to auditory stimulation does not occur at the peak of meditation in Zen experts as measured by the EEG, whereas non-meditating control subjects show marked habituation of $\alpha$ blocking to repetitive stimulation.

SUGITA:

Once I asked a Japanese priest something concerning Zen and he said that mental activity deteriorates if one continues Zen too long, say more than 5 years. This I think is an interesting finding.

FISCHER:

It is not surprising! The proof of the (sensory) pudding is in the (motor) eating. If a Zen priest continues his (sensory) meditatory practices for some time, he is experiencing the universe in a "slanted" way, that is under the auspices of a high sensory to motor ratio.

LOCKER:

Why do you use the term "cerebral" instead of "mental" which would suit better here?

FISCHER:

Since I deal with hallucinations and dreams, a field widely open to attack by other scientists, I use the word "cerebral space" which was the first concrete one before the use of "mental". But, I mean "mental"!

ARLEY:

From the department of pharmacology at the University of Gothenburg a series of papers have been published attempting to localize the action of tranquillizers and, if I may use the term, anti-tranquillizers to specific steps in distinct biochemical reactions within the brain cells, so that we might perhaps hope to plan their use in psychiatry somewhat more systematically. What do you think about the reliability of these results and the possibilities they offer?

FISCHER:

Phenothiazine tranquilizers display over a dozen pharmacological properties and it is therefore nearly impossible to relate their organismic effects when experimenting on the molecular level. By the way, the "exact" mode of action of most of the drugs including that of aspirin is unknown.

LOCKER:

It is interesting enough that you, starting from purely psychological studies, arrive at a triadic scheme, which seems to correspond with the structure of the theory of signs. Here we distinguish between signs which represent themselves and exhibit mutual relations; this sphere is called syntax and would represent your "physical space", if preliminarily supposed as existing independent of the observer. The next sphere (within the theory of signs) relates to signs and the things mapped by them; it includes syntax as a partial area and is called semantics. Your "sensory space" would possibly correspond to it. The third sphere, called pragmatics, refers not only to signs and to the world mapped by them, but also to the user: your "cerebral space".

# Circadian Activity Rhythm in the CNS and its Control

## K. P. RAO

*Abstract*

The scorpion, *H. fulvipes* and the cockroach, *P. americana* exhibit a circadian rhythm in their locomotory activity. A parallel rhythm is recordable in the spontaneous electrical activity of the VNC in both. Blood or extracts of the S. O ganglion or the CTNM taken at different times of the day and applied to the VNC *in vitro* also exhibit a rhythm in their ability to enhance or lower the nervous activity. In the scorpion there is also a cyclic reversal of the sign of the photo-kinetic response. It is suggested that the activity rhythm in the CNS is regulated through the interaction of the non-specific sensory input and the excitability state of the CNS as conditioned by the "neurohormone-like factors".

Several organisms are known to exhibit rhythms of a circadian nature in their locomotor activity and other kinds of physiological activities. A great amount of information is available on the time course and phase relations of such rhythms [1]. However very little is known about the internal regulating mechanisms related to the nervous system which play a role in modulating the motor activity into a circadian rhythm.

We have been studying the physiology and ecology of the scorpion, *Heterometrus fulvipes* for the past several years in our laboratory [5] and it has been demostrated that this scorpion exhibits a diurnal rhythm in its locomotor activity when kept under natural conditions of light and dark, with a period of high activity between 5 p.m. and 10 p.m., with the peak around 8 p.m. When kept in continuous darkess the rhythm persists with a circadian periodicity.

Several investigators have shown a similar diurnal rhythm to occur in the locomotor activity of the cockroach, *Periplaneta americana* [1, 7]. It persists for several weeks in darkness. But a decapitated cockroach becomes arrhythmic.

Since the locomotor activity depends on the motor output from the CNS, we examined the "spontaneous" activity in the VNC of the scorpion, isolated at different times of the day, and this revealed that the level of such activity varied with the time of the day, maximum activity being noticed between about 4 p.m. and 11 p.m. After about 11 p.m. the activity in the VNC decreased and reached extremely low levels in the early hours of the morning. The analysis of the spike heights and frequencies indicated that the increase was not merely due to increased frequency of the same units but also to the appearance of larger units [4]. One could observe a similar diurnal rhythm in the activity of the mesosomatic segmental nerves which supply the legs [8].

Similar changes in the electrical activity of the VNC have been noticed in the cockroach (RAO, 1967, unpublished).

These results indicated that the diurnal changes in the locomotor activity noticed in the organism follow from the changes in the electrical activity of the CNS. That the state of excitability of the CNS also varies with the time of the day is indicated by the quantitative measurements made in our laboratory on the "defensive strike" reflex [3].

The neurosecretory cells in the brain of scorpions have been shown to exhibit diurnal changes in their secretory activity [2]. Therefore it was thought possible that there might be some relation between this cyclic activity of the neurosesecretory cells and the diurnal variation in the motor activity of the CNS. This possibility was investigated by measuring the effects of blood and the extracts of the cephalothoracic nerve mass (CTNM) made at different times of the day on the electrical activity of the isolated ventral nerve cord.

Such an experiment revealed that there was a great increase in the electrical activity of the VNC when either blood or the CTNM extract from the 5 p.m. or the 8 p.m. scorpions was applied. On the other hand there was a marked depression of activity when the blood or CTNM extract from the 11 p.m. or 2 a.m. scorpions is applied to the VNC. These effects are maintained for long periods if the nerve cord is bathed in perfusion fluid containing the active substance [6]. Similar results were obtained in the cockroach (RAO, 1967 unpublished). In the cockroach extracts of the supra-oesophageal ganglion alone were adequate to obtain these effects.

Injection of these extracts into intact scorpion results in modification of the activity associated with "the defensive strike reflex". 8 p.m. blood or CTNM extract enhances the excitability, while the 8 a.m. blood or extract depresses the activity considerably (Table 1).

Now the question would arise as to how merely an increase or decrease in the electrical activity and the excitability of the CNS would result in the natural diurnal behaviour of the animal. Such behaviour consists of emerging from a place of hiding in the evening and returning to a place of hiding in the night (about midnight).

It has been shown in our laboratory that the scorpion exhibits also a rhythm in the photo-kinetic response, which involves a reversal of the sign of the photo-kinetic response. Using a standard choice chamber technique, it has been shown that there is a change from photonegativity during the day to photopositivity in the evening. The change from photonegativity to photopositivity occurs at about 4 p.m. and it reverts back to photonegativity at about 11 p.m. or 12 midnight [3].

Table 1. *Effect of injection of 0.3 to 0.4 ml blood or extract of Cephalothoracic nerve mass from evening scorpions into morning scorpions and vice versa on the "defensive-strike" reflex (an indicator of the excitability of the CNS). Values are mV deflection in the liquid-junction potentiometer measuring the magnitude of the "defensive-strike" of the sting-bearing tail of the scorpion. Ringer injected animals are controls*

| Response in 8 a.m. Scorpions | | | Response in 8 p.m. Scorpions | | |
|---|---|---|---|---|---|
| Untreated or Ringer injected animals | After injection of 8 p.m CTNM extract | After injection of 8 p.m blood | Untreated or Ringer injected animals | After injection of 8 a. m CTNM extract | After injection of 8 a. m. blood |
| $1.14 \pm 0.08$ | $1.90 \pm 0.03$ | $1.75 \pm 0.20$ | $1.84 \pm 0.21$ | $0.75 \pm 0.07$ | $0.75 \pm 0.11$ |
| % of increase (+) from control → | $+67\%$ | $+53.5\%$ | % of decrease (−) from control → | $-59.2\%$ | $-59.2\%$ |

Since it has been shown that the non-specific sensory input plays a role in the maintenance of the tone of the CNS it might be suggested that there is an interaction between the non-specific sensory input and the excitability state of the CNS which tends to maintain the ratio between these two at an approximately constant level. If this supposition is correct, then in the evening when the CNS is excited by the release of hormones the scorpion seeks a situation permitting increased non-specific sensory input by coming out into greater illumination, from its burrow. This is facilitated by the reversal in the sign of the photo-kinetic response.

At about midnight when the activity of the CNS is decreasing, there is a tendency to maintain the ratio by decreasing the non-specific sensory input and this is achieved by the scorpion by getting into areas of lesser illumination such as crevices or burrows.

### References

1. Bünning: The Physiological Clock. Springer Verlag 1967.
2. Habibulla, M.: Dissertation, Sri Venkateswara Univ., Tirupati, India 1962.
3. Ramakrishna, T.: Pers. comm. 1967.
4. Rao, K. P.: Proc. Int. Cong. Zool., Washington, D. C. 1963.
5. — J. Anim. Morph. Physiol. **11**, 133 (1964).
6. —, and T. Gopalakrishnareddy: Nature **213**, 1047 (1967).
7. Roberts, S. K. de F.: J. Cellular Physiol. **67**, 473 (1966).
8. Venkatachari, S.: Personal communication 1967.

*Discussion*

Betz:

I think you have a wonderful system for the study of circadian rhythms in a multicellular animal. Do you already have some analytic data on concentrations of hormones or substances like cyclic AMP in your scorpion's blood.

RAO:

I have some idea of the active fraction; it seems to be heat stable.

WALTER:

Although it is commonly accepted that biological rhythmic events are maintained by external forces, there are some rhythms for which it is difficult to identify such a force. Have you investigated the possibility that the rhythms you observe are an inherent part of the scorpion and do not require an external stimulus?

RAO:

Many biologists believe these are of an inherent nature based on some oscillators in some key cellular reactions.

# The Use of Theoretical Models in the Study of Thermoregulation in Man

J. A. J. Stolwijk

With 5 Figures

*Abstract*

A somewhat unexpected benefit of the development of theoretical models of body temperature regulation has been a significant improvement in communication between research and practice in the field of thermoregulation. A computer program containing the best available insight into and quantitative estimates of physiological responses to environment and exercise can be used by those interested in practical problems to obtain the best quantitative prediction of the responses to any particular combination of environment and work load.

## Introduction

In the study of a complex regulatory system such as the regulation of body temperature in homoiotherms, there are relatively few measurements of important parameters which can be made easily and conveniently. The result tends to be that because a measurement is accessible and in widespread use we begin to give more weight to the measurement than it deserves. Especially in dynamic conditions, a measurement, e.g. of rectal temperature in man, is a notoriously poor indicator of total body heat content, average core temperature, temperature of central thermoreceptors or rate of change of body heat content.

On the other hand, again similar to other physiological regulatory mechanisms, the system is so effective in maintaining body temperature that a sustained thermal load results in only an insignificant displacement of central body temperatures. Thus interpretation of steady state data, in an attempt to gain insight into the mechanism of thermoregulation, allows only very moderate progress.

Since long ago, physiologists have realized either intuitively or explicitly that the study of a complete and intact system can only proceed along certain lines. More recently, the engineering profession has developed a formal approach to the study of complex sytems with feedback control. In general the formal solutions which they have obtained only cover relatively simple systems and then only if certain criteria are met. The approach which systems analysis uses can be applied to quite complex biological control problems, but both the complexity and the non-linear characteristics make it virtually impossible to use the formal theory which systems analysis has developed. It is, perhaps, even unwise to attempt to force simplified biological control concepts into the constraints which apply to control systems theory.

It is perhaps superfluous to point out the major difference between the use of the approaches of systems analysis in engineering and in biology. The ultimate aim in engineering is a complete and general mathematical prediction of the behavior of a complex system consisting of components with known characteristics and known interrelationships. In biological research, we deal with a complex system of which the behavior can be investigated, and we attempt to gain insight into characteristics of components and the nature of their interrelationships. The number of such anatomical and functional components is large, and their interrelationships are not all known in a quantitative way or even a qualitative sense.

Embedded in the literature we can find many isolated observations on the characteristics or relationships of a single component. In intuitive evaluation of complex systems, such observations are of small value—even if they are recollected. The great contribution which the approach of systems analysis can make is to provide a rigorous framework which allows us to transfer such isolated observations with all its quantitative consequences into a structure where it is combined with many more similar observations. The completed mathematical structure then yields a predictive capability for the whole system which is very enlightening and has recently been recognized to be of considerable value in evaluation of applied problems.

### Development of a Quantitative Model

A model of a complex system, such as thermoregulation, develops gradually and is never complete: it should be a continuing process, combined with a program of experimental work in the same field.

In our experience, after a series of small additions and refinements, it usually becomes necessary to re-formulate the mathematical description as insight improves or as new data become available. The initial definition is conveniently expressed in the form of a block diagram. The simplest division of the system is into a controlling system and a controlled system as shown in Figure 1. Both blocks of Figure 1 can be subdivided into a number of components. Obviously the final diagram will still represent a very considerable simplification as compared with the real system. A model must always be simpler than the system it represents if it is to fulfill a major requirement of a model: it must be easier to understand and study than the systems it represents. The development which such a model undergoes is to a considerable intent related to better insight as to which components need to be represented accurately and in detail, and which can be lumped into an equivalent component. Thus it can be important to differentiate the hands from the arms, but it will usually not be necessary to distinguish the left hand from the right hand.

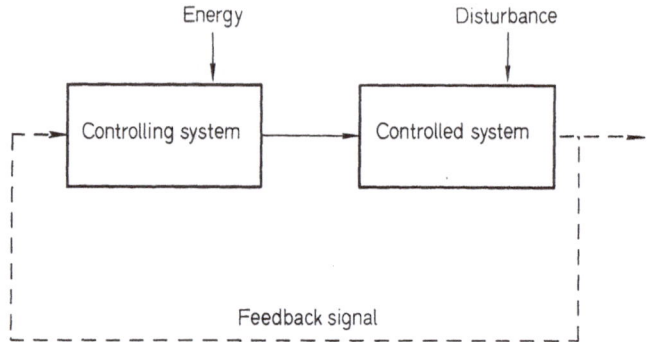

Fig. 1. Block diagram of simple control system. Broken lines represent signal paths; drawn lines depict power transfer

## The Controlled System in Human Thermoregulation

The model presented here is based on a standard man of 73.5 kg weight and a total body surface area of 1.85 m². The complex human form has to be reduced to a simpler analog. We chose to represent, as separate cylindrical components, the head, the trunk, the arms, the legs, the hands and feet.

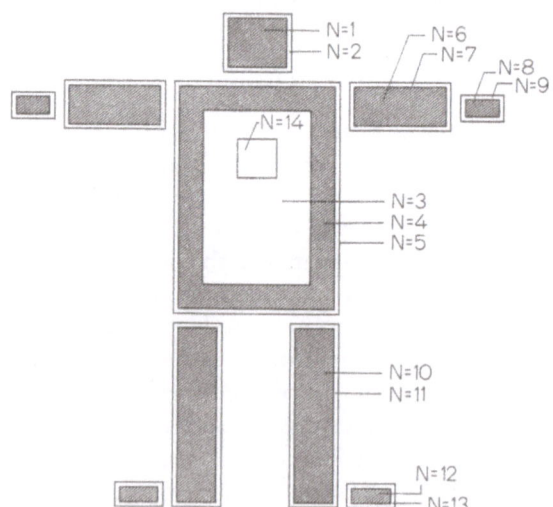

Fig. 2. Diagrammatic representation of the controlled system

This choice is the result of studies with simpler models and the discrepancies between their predictions and experimental data. There is thus a total of 6 cylinders; in addition a central blood compartment is assumed which exchanges heat convectively with all of the compartments, via the respective blood flows. A simplified diagram of the controlled system is given in

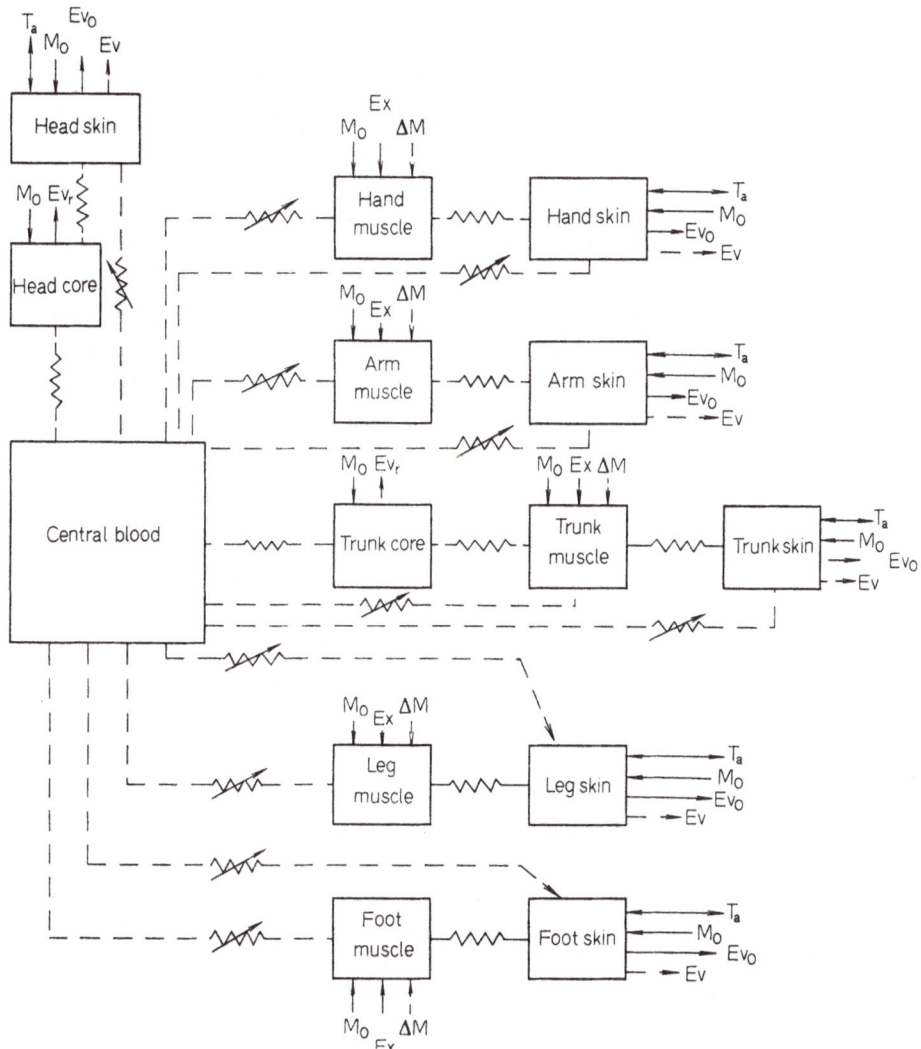

Fig. 3. Interrelations between components of the controlled system factors which are under the influence of the controlling system are given in broken lines. Solid lines indicate physical relationships not under the influence of the controller. ($M_0$: basal metabolism, $\Delta M$: shivering, Ex exercise, Ev evaporative heat loss: $Ev_0$ basal evaporative loss, $Ev_r$ respiratory evaporative loss, Ev sweating)

Figure 2. Each of the cylinders is divided up into two or more layers. The length and radius of each of the concentric rings making up the cylinders is chosen so that weight, surface area and radial thickness correspond as closely as possible to anatomical data. The actual values correspond closely to those given in [2] and the sources upon which the estimates are based are also given in the same contribution.

It is clear that each of the compartments consists of a combination of tissues with varying characteristics. In Figure 2, the head consists of two compartments: the head skin and the head core. The head core consists of the cranium and its contents. The hand cylinder consists of skin and muscle. Obviously bone and tendons are considered to be included in the muscle mass. This lumping means that the heat capacitance and the weight of hand muscle will be increased by that of the added components. Similar considerations apply to the other cylinders and their compartments. The interrelations between the various components are, naturally, also of a distributed nature. The lumped interrelations, which will be accounted for in the model, are graphically represented in Figure 3. It will be noticed that the majority of interrelations are given in broken lines, indicating that they are under control of the controlling system. The natural consequence is that the controlling system is capable of modifying the controlled system by changes in many of the parameters. Since this results in an extremely complex non-linear system, the formal approaches of control theory are no longer applicable. It should be pointed out that most of the parameters and their bounds are available in the literature or their value can be estimated with reasonable accuracy.

The important values of physical characteristics and basal interrelationships of the compartments of Figures 2 and 3 are given in Table 1. A list of symbols with their descriptions and dimensions is given in Table 2.

The heat flow rate into or out of any compartment N is given by the relationship:

$$F(N) = Q(N) - E(N) + TD(N) * C1 - TD(N-1) *$$
$$C2 + BF(N) * TB(N) - A * H * (T(N) - TAIR) \tag{1}$$

in which C1 and C2 are constants as given in Table 1, denoting the thermal conductance between adjacent compartments and in which * is the FORTRAN notation for multiplication; A is the surface area of the compartment in $m^2$.

The instantaneous values of $Q(N)$, $E(N)$ and $BF(N)$ are produced by the controlling system. Any combination of compartment temperatures can serve as inputs to the controlling system.

There are relatively few temperatures and rates of heat flow in the model of Figure 3 which are readily accessible to measurement in experiments.

Table 1

| N | Compartment | Heat Cap. C(N) Kcal/°C | Basal Evap. Heat Loss EB(N) Kcal/h | Basal Heat Prod. QB(N) Kcal/h | Basal Blood Flow BFB(N) l/h | Skin Surface Area AREA m² | AREA % | Heat. Cap. Fraction | Wgt. Kg | TSET(N) °C | Initial Temp. T(N) °C | Conduc. between adj. compartments K Kcal/h/°C |
|---|---|---|---|---|---|---|---|---|---|---|---|---|
| 1 | Head Core | 3.88 | 4.5 | 12.42 | 48.0 | | | .0647 | 4.88 | 36.60 | 37.34 | 2.63 |
| 2 | Head Skin | 0.25 | 0.5 | 0.10 | 1.2 | .104 | 5.6 | .00417 | 0.3 | 36.0 | 36.19 | |
| 3 | Trunk Core | 20.55 | 4.5 | 45.2 | 228.0 | | | .3428 | 25.75 | 37.0 | 37.40 | 4.85 |
| 4 | Trunk Muscle | 9.95 | | 4.3 | 10.0 | | | .1661 | 12.0 | 35.0 | 36.67 | 13.5 |
| 5 | Trunk Skin | 1.28 | 2.5 | 0.51 | 1.0 | .514 | 27.6 | .0231 | 1.54 | 35.0 | 35.42 | |
| 6 | Arm Muscle | 3.88 | | 1.56 | 3.0 | | | .0647 | 4.69 | 36.0 | 36.47 | 4.4 |
| 7 | Arm Skin | 0.78 | 1.6 | 0.31 | 0.5 | .322 | 17.3 | .0131 | 0.94 | 36.0 | 34.51 | |
| 8 | Hand Muscle | 1.46 | | 0.58 | 0.5 | | | .0243 | 1.76 | 36.0 | 36.82 | 2.88 |
| 9 | Hand Skin | 0.20 | 0.4 | .080 | 4.0 | .080 | 4.3 | .0033 | 0.24 | 36.0 | 36.56 | |
| 10 | Leg Muscle | 11.55 | | 4.68 | 8.0 | | | .1927 | 14.06 | 36.0 | 36.71 | 4.28 |
| 11 | Leg Skin | 1.77 | 3.5 | 0.71 | 1.0 | .713 | 38.3 | .0295 | 2.14 | 36.0 | 33.47 | |
| 12 | Foot Muscle | 2.85 | | 1.14 | 1.0 | | | .0475 | 3.44 | 36.0 | 36.80 | 2.16 |
| 13 | Foot Skin | 0.30 | 0.6 | 0.12 | 3.0 | .120 | 6.9 | .0051 | 0.36 | 36.0 | 36.11 | |
| 14 | Central Blood | 1.24 | | | | | | .0207 | 1.35 | 36.66 | 37.25 | |
| 15 | Aver. Skin | | | | | | | | | 34.10 | 34.65 | |
| 16 | Aver. Muscle | | | | | | | | | 36.0 | 36.66 | |
| | TOTAL | 59.94 | 18.1 | 71.71 | 309.2 | 1.853 | | .9999 | 73.45 | | | |

It has been realized for a long time that measurement of accessible body temperatures does not allow realistic estimates of body heat content or even change in body heat content. One of the benefits of the controlled system analog outlined above is that quantitative estimates can be made which are based on a wide variety of isolated experimental observations.

Table 2. *Symbols, description and dimensions*

| | | |
|---|---|---|
| T (N) | temperature of compartment N | °C |
| QB (N) | basal metabolic heat production of compartment N | Kcal·h⁻¹ |
| BFB (N) | basal blood flow to compartment N | l·h⁻¹ |
| EB (N) | basal evaporative heat loss from compartment N | Kcal·h⁻¹ |
| Q (N) | total metabolic heat production in compartment N | Kcal·h⁻¹ |
| BF (N) | total blood flow to compartment N | l·h⁻¹ |
| E (N) | total evaporative heat loss from compartment N | Kcal·h⁻¹ |
| C (N) | heat capacitance of compartment N | Kcal·°C⁻¹ |
| F (N) | heat flow rate into or out of compartment N | Kcal·h⁻¹ |
| TSET (N) | set point temperature for compartment N | °C |
| ERROR(N) | difference between T (N) and TSET (N) | °C |
| SIGN(N) | signal from detectors into controlling system | |
| TB (N) | temperature difference between compartment N and central blood compartment | °C |
| TD (N) | temperature difference between compartments N and N + 1 | °C |
| TAIR | environmental temperature | °C |
| H | environmental heat transfer coefficient | Kcal·m⁻² h⁻¹·°C⁻¹ |
| WORK | total internal heat production due to muscular work | Kcal·h⁻¹ |
| SWEAT | total evaporative heat loss due to sweating | Kcal·h⁻¹ |
| CHILL | total of increased heat production due to shivering | Kcal·h⁻¹ |
| DILAT | total increase in skin blood flow due to vasodilatation | l·h⁻¹ |
| STRIC | total decrease in skin blood flow due to vascoconstriction | l·h⁻¹ |

## The Controlling System in Human Thermoregulation

If we proceed under the assumption that the controlling system relies basically on temperature signals for its input, we can study the effect of any and all internal temperatures on the controller.

The controlled system, as outlined here, is an open loop system. The controlling system closes the control loop, but it does it with a multiplicity of connections, which results in a multiple loop control system.

A number of neural temperature signals are integrated at one or more sites in the nervous system and a number of controller actions are initiated by efferent neural pathways; these controller outputs in turn act upon the controlled system in a number of locations which are not necessarily identical with the locations from which the temperature signals arise. This arrangement can be schematically represented as in Figure 4. The completed system is of such complexity that intuitively only qualitative estimates of input-output relationships are possible at best, and quantitative estimates are out of the question.

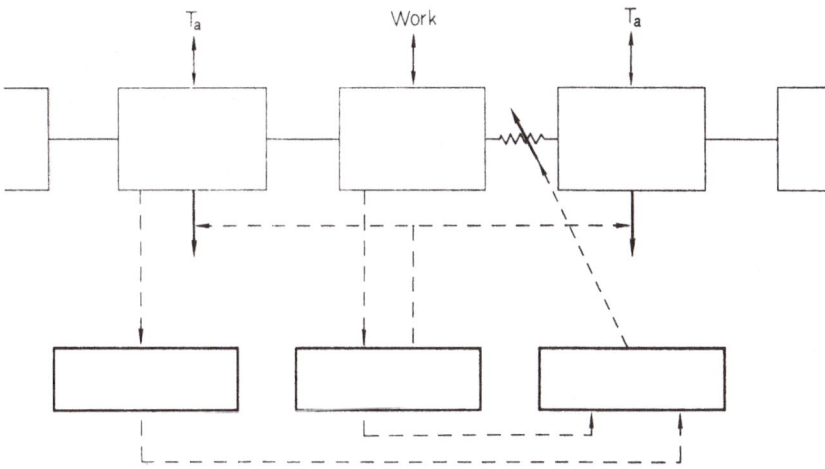

Fig. 4. Schematic diagram of multiple feedback loops. Heavy lines: components or actions of controlling system, light lines: controlled system, broken lines: signal pathways

If the controlling system is efficient and effective, the expectation is that a disturbance will result in counter-action without any of the temperatures showing much displacement. The small displacements which do occur are almost impossible to interpret because of the multiple connections: a given relationship between a temperature displacement and a controller action is first as likely to be fortuitous as real. The reason for this of course is found in the strong connections between the various temperatures via the passive links, especially in or near steady states.

There is only one way to further our knowledge of the thermoregulatory system: by breaking the strong and multiple connections in such a way that isolated parts of the controller can be examined. This can be accomplished by studying the effects of strong transient disturbances in the dynamic phase during which strong links can be temporarily disconnected, or by opening one or more of the loops. In man, the study under dynamic loads is the most suitable; while in animals it is usually possible to open one of the loops. In either case, an exact knowledge of the dynamic responses in all parts of the body is essential since neither approach is useful for steady state studies. In man, steady state studies are made fairly useless because of the close coupling between various body temperatures whereas in animal studies, even with one control loop interrupted, other control loops take over after only a small deviation of their signal temperatures and we no longer are dealing with an isolated component of the controlling system. In either case, simulation of such experiments with mathematical models will yield valuable insight.

In our current version of a thermoregulation model in man, we are assuming that temperature signals come from the skin, from the brain stem and from the muscle. There is no difficulty in deriving signals from other locations, if desired. After integration these signals are used to cause appropriate changes in heat production, evaporative heat loss and blood flow. These values, as modified by the controller system, are then used in the calculation of heat flow rates and temperatures in the controlled system via equations of the type given in (1).

All compartments are given a value TSET, a temperature at which the output from the receptors is effectively considered to be zero. Deviations from the set points are named ERROR as in (2).

$$ERROR (N) = T (N) - TSET (N) \tag{2}$$

ERROR (1) is the deviation of the brain temperature from its setpoint: a negative value is then named SIG (1), a positive value is SIG (2). ERROR (15) is the deviation of average skin temperature; if negative, this becomes SIG (3); if positive SIG (4). Similarly, ERROR (16), the deviation in average muscle temperature, is SIG (5) when negative and SIG (6) when positive. Unless otherwise indicated, SIG (N) are all set to zero. The physiological responses can now be expressed as combinations of signal values SIG (N) and fixed coefficients which can be set to zero or to any other value. The effector actions are represented by the symbols SWEAT, CHILL, DILAT and STRIC. Appropriate fractions of each of these effector actions are assigned to each of the compartments.

Blood flow, evaporative heat loss rates and metabolic heat production are first set to their basal values and subsequently modified by the addition of appropriate fractions of each of the effector actions. Both the head core and trunk core are assigned an additional evaporative heat loss of 2 % of the heat production due to work or shivering, reflecting the increased respiratory rate. Muscle compartments which have a variable metabolic rate are given an addition in blood flow proportional to their extra metabolic rate, thus reflecting the increased blood flow during exercise.

These newly developed values for heat production, evaporative heat loss and blood flow are used in the computation of heat flows during the next iteration interval.

At specifiable intervals, the program can be made to branch to an output sub-routine which can provide values for any and all of the variables and any values derived from them. Such output can be in the form of a graph or a table.

## Application of Models of Thermoregulation in Man

It is evident that almost any experiment or condition can be simulated with the model by applying identical experimental conditions. The coefficients and values of the controlled system are based on careful evaluation of pertinent data in the literature and should not be changed unless there is compelling reason to do so, such as new findings or new measurements of heat capacitances, heat conductances, basal rates of metabolic heat production and of evaporative heat loss, and basal local blood flow rates. Naturally if this model of man is to be modified for other homoiotherms, all these basal values will need readjustment.

The controller of the present model calls upon a number of other biological control systems for its proper operation such as the respiratory control system for supply of oxygen and release of carbon dioxide; the cardiovascular control system for supplying the required blood flow to all regions; the water volume control system for making water available for sweating, etc. As formulated here, no limits have been assigned to the efficacy of any of these control systems, nor is there a limit to the evaporation of secreted sweat. If required, it is relatively simple to put limits on the capabilities of all these systems so that exhaustion, dehydration and circulatory limits can be properly evaluated.

In studies in which the objective is the evalution of practical situations, such as tolerance times, comparisons of stressful conditions, etc., the precise values of the controller coefficients in the thermoregulatory controller are not of primary importance, and in the model, the values given in Table 3 will serve adequately.

Table 3. *Controller coefficients for appled studies with the thermoregulation model*

| Coefficient | | |
|---|---|---|
| HSW | 60 | $Kcal \cdot -h^{-1} \cdot {}^\circ C^{-2}$ |
| WSW | 100 | $Kcal \cdot h^{-1} \cdot {}^\circ C^{-2}$ |
| DIL | 100 | $l \cdot h^{-1} \cdot {}^\circ C^{-1}$ |
| ST | 10 | $l \cdot h^{-1} \cdot {}^\circ C^{-1}$ |
| PC | 60 | $Kcal \cdot h^{-1} \cdot {}^\circ C^{-1}$ |
| CC | 60 | $Kcal \cdot h^{-1} \cdot {}^\circ C^{-2}$ |

If the purpose of the model study is to improve understanding of the thermoregulatory controller itself, then the emphasis should be on variation in value and character of the controller coefficients. It is e.g. noticed in

Table 3 that the controller coefficients which would represent additive integration of different sensors are not used, whereas most of the coefficients used represent integration of the multiplicative or gain control type.

By suitable adjustments of set point temperatures and controller coefficients, it is usually possible to obtain exact simulations of a particular experiment. In fact, it is usually possible to do this with more than one combination of set point temperatures and controller coefficients. The most critical test of the predictive capabilities of a theoretical controller system is the simulation of an experiment which is totally different in experimental design and method from the ones on which the set point temperatures and controller coefficients are based.

An example of such a test is shown in Figure 5 based on a series of experiments performed by PIIRONEN [1]. The observed esophageal temperature is shown in steady state conditions in men working at different work

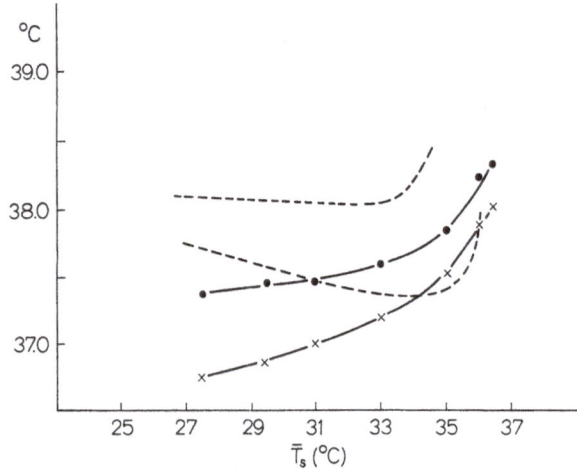

Fig. 5. Comparison of computer predictions of steady state rectal temperature and ear temperature with measured esophageal temperatures in an experiment in which subjects worked at 900 kpm at various average skin temperatures

Work level — 900 Kpm, ● $T_r$, × $T_e$,
———— Computer predictions, - - - - - - Experimental results (2 subjects)

rates in environments which were so manipulated as to produce given average skin temperatures. Fig. 5 gives the computer prediction for rectal temperature and head core temperature at a work level of 900 kgm/min for skin temperatures between 27 °C and 36.5 °C. In the same figure are also given the experimentally determined values of esophageal temperatures for two of PIIRONEN's subjects.

In the case of applied problems in body temperature regulation, predictions of this accuracy are quite useful. The model then presents itself as a vehicle to transfer the knowledge, experience and insight of research workers in the area of thermoregulation to workers in applied fields in such a way as to enable the latter to obtain quantitative and particular answers to quantitative and particular problems. Thus the model becomes an extremely powerful means of communication between research and application.

Similarly, the communication between research workers should improve by the use of this type of model: concepts and ideas can be transmitted in quantitative form, and, even more important, the quantitative consequences of such concepts can be evaluated and discussed in a constructive manner.

## References

1. PIIRONEN, P., and E. ÄIKÄS: Techn. Report No. AF-EOAR-63-113, Institute Occupational Health, Helsinki (Finland) 1964.
2. STOLWIJK, J. A. J., and J. D. HARDY: Pflügers Arch. 291, 129 (1966).

# On the Mechanism of the Calorigenic Action
# of Catecholamines

O. Strubelt

With 2 Figures

*Abstract*

The increase in metabolic rate after injection of adrenaline, noradrenaline and isoprenaline was the same in anesthetized as in awake rats. Partial or total paralysation of skeletal musculature had no influence on the calorigenic action of the catecholamines. The contribution of increased cardiac work and augmented body temperature to sympathomimetic hypermetabolism is minimal. Hence, calorigenic action of catecholamines is a real increase of basal metabolism. This hypermetabolism is a consequence of the metabolic cycles induced by sympathomimetic glycogenolysis and lipolysis and is probably mediated by the formation of cyclic 3′,5′-AMP.

In spite of many investigations, the cause of catecholamine-induced calorigenesis is still a matter for discussion [1, 3]. Most investigators have failed to demonstrate calorigenic action of catecholamines in isolated tissues [2]. In our experiments, adrenaline did not increase oxygen consumption of rat liver slices. Hence, catecholamines probably do not influence cellular respiration directly. Many authors, therefore, have postulated that catecholamine-induced hypermetabolism is not a real increment of basal metabolic rate but the consequence of increased organ activities [1, 2]. This hypothesis, however, is neither proved nor refuted.

First we found that increments of motility do not contribute to catecholamine-induced hypermetabolism. 0.3 mg/kg of adrenaline, noradrenaline or isoprenaline were injected into awake rats and in rats anesthetized with urethane (1.2 g/kg i.m.). Ambient temperature was 28 °C. Oxygen consumption was recorded continuously during 120 min with Noyon's diaferometer. The integrated increase of oxygen consumption over basal values was the same in awake and anesthetized rats.

To exclude increased involuntary activity of skeletal musculature as a source of sympathetic hypermetabolism, we performed experiments in awake rats with the spinal cord destroyed from the first dorsal vertebra downwards. Although 70 to 80% of the musculature was paralysed, calorigenic actions of catecholamines were not smaller in these pithed rats than in awake control animals. The total paralysation of rats by curare was also without influence on calorigenic action of adrenaline (Fig. 1).

The possible contribution of cardio-acceleration to the calorigenic action of catecholamines was evaluated indirectly. Bilateral vagotomy in the rat induced a lasting increase of blood pressure and heart rate and augmented by 65% the cardiac index of MEESMANN [4] which is said to be a suitable indicator of cardiac oxygen consumption. 0.3 mg/kg adrenaline—a highly calorigenic dose (see Fig. 1)—increased cardiac index only by 30%. Bilateral vagotomy, nevertheless, did not increase oxygen consumption of anesthetized rats. Thus, increased cardiac work does not contribute essentially to the calorigenic effects of catecholamines in the rat.

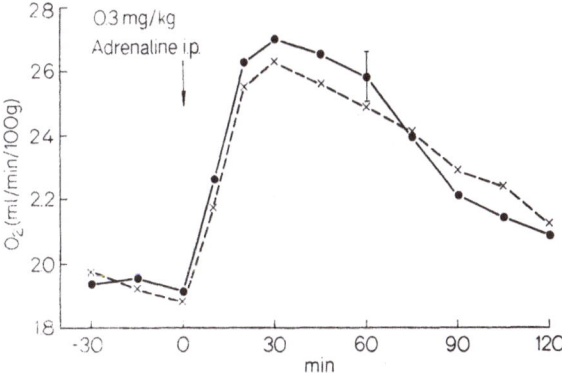

Fig. 1. Mean oxygen consumption of 12 anesthetized (●———●) and 8 anesthetized, curarised, and artificially respirated rats (×——×) after 0.3 mg/kg adrenaline i.p.

Catecholamine-induced restriction of skin blood-flow is said to increase body temperature and thereby to augment metabolism [1]. Increments of body temperature after calorigenic doses of catecholamines ranged in our experiments from 0.2 to 1.2°C. They did not precede, but rather followed the increase of metabolism with a delay, and therefore are not the cause but the consequence of energy release. In further experiments we raised ambient temperature from 28° to 32.5°C. This effect increased the body temperature of anesthetized rats as compared with control animals (28 °C) by 1.2 °C, but did not augment oxygen consumption (Fig. 2). The rise of oxygen consumption after adrenaline was also the same in both groups; again hyperthermia proved unimportant for the calorigenic action of catecholamines.

Our experiments demonstrate that catecholamine-induced hypermetabolism is a true increment of basal metabolic rate. The difficulties with experiments in vitro probably result from the fact that sympathomimetic calorigenesis is a consequence of metabolic events induced by sympathetic glycogenolysis and lipolysis, namely lactic acid and fatty acid cycle. These cycles which increase metabolic rate by consumption of ATP depend on the cooperation of musculature and liver or adipose tissue and liver, respectively, and therefore can take place only in the whole organism.

Sympathetic glycogenolysis and lipolysis are mediated by the formation of cyclic 3′,5′-AMP [5, 6]. Therefore, this compound should augment metabolic rate, too. However, in our experiments extrinsic 3′,5′-AMP (50 and 100 mg/kg i.v.) failed to augment oxygen consumption of rats, perhaps as

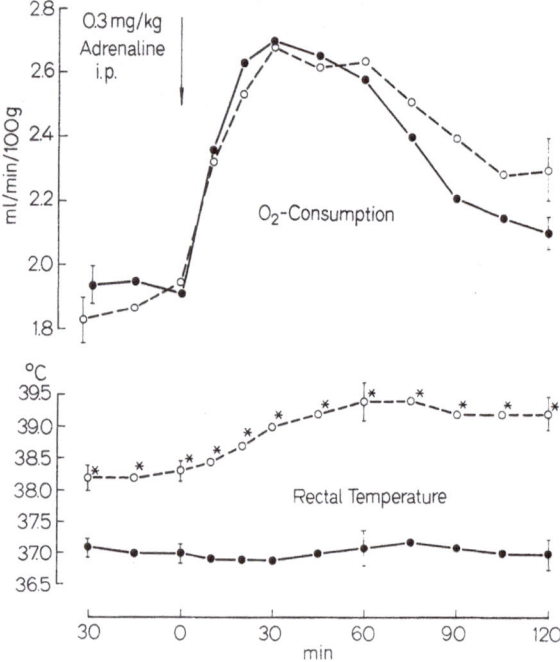

Fig. 2. Oxygen consumption and rectal temperature of anesthetized rats after 0.3 mg/kg adrenaline i.p. at ambient temperatures of 28 °C (●———●) and 32.5°C (○――○). Means of 12 and 8 experiments, respectively. ∗ = significant difference to the simultaneous value of the other group

a consequence of its low permeation into tissues. Theophylline, on the other hand, which inhibits destruction of 3′,5′-AMP [5], proved to be a potent calorigenic agent; the very slow onset of its action is compatible with the idea that it exerts calorigenesis by accumulation of endogenous 3′,5′-AMP.

### References

1. ELLIS, S.: Pharmacol. Rev. **8**, 485 (1956).
2. GRIFFITH, F. R. JR.: Physiol. Rev. **31**, 151 (1951).
3. HOLTZ, P., u. D. PALM: Ergebn. Physiol. **58** (1966).

4. MEESMANN, W.: Med. Welt (Stuttg.) 1778 (1965).
5. SUTHERLAND, E. W., and G. A. ROBISON: Pharmacol. Rev. **18**, 145 (1966).
6. WEISS, B., J. J. DAVIS and B. B. BRODIE: Biochem. Pharmacol. **15**, 1553 (1966).

*Discussion*

SMITH:

Which were the environmental circumstances under which you measured the metabolic rates?

STRUBELT:

I used the indifference temperature, and this in our rats was 28 °C.

SMITH:

WILL found that the effect of catecholamines in raising the calorigenic activity is most strongly elicited in an animal that is slightly if not considerably cooled. Under your conditions the thermogenesis which you followed is undoubtedly ascribable to non-shivering thermogenesis.

STRUBELT:

Yes, the thermogenesis recorded is not a consequence of increased activity but rather of increased BMR.

SMITH:

Did you calculate the cardiac work?

STRUBELT:

In the rat the calorigenic effect of the cardiac work is only 2.5% of the total metabolic increment induced by catecholamines.

AMMON:

Catecholamines activate glycogenolysis with elevation of blood glucose and lipolysis with increase of FFA, but nothing is known about whether they also exert an influence on the further breakdown of glucose and FFA. Do you know whether an increased glucose and FFA blood level does enhance total metabolism?

STRUBELT:

F. R. GRIFFITH (Physiol. Rev. **31**, 151 (1951)) has shown that hyperglycaemia does not increase total metabolism; I can confirm these results. As to the FFA, there are some papers where an increase could be shown in the metabolic rate of isolated organs—heart, liver (NESTEL, P. J., and D. STEINBERG: J. Lipid Res. **4**, 461 (1963)); CHALLONER, D. R., and D. STEINBERG: Amer. J. Physiol. **210**, 280 (1966)); however it is not possible to bring about such an increase in the whole organism, too. This is probably a consequence of the high toxicity of FFA.

HORWITZ:

(1) Under the conditions of your experiment, does the blood lactate increase?
(2) Is the magnitude of this increase accountable for the observed calorigenesis?

STRUBELT:

An increase in the blood lactate after catecholamines has been shown by several authors. L. LUNDHOLM (Acta physiol. scand. **19**, Suppl. 67, 1949) found a good correlation between increase in blood lactate and in total metabolism after application of adrenaline.

# Some Epicritical Remarks (Instead of a Summary)

A. LOCKER

In dealing with the most outstanding implications of the foregoing discussion, the following points deserve some attention: (1) the relationship between the continuous and the discrete system (or model) description as well as between the deterministic and the stochastic description, (2) the problem—closely connected with the one mentioned above—of the ultimately valid form of description of "reality" or, in other words, of the ultimately possible theory, (3) the problem of optimization and (4) the problem of different ways of looking at the world.

The formulation of a certain *correspondence principle* for two possible descriptions of a given system [4] in terms of a continuous and discrete description renders the two forms equally valid. Therefore, it seems hopeless to search for the truly conformable representation. On the contrary, for the purpose of serving certain prescribed needs, the selection between the two depends on our own decision. Thus, the correspondence principle seems to be a kind of mapping. The epistemological interest, however, centers around the problem as to whether or not the system conceived may exist without or independently of at least one of its possible representations; the answer inevitably will be that the system does exist within the human mind and by dint of its creative ability.

Concerning the problem of an *adequate description of "reality"* by means of deterministic or stochastic models, the question was raised during the symposium [1] as to whether or not a scientifically valid description has to be based on quantum physics. We clearly have to distinguish between two facts: (1) By applying quantum physics within the range of reality where the separation of $\varepsilon_1(t)$ and $\varepsilon_2(t)$ makes sense, we have to take into consideration the complementary 3-valued logic with the consequence that an equivalence relation between thinking and being, as pre-supposed by classical logic, no longer holds true. Nevertheless, it may be shown that this fact still remains compatible with critical gnoseology. Moreover, it does not suffice to support the exclusiveness of a probabilistic concept; rather, an epistemologically satisfactory explanation of nature has to include both possibilities, viz. the classical deterministic as well as the quantum physical and classical indeterministic one. Only the two together would constitute a generally valid framework for a universal explanation of nature. (2) By accepting this, one is easily able to avoid the pitfall of postulating the quantum physical description as the only scientifically admissible one. On the contrary, depending on the level of "reality" under consideration, a reducibility of higher levels to the pretended ultimate one, i.e. the quantum physical level, may become scientifically meaningless. The increase of (phenomenological) complexity reflects itself at different levels, leading to the occurrence of new properties, the explanation of which not only depends on prime elements (e.g. quantum physical events) but also on new specific system characters [3] as the consequence of the structure and all other determinants which now play a role—irrespective of a theoretically postulated reducibility to quantum physics.

The third problem of some significance, which evoked a rather lively discussion among participants, *optimization*, refers to the need of a certain entity (be it a process, an efficiency, a turnover) to be optimized. However, irrespective of the several mathematical methods we have at hand and which we may apply routinely—such as linear or non-linear programming or even dynamic programming—again certain definite pre-suppositions have to be accepted. For instance, there is PONTRIJAGIN's optimum principle defined in a given system of differential equations to maximize (or minimize) a pay-off function with respect to certain control variables; this clearly reveals as the essential question the introduction of a certain objective. This objective acts as a criterion which is to be achieved optimally or in accordance with which some functions must become optimal. For biology, of course, it has to be postulated that the objective needs to be biologically meaningful. The well known "bottle-neck problem", however, convinces us of the fact that optimization depends on decisions between more or less equal possibilities. If, for instance, in the production of automobiles steel is becoming a critical intermediate it has to be decided whether the steel production or the automobile production has to be promoted. In this situation the production of one of these two has to be temporarily delayed. The objective or the decision criterion either has to be brought to the system from outside—as in systems optimized by man or, as is the case in organisms capable of optimizing their own activities, we have to assume that it is an inherent part of the organism's organization. Incidentally, mention has to be made as to the amount of relativity involved in applying an objective for optimization.

Finally, we may envisage an interesting problem which is referred to more or less clearly in several of the foregoing papers. Aside from the interest centered around the adequate description of nature, the question arises as to the principally possible and intrinsically different but nonetheless intimately interlinked *modes of cognition*. In other words, which different modes of attaining and expressing cognition and of making statements about "reality" should principally be taken into consideration? If we are aware of the necessity of founding cognition on the grounds of the doctrine of being (i.e. ontology) we may gladly accept the proposal made by BENSE [2] to the effect that we have to distinguish between several kinds of ontology: (a) a processual ontology based on the constructive (operational) capability of the human mind, (b) an intensional ontology concerned with the essentials of being (e.g. the "being of the being", etc.) identical with metaphysics in the classical sense and (c) an extensional ontology, considering the being as an extensional manifold to which, therefore, all formal or formalized descriptions of the being (or of its representations such as systems, models, etc.) belong. The main concepts inherent in the above-mentioned ontology are quantitative (metric) as well as qualitative (relational) although the latter already indicate the existing links between extensional and intensional ontology. No matter how highly speculative such a triadic scheme of cognition and ontology might be, it seems to have a fundamental significance and in the topics of some of the foregoing papers apparently there is an implicitly or even explicitly conceivable line of thought which might be accepted as a hint to consider this as significant rather than purely accidental.

## References

1. ARLEY, N.: Comment on the paper by BARTHOLOMAY, A. F., this symposium.
2. BENSE, M.: Grundlagenstudien Kybern. Geisteswiss. **4**, 12 (1963).
3. LANGE, O.: Ganzheit und Entwicklung in kybernetischer Sicht. Berlin: Akademie-Verlag 1967.
4. ROSEN, R.: this symposium.

# Author Index

(Numbers in *italics* indicate the pages where references are listed)

# Subject Index

Satz und Druck: Werk- und Feindruckerei Dr. Alexander Krebs, Weinheim/Bergstr. und Bad Homburg v. d. H.